GAME THEORY AND ECONOMIC BEHAVIOUR
VOLUME ONE

Game Theory and Economic Behaviour

Selected Essays, Volume One

Reinhard Selten

Professor Emeritus of Economics,
Rheinische Friedrich-Wilhelms Universität Bonn, Germany

Edward Elgar
Cheltenham, UK • Northampton, MA, USA

© Reinhard Selten, 1999

All rights reserved. No part of this publication may be reproduced, stored in a retrieval system or transmitted in any form or by any means, electronic, mechanical or photocopying, recording, or otherwise without the prior permission of the publisher.

Published by
Edward Elgar Publishing Limited
Glensanda House
Montpellier Parade
Cheltenham
Glos GL50 1UA
UK

Edward Elgar Publishing, Inc.
6 Market Street
Northampton
Massachusetts 01060
USA

A catalogue record for this book is available from the British Library

Library of Congress Cataloguing in Publication Data

Selten, Reinhard.
    Game theory and economic behaviour : selected essays / Reinhard Selten.
    Includes bibliographical references.
    1. Game theory. 2. Oligopolies. I. Title.
HB144.S43   1999
330'.01'5193—dc21   98–45786
                                                                                               CIP

ISBN 1 85898 872 1 (2 volume set)

Printed and bound in Great Britain by Biddles Ltd, Guildford and King's Lynn

Contents

Acknowledgements	vii
Foreword Alvin E. Roth	ix
Introduction Andreas Ortmann	xi
Bibliography of Reinhard Selten's writings	xxiii
Nobel Prize citation	xxxi

PART I

1. 'In Search of a Better Understanding of Economic Behaviour' in *The Makers of Modern Economics*, Vol. 1, Arnold Heertje (ed.), 1993, Harvester Wheatsheaf, 115–39 — 3

PART II   AXIOMATIC CHARACTERIZATIONS

2. 'Valuation of n-Person Games', *Advances in Game Theory: Annals of Mathematics Studies*, **52**, 1964, 577–626 — 31
3. 'Properties of a Measure of Predictive Success', *Mathematical Social Sciences*, **21**, 1991, 153–67 — 81
4. 'An Axiomatic Approach to Consumers' Welfare', with Eyal Winter, *Mathematical Social Sciences*, **27**, 1994, 19–30 — 96
5. 'An Axiomatic Theory of a Risk Dominance Measure for Bipolar Games with Linear Incentives', *Games and Economic Behavior*, **8** (1), January 1995, 213–63 — 108

PART III   LEARNING

6. 'Evolution, Learning, and Economic Behavior', *Games and Economic Behavior*, **3**, 1991, 3–24 — 161
7. 'Anticipatory Learning in Two-Person Games' in *Game Equilibrium Models I: Evolution and Game Dynamics*, Reinhard Selten (ed.), 1991, Springer-Verlag, 98–154 — 183
8. 'End Behavior in Sequences of Finite Prisoner's Dilemma Supergames: A Learning Theory Approach', with Rolf Stoecker, *Journal of Economic Behavior and Organization*, **7**, 1986, 47–70 — 240
9. 'Experimental Sealed Bid First Price Auctions with Directly Observed Bid Functions', with Joachim Buchta in *Games and Human Behavior: Essays in Honor of Amnon Rapoport*, David V. Budescu, Ido Erev and Rami Zwick (eds), 1998, Lawrence Erlbaum Associates Inc., 79–102 — 264

PART IV POLITICAL AND SOCIAL INTERACTION

10 'The Scenario Bundle Method', Research Conference on Strategic Decision Analysis Focusing on the Persian Gulf, Verein zur Förderung der Arms Control, e.V. SADAC, 5–56, reset with minor revisions 291
11 'Balance of Power in a Parlor Game' in *Game Equilibrium Models IV: Social and Political Interaction*, Reinhard Selten (ed.), 1991, Springer-Verlag, 150–209 326
12 'The Distribution of Foreign Language Skills as a Game Equilibrium', with Jonathan Pool in *Game Equilibrium Models IV: Social and Political Interaction*, Reinhard Selten (ed.), 1991, Springer-Verlag, 64–87 386

Name index 411

Acknowledgements

The author and publishers wish to thank the following who have kindly given permission for the use of copyright material.

Academic Press Inc for articles: 'An Axiomatic Theory of a Risk Dominance Measure for Bipolar Games with Linear Incentives', *Games and Economic Behavior*, **8** (1), January 1995, 213–63; 'Evolution, Learning, and Economic Behavior', *Games and Economic Behavior*, **3**, 1991, 3–24.

Elsevier Science B.V. for articles: 'Properties of a Measure of Predictive Success', *Mathematical Social Sciences*, **21**, 1991, 153–67; 'An Axiomatic Approach to Consumers' Welfare', with Eyal Winter, *Mathematical Social Sciences*, **27**, 1994, 19–30; 'End Behavior in Sequences of Finite Prisoner's Dilemma Supergames: A Learning Theory Approach', with Rolf Stoecker, *Journal of Economic Behavior and Organization*, 7, 1986, 47–70.

Lawrence Erlbaum Associates Inc for excerpt: 'Experimental Sealed Bid First Price Auctions with Directly Observed Bid Functions', with Joachim Buchta in *Games and Human Behavior: Essays in Honor of Amnon Rapoport*, David V. Budescu, Ido Erev and Rami Zwick (eds), 1998, 79–102.

Princeton University Press for article: 'Valuation of n-Person Games', *Advances in Game Theory: Annals of Mathematics Studies*, **52**, 1964, 577–626.

Springer-Verlag for articles: 'Anticipatory Learning in Two-Person Games' in *Game Equilibrium Models I: Evolution and Game Dynamics*, Reinhard Selten (ed.), 1991, 98–154; 'Balance of Power in a Parlor Game', in *Game Equilibrium Models IV: Social and Political Interaction*, Reinhard Selten (ed.), 1991, 150–209; 'The Distribution of Foreign Language Skills as a Game Equilibrium', with Jonathan Pool in *Game Equilibrium Models IV: Social and Political Interaction*, Reinhard Selten (ed.), 1991, 64–87.

Every effort has been made to trace all the copyright holders but if any have been inadvertently overlooked the publishers will be pleased to make the necessary arrangements at the first opportunity.

Foreword
Alvin E. Roth[*]

Two of the most important developments in economics in the latter half of the twentieth century are surely the extraordinary progress in game theory and in experimental economics. Reinhard Selten is one of the pioneers in both of these endeavours, and he has been a leader in each of them throughout his career. This makes him unique: no one else in the world has made such important or such sustained contributions to both fields.

This is all the more noteworthy because game theory and experimental economics have been intertwined since their beginnings. Indeed, Professor Selten cites the 1953 experimental paper by Kalish, Milnor, Nash, and Nering as one of the inspirations for his own first paper, his 1959 paper with Sauermann on an oligopoly experiment. But while several other pioneers of game theory were at least briefly involved in early experiments, only Selten was instrumental in developing both kinds of research to the point where, first game theory and more recently experimental economics too, have become part of mainstream economic research.

The historical connection between game theory and experimental economics is a natural one, because game theory brought to economics a kind of theory that lent itself to experimental investigation, and in some cases demanded it. The reason is that game theory seeks to provide precise models of both individual behaviour and of economic environments. This concern with the 'rules of the game', the institutions and mechanisms by which transactions are made, together with precise assumptions about the behaviour of individuals given the information available to them, gave rise to theories that could be tested in the laboratory.

But theories precise enough to be clearly tested under controlled conditions can also be found wanting. The reason Selten's contributions to game theory and to experimental economics constitute one scientific career, and not two separate ones, is that he has been a leader in developing the theoretical implications of how games might be played by ideally rational players, and also, when these theories fail to be descriptive of observed behaviour, in undertaking the related endeavour of proposing more descriptive theories.

Thus, Selten, the philosopher of ideal rationality who observed that equilibrium alone may not be a sufficient condition to rule out irrational play, was also the scientist who examined the limitations of perfect equilibrium (as in his chain store paradox) and even of any kind of equilibrium (as in his experimental study with Stoecker of the finitely repeated prisoner's dilemma), and he proposed novel theories of boundedly rational behaviour to explain what he observed.

As economics prepares to enter the twenty-first century, we will certainly need to explore further both the implications of rationality and their limitations, both normative theory and descriptive theory, and new methods of both theoretical and empirical research. In this enterprise we will have the proverbial advantage of standing upon the shoulders of giants. As the selected essays in the present two volumes make clear, on Selten's shoulders we will find solid support for both legs.

[*] Alvin E. Roth is Gund Professor of Economics and Business Administration at Harvard.

Introduction[1]
Andreas Ortmann

In 1994, fifty years after John von Neumann and Oskar Morgenstern published *Theory of Games and Economic Behavior* (1944), the Royal Swedish Academy of Sciences awarded the Nobel Prize in Economic Sciences jointly to Professors John F. Nash (Princeton University), John C. Harsanyi (University of California) and Reinhard Selten (University of Bonn) for their path-breaking analyses of equilibria in the theory of non-cooperative games. The Academy cited 'Spieltheoretische Behandlung eines Oligopolmodells mit Nachfrageträgheit' (A1965a,b)[2] and 'Reexamination of the Perfectness Concept for Equilibrium Points in Extensive Games' (A1975) as Professor Selten's prize-winning achievements. In the first article he introduced the most fundamental refinement of the standard game theoretic equilibrium (Nash 1951), now known as subgame perfection. In the second article he introduced another important refinement, now known as 'trembling-hand' perfection. Subgame perfection identifies credible Nash equilibria. Trembling-hand perfection identifies those subgame perfect equilibria that are robust to small probabilities of mistakes that players may make. These two equilibrium refinements have spawned a large body of research (Van Damme 1991) and are classic examples of normative game theory.

Normative game theory builds equilibrium models of strategic interaction under full rationality, drawing whenever possible on an explicit set of axioms and assuming often common knowledge of all players' preferences and strategic possibilities. In its attempt to clarify the concept of ideal rationality, normative game theory does not bother much with actual human behaviour. Descriptive game theory, in contrast, builds models whose explicit task is to explain actual human behaviour. In the introduction to an earlier collection of his articles (Selten B1988b), Professor Selten reiterated his view – stated originally in a comment to Aumann's paper 'What is Game Theory Trying to Accomplish?' (Aumann 1985; Selten C1985) – that a sharp distinction ought to be made between normative (ideal-normative) and descriptive (prescriptive) game theory (Selten B1988b, p. vii). He has labelled this distinction 'methodological dualism'.

Professor Selten was intrigued by, and has pursued work in, both normative and descriptive game theory throughout his career (Güth and Strobel 1996, pp. 5–6). While important parts of his theoretical work, including the work prominently featured in the Nobel Prize citation, represent an elaboration of the consequences of (full or Bayesian) rationality, other parts of his theoretical work have contributed significantly to a descriptive game theory. His contributions to the latter consistently have been informed by his own early experimental work as well as H.A. Simon's work on bounded rationality (Simon 1957). These influences are evidenced in his first paper on oligopoly experiments and his aspiration-level adaptation theory of the firm, both co-written with Professor Heinz Sauermann and published in German in *Zeitschrift für die gesamte Staatswissenschaft* (Selten A1959, Selten A1962). An English version

(C1960) of the first paper is reprinted in these present volumes (Volume Two, Chapter 5); an English and updated version of the second paper has recently become available (Selten A1998c).

Professor Selten's simultaneous pursuit of normative and descriptive game theory and experiments has occasionally confounded his colleagues (Selten B1988b, p. vii). To him, there is nothing contradictory in such a research strategy. Normative and descriptive game theory and experiments are simply different paths towards a better understanding of human behaviour. Normative game theory clarifies the conditions of ideal rationality and thus defines benchmarks. Experiments are one important way to confront the predictions of normative game theory with observed behaviour. Sometimes normative game theory explains observed behaviour in experiments quite well and hence can be descriptive. Sometimes, however, it does not. It is in those situations that descriptive theories may be needed. In the computerized laboratory for experimental economics at the University of Bonn – a research unit of Sonderforschungsbereich (special research unit) 303 – Professor Selten and most of his assistants have, since 1984, worked towards establishing 'a descriptive branch of decision and game theory which takes the limited rationality of human behaviour seriously' (Selten C1994b, p. 319).

As scientific director, Professor Selten remains deeply involved with his laboratory which is funded by the Deutsche Forschungsgesellschaft (Germany's National Science Foundation). He continues to pursue work in normative and descriptive game theory and experimental economics. For recent developments, see *http://www.econ1.uni-bonn.de/labor/people/selten/index.html*.

These selected essays, compiled and categorized in close collaboration with Professor Selten, make accessible a representative selection of his articles. The articles reflect his long-standing simultaneous interest in both normative and descriptive game theory as well as in experiments; it also reflects his diverse interests in economics and other social sciences. Choices had to be made. Professor Selten's work in biology (for example, Selten A1980a, A1983c, A1984, A1988; see also Selten, Kadmon, & Shmida (Selten A1991c), Selten & Shmida (Selten C1991b), and Hammerstein & Selten (Selten C1994a)) is not represented here. Most of this work, however, is readily accessible. There is no overlap between the articles in the present two volumes and an earlier collection of his essays still in print (Selten B1988b).

Following these introductory remarks, the present volume proffers the reader an up-to-date bibliography of Professor Selten's work, and the Nobel Prize citation. Part I, Volume One also proffers a longer essay, 'In Search of a Better Understanding of Economic Behavior'. This article (Selten, C1993b) is a highly readable description of Professor Selten's work through 1992, written by himself.[3] It has 20 sections which are thematically arranged, starting with an account of early influences on his work and remarks about the beginnings of experimental economics in Germany. The thematic discussion of his work makes it clear that Professor Selten, notwithstanding the fact that he is 'easily attracted by the opportunity to shift my interests into unforeseen exciting new directions' (Ch. 1 Volume One, p. 3), revisits certain themes with great persistence and is often inspired by insights from his excursions. Parts II to IV of Volume One collect the results of some of these excursions under the headings Axiomatic Characterizations, Learning and Political and Social Interaction. Parts I to IV, in Volume Two, under the headings Theories of Oligopolistic Competition, Oligopoly Experiments, Bilateral Bargaining and Coalition Bargaining collect the results of some other excursions. Volume Two also reprints a biographical outline that Professor Selten gave before his Nobel Prize lecture.

Brief sketches of the contents of the articles in both volumes follow below. In some cases I provide additional background and references.

Part II, Volume One: Axiomatic Characterizations

What is the value of an n-person game for a player taking part in it? In 'Valuation of n-Person Games' (Ch. 2), a translation of his PhD thesis from 1961, Professor Selten addressed this question. Building on his 1957 master's thesis and a lengthy paper available only in German (Selten A1960), the paper reprinted here presents an axiomatic characterization of the value to a player of an n-person game. Its importance is three-fold. First, by building his characterization on intuitively justifiable postulates, Professor Selten made an end-run around what he already called then 'the difficult question of rational behavior' (Selten C1964, p. 577). Second, the paper provided two characterization theorems. The first stated that the Shapley value is the only value function for n-person constant-sum games to satisfy a set of postulates. The second stated that a modified Shapley value is the only value function for general n-person games to satisfy the same set of postulates plus an additional normalization postulate. (For n-person constant-sum games the two coincide.) Third, Professor Selten employed the extensive form in his analysis. At a time when little work was done on games in extensive form, this mode of tackling the problem enabled him to see the perfectness problem early on and to write the contributions which led to his Nobel Prize award.

Chapter 3 in this part deals with measures of predictive success, m. Such measures are functions whose arguments are the relative frequency of correct predictions, r, and the size of the predicted area relative to the set of all possible outcomes, a. Such measures have been useful in the analysis of experimental results. However, several functional specifications exist such as $m = r - a$ and $m = r/a$. 'Properties of a Measure of Predictive Success' makes the case for the difference measure. It is shown that this measure can be characterized up to increasing monotonic transformations by a subset of six axioms, or up to positive linear transformations by another subset of the same six axioms. Selten (C1987) employed this measure successfully to compare the predictive success of characteristic function theories for three-person bargaining games on four sets of experimental data provided by other authors. A recent example of an application of the difference measure of predictive success has been provided by Hey (1998).

The consumers' surplus is a notoriously difficult construct. In 'An Axiomatic Approach to Consumers' Welfare' (Ch. 4), Professor Selten and Eyal Winter propose a new measure of consumers' welfare that avoids the assumption of utility maximization. Instead, aggregate demand is taken as primitive. A welfare gain function is constructed that is uniquely determined by six axioms.

Harsanyi and Selten (Selten B1988a) proposed a general theory of equilibrium selection in games with multiple equilibria, drawing on the third of three approaches worked out during two decades of joint work on this eminently difficult and important issue. One of the key concepts in their book – a concept that has influenced decisively the theoretical and experimental investigation of the possible tension between the selection principles of risk dominance and pay-off dominance – is a particular definition of risk dominance. 'An Axiomatic Theory of a Risk Dominant Measure for Bipolar Games with Linear Incentives' (Ch. 5) uses a different definition of risk dominance – a variant of the first of their original approaches – and characterizes it with the help of eleven axioms. Bipolar games are normal form games with two pure strategies for each player and two strict equilibria. A normal form game has linear incentives if the pay-off difference of any pair of a player's pure strategies depends linearly on the probabilities in the mixed strategies of the other players. The axiomatic theory of risk dominance proposed here is a generalization of earlier results, but for a smaller class of games.

Part III, Volume One: Learning
'Evolution, Learning, and Economic Behavior' (Ch. 6) is the text of a lecture that Professor Selten delivered at the J.L. Kellogg Graduate School of Management, Northwestern University, Evanston, Illinois. It has quickly become a classic both for content and format. The lecture is structured as a fictitious multi-logue between imaginary (idealized) discussants such as a 'Bayesian', an 'economist', an 'adaptationist' (that is, a biologist who strongly believes in fitness maximization), a 'population geneticist', a 'naturalist' and an 'experimentalist'. This motley group of discussants, moderated by a chairman who unmistakably shares many of the views of Professor Selten (as do the 'naturalist' and the 'experimentalist'), discuss the state of our knowledge about human behaviour. The 'Bayesian' is attacked for his stubborn reliance on optimization models. At the end the chairman summarizes the discussion with a plea for quantitative theories of bounded rationality which are supported by experimental evidence, that is, the descriptive decision and game theory that has been one of Professor Selten's central pursuits ever since he began doing experiments.

Evolutionary or learning models, which have become increasingly popular in game theory over the past few years, are typically backward looking. However, while players lack the full rationality required by standard game theoretic models, they do typically employ finite depths of reasoning (see, for example, Nagel 1995). In 'Anticipatory Learning in Two-Person Games' (Ch. 7) Professor Selten injects an element of anticipation into a learning process for two-person games in normal form and then investigates the stability properties of the thus defined process for various classes of games. A stability criterion is derived. It is shown that this criterion survives anticipation of completely mixed equilibrium points.

Sharp equilibrium predictions can be derived for finite prisoner's dilemma games under complete information. However, experimental results have consistently questioned the validity of these predictions. In response to these experimental results, Kreps and his colleagues (1982) injected incomplete information into finite prisoner's dilemma games to rationalize the observed cooperation. In 'End Behavior in Sequences of Finite Prisoner's Dilemma Games' (Ch. 8) Professor Selten and Rolf Stoecker showed that participants in 25 supergames of 10-period prisoner's dilemma games cooperate until shortly before the end. However, contradicting Kreps et al. (1982), the predicted behaviour of cooperation is not uniform across supergames. Rather, behaviour in their experiment tends to be chaotic in the beginning, is informed by learning thereafter, and increasingly shows final rounds effects.

Selten and Stoecker explain participants' behaviour by way of a simple qualitative learning direction theory. This theory suggests that in each period $t = 1, \ldots, T$ a parameter p_t informing a decision be evaluated *ex post*. Learning direction theory predicts that subjects tend to choose a higher (lower) parameter in period $t + 1$ if this had been more successful in period t. The theory predicts, for example, that in sealed bid first price auctions bids will increase (decrease) if bids in the previous round were not successful (successful). Variants of this theory have been successfully applied in a variety of contexts. See, for example, Mitzkewitz and Nagel 1993, Nagel 1995, Ryll 1995, Cason and Friedman 1997, Selten A1997. Selten (C1996b) presents an eminently readable outline of learning direction theory and a summary of some papers that apply it to explain experimental results. Unfortunately, it is in German and Polish only. A similar outline is currently not available in English.

Learning direction theory is also used in the recent 'Experimental Sealed Bid First Price Auctions with Directly Observed Bid Functions' (Ch. 9). Professor Selten and Joachim Buchta report the results of repeated interaction of groups of three subjects in sealed bid first price auctions. Subjects had to specify piecewise linear bid functions. The results of the experiment are statistically significant and confirm the predictions of learning direction theory about the direction of change in bid functions.

Part IV, Volume One: Political and Social Interaction

'The Scenario Bundle Method' (Ch. 10) paper grew out of a collaboration with political scientist Amos Perlmutter on the occasion of a conference on strategic decision analysis. It was an attempt to construct simple game models of international conflicts, namely those in the Persian Gulf at the beginning of October 1976. The scenario bundle method is a systematic way of answering the very questions that a game theorist has to answer in different contexts: 'Who are the players? What are the motivation factors which determine the players' preferences? What are the strategic possibilities of the players? What are the consequences of various combinations of strategic choices? What are the players' preferences over these consequences?'. These questions are answered for the relevant initial options and the tree of reactive options that each of them generates. (An initial option and the tree of reactive options it generates is called a scenario bundle.) The determination of initial and reactive options is preferably based on expert judgements. Those judgements, by necessity, are about possible future developments. Decision makers are assumed to act purposefully. Backward induction identifies likely and unlikely initial options. The major difference to standard game theoretic modelling and analysis is that scenario bundle analysis does not employ numerical parameters. In a sense, scenario bundle analysis is qualitative game theory that employs the extensive form to clarify the strategy space of the participants in international conflicts and relies on preferences over outcomes.

The paper was never formally published, mainly because the experts invited to the conference misjudged the likelihood of an Iranian revolution; it happened shortly after the conference, making the paper seemingly obsolete. As pointed out by Barry O'Neill in his contribution to the *Handbook of Game Theory*,

> Their first scenario in 1976 had Iraq tempted to invade Kuwait but deterred by the prospect of an Iran/Saudi/U.S. coalition and a Soviet withdrawal of support. In 1990 Iraq really invaded, and the common wisdom was that Saddam went ahead *in spite*

of the Soviet Union's retirement as his patron. The model suggests the opposite possibility, that Soviet absence may have given Iraq less to lose. (O'Neill 1994, p. 999)

Whatever the merits of this suggestion, the approach to the modelling of international relations proposed in 'The Scenario Bundle Method' warrants its belated publication here. (For more details about the history of this paper, see section 11 of Ch. 1, Volume One.)

'Balance of Power in a Parlor Game' (Ch. 11) presents a theoretical solution for a simplified version of a parlour game in which alliances and counteralliances can be built and provinces can be won and lost in war. The theoretical model is an infinite game with perfect information, with behavioural strategies that depend on the current province distribution and position. Eleven types of province distributions are identified of which five are stable. (For more details about the history of this paper, see section 11 of Ch. 1, Volume One.)

'The Distribution of Foreign Language Skills as a Game Equilibrium' (Ch. 12), written jointly with Jonathan Pool, is one of Professor Selten's seven contributions to the four-volume *Game Equilibrium Models*. The paper reflects Professor Selten's interest in the Esperanto movement, as it explores the distribution of native and auxiliary languages, that is, languages like Esperanto that have no native speakers. The distribution of languages is modelled as the outcome of a non-cooperative normal form game whose pay-offs are the net benefits players derive from the 'communicative benefits' and 'learning costs' of the different languages. The model in the article reprinted here is more general than the model in Selten (B1995) which – though published later – was written earlier. The earlier version discussed a symmetric two-country case and explored in detail equilibrium points and comparative statics. Unfortunately, the earlier version is at this point available in Esperanto and German only.

Volume Two starts with the autobiographical outline that Professor Selten gave before his Nobel Prize lecture. In these sometimes very personal remarks, Professor Selten traces his intellectual independence back to the particular circumstances of his childhood in Breslau, now Wroclaw (Poland): as 'a half-Jewish boy under the Hitler regime . . . , I had to learn to trust my own judgement rather than official propaganda or public opinion' (C1994b, p. 313). He also acknowledges the important influence that mathematician Ewald Burger and economist Heinz Sauermann had on his intellectual development and enumerates his key collaborators and students over time.

Part I, Volume Two: Theories of Oligopolistic Competition
'Multistage Game Models and Delay Supergames' (Ch. 1) is Professor Selten's Nobel lecture of 9 December 1994. Kreps and Scheinkman (1983) have shown that capacity-constrained Bertrand competition can lead to Cournot outcomes. Their model employs a two-stage game where (long-term) capacities are selected in the first period and (short-term) prices in the second. Models of this kind can be solved by subgame perfection (Selten A1965a). Professor Selten asks, 'What does it mean in the model of Kreps and Scheinkman that capacity decisions are long term and price decisions are short term?' (p. 322). He proposes that it may be the differential delay for capacity adjustments on the one hand, and price adjustments on the other, to become effective.

Accordingly, a delay supergame models decisions on strategic variables as being made at the same time, say t, but becoming effective at different times, say $t + 1$ and $t + 10$. Players have full information about the history of play up to, but not including t. A delay supergame is bounded if the number of periods has an upper bound.

Professor Selten shows that every subgame perfect equilibrium of a bounded multistage game always induces a subgame perfect equilibrium for every one of its bounded or unbounded delay supergames. Calling a multistage game or a delay supergame determinate if the set of all its subgame perfect equilibria yields the same pay-off in the game and each of its subgames, he also shows that every bounded delay supergame of a determinate bounded multistage game is itself determinate. In a concluding discussion, he relates this exercise in normative game theory to his earlier experimental work – for example, his work with Stoecker on finite prisoner's dilemma games (Selten A1986). Stressing that observed behaviour tends to be boundedly rational, but acknowledging that under certain circumstances experience with a game can teach a lesson or two, he submits the following intriguing hypothesis:

Learning in a modified delay supergame with reduced payoff information converges to behavior in agreement with the predictions derived from a subgame perfect equilibrium of the underlying multistage game, the more so, the less the players know about other players' payoffs. (p. 347)

'Restabilizing Responses, Inertia Supergames and Oligopolistic Equilibria' (Ch. 2) and 'Oligopolistic Economies as Games of Limited Information' (Ch. 3), jointly written with Thomas Marschak, propose a model of oligopolistic general equilibrium. It 'took its point of departure from the German tradition of conjectural oligopoly theory. . . . Our theory describes oligopolisitic competition by a specific convolution which combines aspects of kinked demand curve theory and limit pricing'. (For more details see section 12 of Ch. 1, Volume One.)

'Elementary Theory of Slack-Ridden Imperfect Competition' (Ch. 4) builds on the well-documented existence of organizational slack. In fact, the point of departure for the behavioural theory of the firm developed here is the assumption that the separation of ownership and management allows slack to grow as long as profits are non-negative. Organizational slack is assumed to be so inescapable that it is considered a component of marginal cost. However, it is acknowledged that slack has both negative aspects ('inefficiency') and positive aspects ('consumption at the work place'). A symmetric, linear Cournot model is used to trace out the disciplining force of free and efficient entry, the role of fixed costs and the composition of organizational slack, and the welfare implications of different scenarios.

Part II, Volume Two: Oligopoly Experiments

'An Experiment in Oligopoly' (Ch. 5) is the English translation of an article originally published jointly with Professor Sauermann in 1959 (Selten A1959). The German article was Professor Selten's first publication. The experiments are based on a modified Cournot model that features three firms which produce a homogeneous good but have different cost functions and capacity limits. Cost functions and capacity limits cannot be adjusted. It is important to recall that in the 1950s, experiments were truly innovative enterprises. To wit, 'At this time experiments in economics were almost unheard of. It was a standard assertion that experiments are impossible in economics' (Ch. 1, Volume One, p. 5). The paper gives a good impression of the daunting logistics, and very different conventions (the experiments lasted about four hours!) of that pioneer phase of laboratory experimentation.

One of the interesting features of the 13 sessions reported here was that each firm had to document in writing the reasons for the decisions made. On the basis of these minutes, Sauermann and Selten classified eleven essential 'motive groups' such as 'good by experience' (by a far margin the most frequent) and 'optimization by table sheet' (the second most frequent), with the former being clearly an adaptive strategy and the second reflecting best replies to predicted levels of competitors' output. We see thus already in the interpretation of these data, a concern for the failure of normative theory to describe observed behaviour in the laboratory. We also see, in one of the 'four postulates which should be satisfied by an explicative theory of our oligopoly model', a concern that learning effects be incorporated in such a descriptive theory.

'Duopoly Strategies Programmed By Experienced Players' (Ch. 6), with Michael Mitzkewitz and Gerald R. Uhlich, is one of Professor Selten's more recent publications (Selten A1997); it had, however, a long gestation period. (The first version circulated as a discussion paper in 1988.) The authors investigate experimentally the finite supergame of an asymmetric Cournot duopoly. Their goal is to understand how cooperation can evolve in such a scenario. To that end the authors employ the strategy method which was first proposed in Selten (C1967e) and in a sense is a refinement of the 'motive groups' approach discussed above. (For a discussion of the strategy method see section 19 of Ch. 1, Volume One.) The strategy method requires participants, after they have gained experience with the game by playing it, to write down 'strategy programs'. These strategy programs are then 'played' against each other, evaluated and revised, and so on.

Specifically, the 24 participants of a seminar played three 20-period supergames of the asymmetric Cournot duopoly game, with one week between each. After these three rounds of 'play', or training, participants had to write a first strategy program which then was run in a tournament-like setting against the other participants' programs. This procedure was repeated three times. Among the many interesting results of this paper are: participants do not implement the subgame perfect Cournot solution for finitely repeated games under complete information; strategies change over both the three training rounds and tournament rounds, with first round outcomes being scattered and often not even resulting in Cournot pay-offs, the second outcomes being less scattered and moving on average away from Cournot pay-offs towards pay-offs on the Pareto efficient frontier, a trend that is also enhanced in the third round outcomes; and strategies in initial, main, and end phase periods differ significantly.

The authors extract 13 characteristics from the structure of the programmed strategies that were used in the final tournament round. Contrary to what many oligopoly theories assume, one of the characteristics suggests that few participants tried to optimize against the predicted behaviour of opponents. Rather, decisions were guided by 'ideal points' – that is, decisions were made that reacted to or signalled moves towards Pareto efficient outcomes that were informed by equity considerations, as well as prominence and prior experience. Professor Selten and his colleagues also construct a measure of typicity and show that there is a strongly significant positive correlation between the typicity and the success of final strategies. Thus, strategies with many typical characteristics are indeed smart strategies even if they do not engage in complex computations that try to predict opponents' behaviour. Similar to the results reported in Selten and Stoecker (Selten A1986), there are final rounds effects here too. While the particular circumstances of this experiment were unconventional (seminar participants, no monetary pay-offs, strategy method, one week intervals between each supergame, verbal communication, and so on), the results of this paper provide food for thought about the value of established normative theories, the need for descriptive theories, and the important role of experiments in all this.

Part III, Volume Two: Bilateral Bargaining
'Game Theoretical Analysis of Wage Bargaining in a Simple Business Cycle Model', with Werner Güth, conceptualizes wage bargaining as a repeated bilateral bargaining game between a union and an employers' association. Both bargainers are assumed to be rational and to have complete knowledge. While they are assumed to pursue long-run goals, they can commit themselves for one period only. Bargaining results of the current period affect the functional distribution, the level of present and future national income, and future bargaining positions. Thus, players face a sequence of interdependent, yet separate bargaining games. The solution for this dynamic bargaining game draws on Nash's bargaining theory and the idea of subgame consistency proposed in Selten (C1979c).

In 'Equilibrium Point Selection in a Bargaining Game with Opportunity Costs' (Ch. 8) Professor Selten and Ulrike Leopold analyse a simple two-person bargaining problem with two agreements. One of the agreements favours player 1, the other player 2. Bargaining, whether successful or not, comes with opportunity costs that can be avoided by not bargaining at all. The article analyses the selection of one of the three resultant pure strategy equilibria. The selected equilibrium is a function of the given parameter constellation. The solution employs the risk dominance concept in Harsanyi and Selten (Selten B1988a).

Harsanyi and Selten's equilibrium selection theory (Selten B1988a) is also employed in 'Original or Fake – A Bargaining Game with Incomplete Information' (Ch. 9), with Werner Güth. The potential buyer does not know whether the artwork is an original or a fake; it is common knowledge that the seller knows. Selten and Güth show that this bilateral bargaining game has three types of solutions which emerge as a function of the probability of the fake and the buyer's value of the original.

'Bargaining Experiments with Incomplete Information' (Ch. 10) is the result of a collaboration with Austin Hoggatt and three of his research assistants. The experiments were based on a semi-numerical example that Professor Selten, together with Harsanyi, had worked out as an illustration of a theory of two-person bargaining under incomplete information (Selten A1972). The experiments themselves are interesting for the simple reason that they were computerized. They also produced some interesting results. For example, the authors report significant learning effects over the five rounds of bargaining that were not addressed by the theory. The partial lack of agreement between theory and experimental data led the authors to construct 'robots', that is, computer programs based on observations of the participants' modal behaviour. The paper reports results of robot interaction. The paper is a wonderful early illustration of Professor Selten's major themes and frequent mode of attack: normative theory is employed as a benchmark that informs the design and implementation of experiments whose results (casting doubt on utility theory) then motivate the search for more realistic (descriptive) theories. (For more details see section 10 of Ch. 1, Volume One.)

Part IV, Volume Two: Coalition Bargaining
Three-person games in characteristic function form are simple environments in which coalition formation has been studied, both theoretically and experimentally. A characteristic function game is defined by a set of players, a set of permissible coalitions, and a functional value assigned to each of the permissible coalitions. This functional value allows an evaluation of different options and the gains to be had from each. Equity considerations reflect a splitting of the pie that is tied to some standard of comparison. For example, as regards a quota cartel, a standard of comparison may be defined by firms' relative capacities or their sales in the past. The equity principle requires that the pie is split relative to the weights defined by the relevant standard of comparison. In 'Equity and Coalition Bargaining in Experimental Three-Person Games' (Ch. 11) Professor Selten uses equity considerations to propose a new descriptive theory – a revised version of his theory of equal division pay-off bounds (Selten C1983b) – of coalition bargaining in three-person characteristic function games. The important starting point of this undertaking is his belief that participants in experiments rely on easily accessible cues such as obvious ordinal power comparisons and equitable shares to determine lower pay-off bounds, that is, they are boundedly rational. Comparing four sets of data from experiments conducted by other experimenters, he finds the explanatory power of his proposed descriptive theory to be superior to two well-known competitors. (The history of the 'equal division pay-off bounds' theory and an important extension by one of Professor Selten's students, Gerald Uhlich, are discussed in section 6 of Ch. 1, Volume One.)

'A Demand Commitment Model of Coalition Bargaining' (Ch. 12) presents a finite model of characteristic function bargaining which yields quota agreements in the case of three-person quota games without the grand coalition as the only subgame perfect outcomes. In this 'demand commitment model' players may temporarily commit themselves to pay-off demands or they may form coalitions with players who are already committed. (For details see section 16 of Ch. 1, Volume One.)

'Demand Commitment Bargaining in Three-Person Quota Game Experiments' (Ch. 13), with Bettina Kuon, takes an earlier and simpler version of the finite model of characteristic function bargaining, which allows for other equilibria, to the laboratory. The experimental results suggest that the laboratory procedure implicit in the demand commitment model are favourable to quota agreements. Experience with the same game contributes significantly to quota agreements. The difference measure of predictive success suggests that quota agreements do not perform better than equal division pay-off bounds.

Notes
1. I am very grateful for the numerous comments that Professor Selten provided on my earlier drafts of this introduction. The usual caveat applies.

2. For references to Professor Selten's publications see the Bibliography following this introduction: A denotes journal articles, B and C denote books and chapters in books, and D discussion and working papers, respectively. All other references can be found at the end of this introduction.
3. See also the excellent article by Van Damme and Weibull (1995).

References

Aumann, R.J. (1985), 'What is Game Theory Trying to Accomplish?', in *Frontiers of Economics*, K.J. Arrow and S. Honkapohja (eds), pp. 28–76, Oxford: Basil Blackwell.

Cason, T. and D. Friedman (1997), 'Price Formation in Single Call Markets', *Econometrica*, **65** (2): 311–45.

Güth, W. and M. Strobel (1996), 'Interview with Elisabeth and Reinhard Selten', in *Understanding Strategic Interaction: Essays in Honor of Reinhard Selten*, W. Albers, W. Güth, P. Hammerstein, B. Moldovanu, and E. van Damme (eds), pp. 1–7, Berlin–Heidelberg–New York: Springer-Verlag.

Hey, J. (1998), 'An Application of Selten's Measure of Predictive Success', *Mathematical Social Sciences*, **35**: 1–15.

Kreps D.M. and J.A. Scheinkman (1983), 'Quantity Precommitment and Bertrand Competition Yield Cournot Outcomes', *Bell Journal of Economics*, **14**: 326–37.

Kreps D., P. Milgrom, J. Roberts and R. Wilson (1982), 'Rational Cooperation in the Finitely Repeated Prisoner's Dilemma', *Journal of Economic Theory*, **27**: 245–52.

Mitzkewitz, M. and R. Nagel (1993), 'Envy, Greed, and Anticipation in Ultimatum Games with Incomplete Information', *International Journal of Game Theory*, **22**: 171–98.

Nagel, R. (1995), 'Unravelling in Guessing Games: An Experimental Study', *American Economic Review*, **85** (5): 1013–26.

Nash, J, (1951), 'Non-Cooperative Games', *Annals of Mathematics*, **54**: 286–95.

O'Neill, B. (1994), 'Game Theory Models of Peace and War', in *Handbook of Game Theory with Economic Applications, Volume II*, R.J. Aumann and S. Hart (eds), pp. 995–1053, Amsterdam: Elsevier.

Ryll, W. (1995), *Litigation and Settlement in a Game with Incomplete Information: An Experimental Study.* Lecture Notes in Economics and Mathematical Systems **440**, Berlin–Heidelberg–New York: Springer-Verlag.

Simon, H.A. (1957), *Models of Man*, New York: Wiley.

Van Damme, E. and J. Weibull (1995), 'Equilibrium in Strategic Interaction: The Contributions of John C. Harsanyi, John F. Nash and Reinhard Selten', *Scandinavian Journal of Economics*, **97** (1), 15–40.

Von Neumann, J. and O. Morgenstern (1944), *Theory of Games and Economic Behavior*, Princeton, NJ: Princeton University Press.

Bibliography of Reinhard Selten's writings

Books (B)

(1970), *Preispolitik der Mehrproduktenunternehmung in der statischen Theorie*, Berlin–Heidelberg–New York: Springer-Verlag.

(1974), *General Equilibrium with Price-Making Firms* (with Thomas Marschak), Lecture Notes in Economics and Mathematical Systems, Berlin–Heidelberg–New York: Springer-Verlag.

(1988a), *A General Theory of Equilibrium Selection in Games* (with John C. Harsanyi), Cambridge, MA: MIT Press.

(1988b), *Models of Strategic Rationality, Theory and Decision Library, Series C: Game Theory, Mathematical Programming and Operations Research*, Dordrecht–Boston–London: Kluwer Academic Publishers.

(1995), *Enkonduko en la Teorion de Lingvaj Ludoj – Cu mi lernu Esperanton* (with Jonathan Pool), Berlin-Paderborn: Akademia Libroservo, Institut für Kybernetik.

Journal articles (A)

(1959), 'Ein Oligopolexperiment' (with Heinz Sauermann), *Zeitschrift für die gesamte Staatswissenschaft*, **115**: 427–71; reprinted in Heinz Sauermann (ed.), *Beiträge zur experimentellen Wirtschaftsforschung*, Tübingen: J.C.B. Mohr (Paul Siebeck), 1967, pp. 9–59.

(1960), 'Bewertung strategischer Spiele', *Zeitschrift für die gesamte Staatswissenschaft*, **116**: 221–81.

(1962) 'Anspruchsanpassungstheorie der Unternehmung' (with Heinz Sauermann), *Zeitschrift für die gesamte Staatswissenschaft*, **118**: 577–97.

(1963), 'Dynamische Theorie der Built-in Flexibility' (with Rudolf Richter), *Zeitschrift für die gesamte Staatswissenschaft*, **119**: 555–78.

(1965a), 'Spieltheoretische Behandlung eines Oligopolmodells mit Nachfrageträgheit – Teil I: Bestimmung des dynamischen Preisgleichgewichts, *Zeitschrift für die gesamte Staatswissenschaft*, **121**: 301–24.

(1965b), 'Spieltheoretische Behandlung eines Oligopolmodells mit Nachfrageträgheit – Teil II: Eigenschaften des dynamischen Preisgleichgewichts, *Zeitschrift für die gesamte Staatswissenschaft*, **121**: 667–89.

(1972), 'A Generalized Nash Solution for Two-Person Bargaining Games with Incomplete Information' (with John C. Harsanyi), *Management Science*, **18** (5) Part 2 (January): 80–106.

(1973), 'A Simple Model of Imperfect Competition where 4 are Few and 6 are Many', *International Journal of Game Theory*, **2** (3): 141–201.

(1975), Reexamination of the Perfectness Concept for Equilibrium Points in Extensive Games', *International Journal of Game Theory*, **4** (1): 25–55; reprinted in H.W. Kuhn (ed.), *Classics in Game Theory*, Princeton: Princeton University Press, 1997, pp. 317–54.

(1977), 'Oligopolistic Economies as Games of Limited Information' (with Thomas Marschak), *Zeitschrift für die gesamte Staatswissenschaft*, **133** (October): 385–410.
(1978a), 'Restabilizing Responses, Inertia Supergames and Oligopolistic Equilibria' (with Thomas Marschak), *Quarterly Journal of Economics*, **92** (1) (February): 71–93.
(1978b), 'The Chain Store Paradox', *Theory and Decision*, **9** (2) (April): 127–59.
(1980a), 'A Note on Evolutionarily Stable Strategies in Asymmetric Animal Conflicts', *Journal of Theoretical Biology*, **83**, 93–101.
(1980b), 'Zum Selbstverständnis der experimentellen Wirtschaftsforschung im Umkreis von Heinz Sauermann' (with Reinhard Tietz), *Zeitschrift für die gesamte Staatswissenschaft*, **136** (1) (March), 12–27.
(1980c), 'Was ist eigentlich aus der Spieltheorie geworden?' *IHS–Journal*, **4**: 147–61.
(1982), 'Game Theoretical Analysis of Wage Bargaining in a Simple Business Cycle Model' (with Werner Güth), *Journal of Mathematical Economics*, **10** (2–3) (September): 177–95.
(1983a), 'Equilibrium Point Selection in a Bargaining Situation with Opportunity Costs' (with Ulrike Leopold), *Economie Appliquée*, **36** (4): 611–48.
(1983b), 'A Model of Oligopolistic Size, Structure, and Profitability', *European Economic Review*, **22** (June): 33–57.
(1983c), 'Evolutionary Stability in Extensive Two-Person Games', *Mathematical Social Sciences*, **5** (3): 269–363.
(1984), 'Gaps in Harley's Argument on Evolutionarily Stable Learning Rules and in the Logic of "Tit for Tat"' (with Peter Hammerstein), *Behavioral and Brain Sciences*, **7** (1): 115–16.
(1986), 'End Behavior in Sequences of Finite Prisoner's Dilemma Supergames' (with Rolf Stoecker), *Journal of Economic Behavior and Organization*, **7** (1): 47–70.
(1988), 'Evolutionary Stability in Extensive Two-Person Games, Correction and Further Development', *Mathematical Social Sciences*, **16** (3) (December): 223–66.
(1990a), 'Bounded Rationality', *Journal of Institutional and Theoretical Economics*, **146** (4) (December): 649–58.
(1990), 'Alternating Bid Bargaining with a Smallest Money Unit' (with Eric van Damme and Eyal Winter), *Games and Economic Behavior*, **2**: 188–201.
(1991a), 'Evolution, Learning, and Economic Behavior', 1989 Nancy Schwartz Memorial Lecture, *Games and Economic Behavior*, **3** (1) (February): 3–24.
(1991b), 'Properties of a Measure of Predictive Success', *Mathematical Social Sciences*, **21** (2) (April): 153–67.
(1991c), 'Within-Plant Foraging Behavior of Bees and Its Relationship to Nectar Distribution in Anchusa Strigosa', (with Ronen Kadmon and Avi Shmida), *Israel Journal of Botany*, **40**: 283–94.
(1993), 'Demand Commitment Bargaining in Three-Person Quota Game Experiments' (with Bettina Kuon), *International Journal of Game Theory*, **22**: 261–77.
(1994a), 'An Axiomatic Approach to Consumers' Welfare' (with Eyal Winter), *Mathematical Social Sciences*, **27** (1) (February): 19–30.
(1994b), 'New Challenges to the Rationality Assumption: Comment', *Journal of Institutional and Theoretical Economics*, **150** (1) (March): 42–4.
(1995), 'An Axiomatic Theory of a Risk Dominance Measure for Bipolar Games with Linear Incentives', *Games and Economic Behavior*, **8** (1) (January): 213–63.

(1996), 'The Work of John Nash in Game Theory' (with Peter Hammerstein), *Journal of Economic Theory*, **69**: 161–65.
(1997), 'Duopoly Strategies Programmed by Experienced Players' (with Michael Mitzkewitz and Gerald R. Uhlich), *Econometrica*, **65** (3) (May): 517–55.
(1998a), 'An Experimental Solidarity Game' (with Axel Ockenfels), *Journal of Economic Behavior and Organization*, **34** (4): 517–40.
(1998b), 'Multistage Game Models and Delay Supergames', *Theory and Decision*, **44** (1) (January): 1–36. Previously published in: Tore Frängsmyr (ed.), *Les Prix Nobel 1994*, 320–48.
(1998c), 'Aspiration Adaptation Theory', *Journal of Mathematical Psychology*, **42**: 191–214.
(1998d), 'Axiomatic Characterization of the Quadratic Scoring Rule', *Experimental Economics*, **1** (1), 43–62.
(1998c), 'Features of Experimentally Observed Bounded Rationality', *European Economic Review (Papers and Proceedings)*, **42** (3–5): 413–36.

Chapters in books (C)
(1960), 'An Experiment in Oligopoly' (with Heinz Sauermann), 85–114, in *General Systems, Yearbook of the Society for General Systems Research*, **5**, Ann Arbor, MI: Society for General Systems (translation of 1959 journal article with Heinz Sauermann).
(1964), 'Valuation of n-Person Games', 555–78, in *Advances in Game Theory, Annals of Mathematics Studies*, **52**, Princeton, NJ: Princeton University Press.
(1967a), 'Der Rangsummentest – Beschreibung und Signifikanztafeln' (with Reinhard Tietz), 353–75, in *Operations Research Verfahren III*, Rudolf Henn (ed.), Meisenheim: Verlag Anton Hain.
(1967b), (i) 'Zur Entwicklung der experimentellen Wirtschaftsforschung' (with Heinz Sauermann), 1–8; (1967c), (ii) 'Investitionsverhalten im Oligopolexperiment', 60–102; (1967d), (iii) 'Ein Oligopolexperiment mit Preisvariation und Investition', 103–35; and (1967e), (iv) 'Die Strategiemethode zur Erforschung des eingeschränkten rationalen Verhaltens im Rahmen eines Oligopolexperiments', 136–68; all in *Beiträge zur experimentellen Wirtschaftsforschung*, Heinz Sauermann (ed.), Tübingen: J.C.B. Mohr (Paul Siebeck).
(1968), 'Psychological Variables and Coalition Forming Behavior' (with Klaus G. Schuster), 221–40, in *Risk and Uncertainty*, Proceedings of a Conference held by the International Economic Association, Karl Borch and Jan Mossin (eds), London–Melbourne–Toronto–New York: Macmillan/St Martin's Press.
(1970a), (i) 'Ein Marketexperiment', 33-98; (1970b), (ii) 'Psychologische Faktoren bei Koalitionsverhandlungen' (with Klaus G. Schuster), 99–135; (1970c), (iii) 'Experiences with the Management Game SINTO-Market (with Otwin Becker)', 136–50; (1970d), (iv) 'Drei experimentelle Oligopolspielserien mit kontinuierlichem Zeitablauf' (with Claus C. Berg), 162–221; (1970e), (v) 'Ein Gerät zur optischen und akustischen Anzeige von Entscheidungszeitpunkten in Oligopolexperimenten mit kontinuierlicher Zeit' (with Claus C. Berg), 222–9; all in *Beiträge zur experimentellen Wirtschaftsforschung*, **2**, Heinz Sauermann (ed.), Tübingen: J.C.B. Mohr (Paul Siebeck).

(1971), 'Anwendungen der Spieltheorie auf die politische Wissenschaft', 287–320, in *Politik und Wissenschaft*, Hans Maier, Klaus Ritter und Ulrich Matz (eds), München: C.H. Beck.

(1972a), '(i) 'Security Equilibria' (with Reinhard Tietz), 103–22; (1972b), (ii) 'A Formal Theory of Security Equilibria' (with Reinhard Tietz), 185–202 in *The Future of the International Strategic System*, Richard Rosecrance (ed.), San Francisco–Scranton–London–Toronto: Chandler Publishing Company.

(1972c), 'Equal Share Analysis of Characteristic Function Experiments', 130–65, in *Beiträge zur experimentellen Wirtschaftsforschung – Contributions to Experimental Economics*, **3**, Heinz Sauermann (ed.), Tübingen: J.C.B. Mohr (Paul Siebeck).

(1975), 'Bargaining under Incomplete Information – A Numerical Example', 203–32, in *Dynamische Wirtschaftsanalyse*, Otwin Becker and Rudolf Richter (eds), Tübingen: J.C.B. Mohr (Paul Siebeck).

(1976), 'Wirtschaftswissenschaften und Mathematik', in *Mathematisierung der Einzelwissenschaften*, B. Boos and K. Krickeberg (eds), *Interdisciplinary Systems Research* **24**, Basel: Birkhäuser Verlag.

(1977), 'A Simple Game Model of Kidnapping', 139–56, in *Mathematical Economics and Game Theory*, R. Henn and O. Möschlin (eds), *Lecture Notes in Economics and Mathematical Systems* **141**, Berlin–Heidelberg–New York: Springer-Verlag.

(1978a), 'Bargaining Experiments with Incomplete Information' (with A. Hoggatt, D. Crocket, S. Gill, J. Moore), 127–78, in *Contributions to Experimental Economics*, **7**, Heinz Sauermann (ed.), Tübingen: J.C.B. Mohr (Paul Siebeck).

(1978b), 'The Equity Principle in Economic Behavior', 289–301, in *Decision Theory and Social Ethics: Issues in Social Choice*, H.W. Gottinger and W. Leinfellner (eds), Dordrecht: D. Reidel Publishing Company.

(1978c), 'Macht Einigkeit stark? Spieltheoretische Analysen einer Verhandlungssituation' (with Werner Güth), 197–217, *Schriften des Vereins für Socialpolitik*, Band 98, Neuere Entwicklungen in den Wirtschaftswissenschaften.

(1979a), 'Experimentelle Wirschaftsforschung', 41–61, in *Rheinisch-Westfälische Akademie der Wissenschaften*, Vorträge N 287, Opladen: Westdeutcher Verlag.

(1979b), 'Limited Rationality and Structural Uncertainty', 476–83, in *Wittgenstein, The Vienna Circle and Critical Rationalism*, Proceedings of the 3rd International Wittgenstein Symposium, August 1978, Kirchberg am Wechsel, Austria.

(1979c), 'Coalition Probabilities in a Non-Cooperative Model of Three-Person Quota Game Bargaining', 90–106, in *Mathematical Systems in Economics – Entscheidungen in kleinen Gruppen*, Wulf Albers, Günter Bamberg and Reinhard Selten (eds), Meisenheim: Verlag Anton Hain.

(1979d), 'Oligopoltheorie', 293–301, in *Handwörterbuch der Mathematischen Wirtschaftswissenschaften*, Wiesbaden: Gabler-Verlag.

(1980a), 'Auswahl eines Gleichgewichtspunktes in einem einfachen Verhandlungsproblem mit Opportunitätskosten' (with Ulrike Leopold), in *Methods of Operations Research*, **38**, Rudolf Henn et al. (eds), Königstein/Taunus: Verlagsgruppe Athenaeum/Hain/Scriptor/Hanstein.

(1980b), 'Oligopoltheorie', 667–78, in *Handwörterbuch der Wirtschaftswissenschaften*, in W. Albers et al. (eds), Stuttgart-New York: Gustav Fischer.

(1981), 'A Noncooperative Model of Characteristic Function Bargaining', 131–51, in *Essays in Game Theory and Mathematical Economics in Honor of Oskar Morgenstern*; in *Gesellschaft, Recht, Wirtschaft*, Band A, V. Bohm and H. Nachtkamp (eds), Mannheim–Wien–Zürich: Wissenschaftsverlag Bibliographisches Institut.

(1982a), 'Equilibrium Point Selection in a Class of Market Entry Games', 101–16, in *Games, Economics, and Time Series Analysis* (with Werner Güth), Wien–Würzburg: Physica-Verlag.

(1982b), 'Subjunctive Conditionals in Decision and Game Theory' (with Ulrike Leopold), 191–200, in *Philosophy of Economics*, W. Stegmueller, W. Balzer and W. Spohn (eds), Berlin–Heidelberg–New York: Springer-Verlag.

(1982c), 'Einführung in die Theorie der Spiele mit unvollständiger Information', 81–147, in *Schriften des Vereins für Socialpolitik*, Band 126, Information in der Wirtschaft.

(1983a), (i) 'Comparison of Two Theories for Characteristic Function Experiments' (with Wilhelm Krischker), 259–64; (1983b), (ii) 'Equal Division Payoff Bounds for Three-Person Characteristic Function Experiments', 265–75; both in *Aspiration Levels in Bargaining and Economic Decision Making*, R. Tietz (ed.), *Lecture Notes in Economics and Mathematical Systems* **213**, Berlin–Heidelberg–New York: Springer-Verlag.

(1983c), 'Towards a Theory of Limited Rationality', 409–12, in *Decision Making under Uncertainty*, R. Scholz (ed.), Amsterdam: North-Holland.

(1984a), 'Are Cartel Laws Bad for Business?', 86–117, in *Operations Research and Economic Theory*, H. Hauptmann, W. Krelle and K.C. Mosler (eds), Berlin–Heidelberg–New York: Springer-Verlag.

(1984b) 'Formale Konzepte eingeschränkt rationalen Verhaltens' (with Angela Klopstech), 11–34, in *Normengeleitetes Verhalten in den Sozialwissenschaften*, H. Todt (ed.), Berlin: Duncker & Humblot.

(1985), 'Comment on R.J. Aumann: What is Game Theory Trying to Accomplish?', 77–87, in *Frontiers of Economics*, K.J. Arrow and S. Honkapohja (eds), Oxford–New York: Basil Blackwell.

(1986a), 'Elementary Theory of Slack-Ridden Imperfect Competition', 126–44, in *New Developments in the Analysis of Market Structure*, J. Stiglitz and G. Mathewson (eds), Cambridge, MA: MIT Press.

(1986b), 'Institutional Utilitarianism', 251–63, in *Guidance, Control, and Evaluation in the Public Sector*, F. Kaufmann, G. Majone, V. Ostrom with assistance of W. Wirth (eds), Berlin–New York: Walter de Gruyter-Verlag.

(1987), 'Equity and Coalition Bargaining in Experimental Three-Person Games', 42–98, in *Laboratory Experimentation in Economics – Six Points of View*, Alvin E. Roth (ed.), Cambridge: Cambridge University Press.

(1988), 'Order of Strength and Exhaustivity as Additional Hypotheses in Theories for 3-Person Characteristic Function Games' (with Gerald R. Uhlich), 235–50, in *Bounded Rational Behavior in Experimental Games and Markets*, R. Tietz, W. Albers and R. Selten (eds), *Lecture Notes in Economics and Mathematical Systems* **314**, Berlin–Heidelberg–New York: Springer-Verlag.

(1989), 'On the Time Aspect of International Negotiations and the Probability for Reaching an Agreement: An Incomplete Information Approach' (with Werner

Güth), 319–33, in *Processes of International Negotiations*, Frances Mautner-Markhof (ed.), Boulder, CO: Westview Press.

(1991a), 'Anticipatory Learning in Two-Person Games', 98–153; (1991b), 'Pollinator Foraging and Flower Competition in a Game Equilibrium Model' (with Avi Shmida), 195–246; both in *Game Equilibrium Models I*, Reinhard Selten (ed.), Berlin–Heidelberg–New York: Springer-Verlag.

(1991c), (i) 'Original or Fake – A Bargaining Game with Incomplete Information' (with Werner Güth), 186–224; (1991d), (ii) 'A Game Equilibrium Model of Thin Markets' (with Myrna H. Wooders), 242–80; both in *Game Equilibrium Models III*, Reinhard Selten (ed.), Berlin–Heidelberg–New York: Springer-Verlag.

(1991e), (i) 'Majority Voting in the Concorcet Paradox as a Problem of Equilibrium Selection' (with Werner Güth), 7–40; (1991f), (ii) 'The Distribution of Foreign Language Skills as a Game Equilibrium' (with Jonathan Pool), 64–87; (1991g), (iii) 'Balance of Power in a Parlor Game', 150–209; all in *Game Equilibrium Models IV*, Reinhard Selten (ed.), Berlin–Heidelberg–New York: Springer-Verlag.

(1991h), 'Evolutorische Spieltheorie', 261–78, in *Der Evolutionsgedanke in den Wissenschaften*, Günther Patzig (ed.), Göttingen: Vandenboeck & Ruprecht.

(1992a), (i) 'A Demand Commitment Model of Coalition Bargaining', 245–82; (1992b), (ii) 'John C. Harsanyi, System Builder and Conceptual Innovator', 419–32; both in *Rational Interaction*, Reinhard Selten (ed.), Berlin–Heidelberg–New York: Springer-Verlag.

(1992c), 'Application of Bargaining I-Games to Cold-War Competition' (with John P. Mayberry), 133–52, in *Game-Theoretic Models of Cooperation and Conflict*, John P. Mayberry (ed.), Boulder, CO: Westview Press.

(1993a), 'Wirtschaftliche und kulturelle Evolution', 38–56, in *Makro, Geld & Institutionen*, Ulrich Schlieper und Dieter Schmidtchen (eds), Tübingen: J.C.B. Mohr.

(1993b), 'In Search of a Better Understanding of Economic Behavior', 115–39, in Arnold Heertje (ed.), *Makers of Modern Economics*, New York–London–Toronto–Sidney–Tokyo–Singapore: Simon & Schuster.

(1994a), 'Game Theory and Evolutionary Biology' (with Peter Hammerstein), 929–93, in *Handbook of Game Theory*, Volume 2, R.J. Aumann and S. Hart (eds), Amsterdam–New York: Elsevier Science B.V.

(1994b), (i) 'Biographical Outline Preceding the Nobel Prize Lecture', 313–19; (1994c), (ii) 'Multistage Game Models and Delay Supergames', 320–48, in *Les Prix Nobel 1994*, Tore Frängsmyr (ed.), Stockholm: Almqvist & Wiksell.

(1996a), 'Ansprache zur Ehrenpromotion an der Universität Graz', in Grazer Universitätsreden, Graz: Verlag Jos. A. Kienreich.

(1996b), 'Lernrichtungstheorie, Vortrag aus Anlass der Ehrenpromotion an der Universität Breslau, Wydawnictwo Akademii Ekonomicznej im Wroclaw: Oskara Langegowe Wroclawiu.

(1997), 'Descriptive Approaches to Cooperation', 289–326, in *Cooperation: Game-Theoretic Approaches*, Sergiu Hart and Andreu Mas-Colell (eds), NATO ASI Series, Series F: Computer and Systems Sciences, 155, Springer-Verlag, Berlin–Heidelberg–New York: Springer-Verlag.

(1998), 'Game Theory, Experience, Rationality', 9–34, in *Game Theory, Experience, Rationality*, W. Leinfellner and E. Köhler (eds) (Yearbook of Vienna Circle Institute), Dordrecht–Boston–London: Kluwer Academic Publishers.

(1998b), 'Experimental Sealed Bid First Price Auctions with Directly Observed Bid Functions' (with Joachim Buchta), 79–102, in *Games and Human Behavior: Essays in Honor of Amnon Rapoport*, David V. Budescu, Ido Erev and Rami Zwick (eds), Mawhaw, NJ: Lawrence Erlbaum Associates, Inc.

Discussion and working papers (D)

(1977), 'The Scenario Bundle Method', Research Conference on Strategic Decision Analysis Focusing on the Persian Gulf, Verein zur Förderung der Arms Control, e.V., SADAC, 5–56.

(1981), 'Strategic Aspects of IIASA's Food and Agricultural Model' (with Werner Güth), WP-81-9, International Institute for Applied Systems Analysis, A-2361 Laxenburg, Austria.

(1985), 'A Commodity Flow Model of Flexible Exchange Rates with Overshooting in Response to Inflation', Universität Bonn, Sonderforschungsbereich 303 discussion paper A-155.

(1995a), 'Cyclic Games, an Introduction and Examples' (with Myrna Wooders), Universität Bonn, Sonderforschungsbereich 303 discussion paper B-334.

(1995b), 'Money Does not Induce Risk Neutral Behavior, but Binary Lotteries do Even Worse' (with Abdolkarim Sadrieh and Klaus Abbink), Universität Bonn, Sonderforschungsbereich 303 discussion paper B-343.

Other Publications

(1998), 'Das Zahlenwahlspiel – Ergebnisse und Hintergrund' (with Rosemarie Nagel), *Spektrum der Wissenschaft*, February, 16–22.

Nobel Prize citation

The Royal Swedish Academy of Sciences has decided to award the Bank of Sweden Prize in Economic Sciences in Memory of Alfred Nobel, 1994, jointly to

Professor John C. Harsanyi, University of California, Berkeley, CA, USA
Dr John F. Nash, Princeton University, Princeton, NJ, USA
Professor Dr Reinhard Selten, Rheinische Friedrich-Wilhelms-Universität Bonn, Germany

for their pioneering analysis of equilibria in the theory of non-cooperative games.

Games as the foundation for understanding complex issues
Game theory emanates from studies of games such as chess or poker. Everyone knows that in these games, players have to think ahead – devise a strategy based on expected countermoves from the other player(s). Such strategic interaction also characterizes many economic situations, and game theory has therefore proved to be very useful in economic analysis.

The foundations for using game theory in economics were introduced in a monumental study by John von Neumann and Oskar Morgenstern entitled *Theory of Games and Economic Behavior* (1944). Today, 50 years later, game theory has become a dominant tool for analysing economic issues. In particular, non-cooperative game theory i.e. the branch of game theory which excludes binding agreements, has had great impact on economic research. The principal aspect of this theory is the concept of equilibrium, which is used to make predictions about the outcome of strategic interaction. John F. Nash, Reinhard Selten and John C. Harsanyi are three researchers who have made eminent contributions to this type of equilibrium analysis.

John F. Nash introduced the distinction between cooperative games, in which binding agreements can be made, and non-cooperative games, where binding agreements are not feasible. Nash developed an equilibrium concept for non-cooperative games that later came to be called Nash equilibrium.

Reinhard Selten was the first to refine the Nash equilibrium concept for analysing dynamic strategic interaction. He has also applied these refined concepts to analyses of competition with only a few sellers.

John C. Harsanyi showed how games of incomplete information can be analysed, thereby providing a theoretical foundation for a lively field of research – the economics of information – which focuses on strategic situations where different agents do not know each others' objectives.

Strategic interaction
Game theory is a mathematical method for analysing *strategic interaction*. Many classical analyses in economics presuppose such a large number of agents that each of them can disregard the others' reactions to their own decision. In many cases, this

assumption is a good description of reality, but in other cases it is misleading. When a few firms dominate a market, when countries have to make an agreement on trade policy or environmental policy, when parties on the labour market negotiate about wages, and when a government deregulates a market, privatizes companies or pursues economic policy, each agent in question has to consider other agents' reactions and expectations regarding their own decisions, i.e. strategic interaction.

As far back as the early nineteenth century, beginning with Auguste Cournot in 1838, economists have developed methods for studying strategic interaction. But these methods focused on specific situations and, for a long time, no overall method existed. The game-theoretic approach now offers a general tool box for analysing strategic interaction.

Game theory

Whereas mathematical probability theory ensued from the study of pure gambling without strategic interaction, games such as chess, cards etc. became the basis of game theory. The latter are characterized by strategic interaction in the sense that the players are individuals who think rationally. In the early 1900s, mathematicians such as Zermelo, Borel and von Neumann had already begun to study mathematical formulations of games. It was not until the economist Oskar Morgenstern met the mathematician John von Neumann in 1939 that a plan originated to develop game theory so that it could be used in economic analysis.

The most important ideas set forth by von Neumann and Morgenstern in the present context may be found in their analysis of two-person zero-sum games. In a zero-sum game, the gains of one player are equal to the losses of the other player. As early as 1928, von Neumann introduced the minimax solution for a two-person zero-sum game. According to the minimax solution, each player tried to maximize his gain in the outcome which is most disadvantageous to him (where the worst outcome is determined by his opponent's choice of strategy). By means of such a strategy, each player can guarantee himself a minimum gain. Of course, it is not certain that players' choice of strategy will be consistent with each other. Von Neumann was able to show, however, that there is always a minimax solution, i.e. a consistent solution, if so-called mixed strategies are introduced. A mixed strategy is a probability distribution of a player's available strategies, whereby a player is assumed to choose a certain 'pure' strategy with some probability.

John F. Nash

John Nash arrived at Princeton University in 1948 as a young doctoral student in mathematics. The results of his studies are reported in his doctoral dissertation entitled 'Non-cooperative Games' (1950). The thesis gave rise to *Equilibrium Points in n-person Games* (Proceedings of the National Academy of Sciences of the USA 1950), and to an article entitled 'Non-cooperative Games' (Annals of Mathematics 1951).

In his dissertation, Nash introduced the distinction between cooperative and non-cooperative games. His most important contribution to the theory of non-cooperative games was to formulate a universal solution concept with an arbitrary number of players and arbitrary preferences, i.e., not solely for two-person zero-sum games. This solution concept later came to be called Nash equilibrium. In a Nash equilibrium, all of the players' expectations are fulfilled and their chosen strategies are optimal. Nash proposed two interpretations of the equilibrium concept: one based on rationality and the other on statistical populations. According to the rationalistic interpretation, the players are perceived as rational and they have complete information about the structure of the game, including all of the players' preferences regarding possible outcomes, where this information is common knowledge. Since all players have complete information about each others' strategic alternatives and preferences, they can also compute each others' optimal choice of strategy for each set of expectations. If all of the players expect the same Nash equilibrium, then there are no incentives for anyone to change his strategy. Nash's second interpretation – in terms of statistical populations – is useful in so-called evolutionary games. This type of game has also been developed in biology in order to understand how the principles of natural selection operate in strategic interaction within and among species. Moreover, Nash showed that for every game with a finite number of players, there exists an equilibrium in mixed strategies.

Many interesting economic issues, such as the analysis of oligopoly, originate in non-cooperative games. In general, firms cannot enter into binding contracts regarding restrictive trade practices because such agreements are contrary to trade legislation. Correspondingly, the interaction among a government, special interest groups and the general public concerning, for instance, the design of tax policy is regarded as a non-cooperative game. Nash equilibrium has become a standard tool in almost all areas of economic theory. The most obvious is perhaps the study of competition between firms in the theory of industrial organization. But the concept has also been used in macroeconomic theory for economic policy, environmental and resource economics, foreign trade theory, the economics of information, etc. in order to improve our understanding of complex strategic interactions. Non-cooperative game theory has also generated new research areas. For example, in combination with the theory of repeated games, non-cooperative equilibrium concepts have been used successfully to explain the development of institutions and social norms. Despite its usefulness, there are problems associated with the concept of Nash equilibrium. If a game has several Nash equilibria, the equilibrium criterion cannot be used immediately to predict the outcome of the game. This has brought about the development of so-called refinements of the Nash equilibrium concept. Another problem is that when interpreted in terms of rationality, the equilibrium concept presupposes that each player has complete information about the other players' situation. It was precisely these two problems that Selten and Harsanyi undertook to solve in their contributions.

Reinhard Selten

The problem of numerous non-cooperative equilibria has generated a research programme aimed at eliminating 'uninteresting' Nash equilibria. The principal idea has been to use stronger conditions not only to reduce the number of possible equilibria, but also to avoid equilibria which are unreasonable in economic terms. By introducing the concept of subgame perfection, Selten provided the foundation for a systematic endeavour in 'Spieltheoretische Behandlung eines Oligopolmodells mit Nachfrageträgheit' (Zeitschrift für die gesamte Staatswissenschaft' **121**, 301–24 and 667–89, 1965).

An example might help to explain this concept. Imagine a monopoly market where a potential competitor is deterred by threats of a price war. This may well be a Nash equilibrium – if the competitor takes the threat seriously, then it is optimal to stay out of the market – and the threat is of no cost to the monopolist because it is not carried out. But the threat is not credible if the monopolist faces high costs in a price war. A potential competitor who realises this will establish himself on the market and the monopolist, confronted with a *fait accompli*, will not start a price war. This is also a Nash equilibrium. In addition, however, it fulfils Selten's requirement of subgame perfection, which thus implies systematic formalization of the requirement that only credible threats should be taken into account.

Selten's subgame perfection has direct significance in discussions of credibility in economic policy, the analysis of oligopoly, the economics of information etc. It is the most fundamental refinement of Nash equilibrium. Nevertheless, there are situations where not even the requirement of subgame perfection is sufficient. This prompted Selten to introduce further refinement, usually called the 'trembling hand' equilibrium, in 'Re-examination of the Perfectness Concept for Equilibrium Points in Extensive Games' (*International Journal of Game Theory* **4**, 25–55, 1975). The analysis assumes that each player presupposes a small probability that a mistake will occur, that someone's hand will tremble. A Nash equilibrium in a game is 'trembling-hand perfect' if it is robust with respect to small probabilities of such mistakes. This and closely related concepts, such as sequential equilibrium (Kreps and Wilson, 1982), have turned out to be very fruitful in several areas, including the theory of industrial organization and macroeconomic theory for economic policy.

John C. Harsanyi

In games with complete information, all of the players know the other players' preferences, whereas they wholly or partially lack this knowledge in games with incomplete information. Since the rationalistic interpretation of Nash equilibrium is based on the assumption that the players know each others' preferences, no methods had been available for analysing games with incomplete information, despite the fact that such games best reflect many strategic interactions in the real world.

This situation changed radically in 1967–68 when John Harsanyi published three articles entitled 'Games with Incomplete Information Played by Bayesian Players' (*Management Science* **14**, 159–82, 320–34 and 486–502). Harsanyi's approach to games with incomplete information may be viewed as the foundation for nearly all economic analysis involving information, regardless of whether it is asymmetric, completely private or public.

Harsanyi postulated that every player is one of several 'types', where each type corresponds to a set of possible preferences for the player and a (subjective) probability distribution over the other players' types. Every player in a game with incomplete information chooses a strategy for each of his types. Under a consistency requirement on the players' probability distributions, Harsanyi showed that for every game with incomplete information, there is an equivalent game with complete information. In the jargon of game theory, he thereby transformed games with incomplete information into games with imperfect information. Such games can be handled with standard methods.

An example of a situation with incomplete information is when private firms and financial markets do not exactly know the preferences of the central bank regarding the tradeoff between inflation and unemployment. The central bank's policy for future interest rates is therefore unknown. The interactions between the formation of expectations and the policy of the central bank can be analysed using the technique introduced by Harsanyi. In the most simple case, the central bank can be of two types, with adherent probabilities. Either it is oriented towards fighting inflation and thus prepared to pursue a restrictive policy with high rates, or it will try to combat unemployment by means of lower rates. Another example where similar methods can be applied is regulation of a monopoly firm. What regulatory or contractual solution will produce a desirable outcome when the regulator does not have perfect knowledge about the firm's costs?

Other contributions of the Laureates
In addition to his contributions to non-cooperative game theory, John Nash has developed a basic solution for cooperative games, usually referred to as Nash's bargaining solution, which has been applied extensively in different branches of economic theory. He also initiated a project that subsequently came to be called the Nash programme, a research programme designed to base cooperative game theory on results from non-cooperative game theory. In addition to his prizewinning achievements, Reinhard Selten has contributed powerful new insights regarding evolutionary games and experimental game theory. John Harsanyi has also made significant contributions to the foundations of welfare economics and to the area on the boundary between economics and moral philosophy. Harsanyi and Selten have worked closely together for more than 20 years, sometimes in direct collaboration.

Through their contributions to equilibrium analysis in non-cooperative game theory, the three laureates constitute a natural combination: **Nash** provided the foundations for the analysis, while **Selten** developed it with respect to dynamics, and **Harsanyi** with respect to incomplete information.

PART I

# Reinhard Selten

In search of a better understanding of economic behaviour

The title of this chapter tries to express a leading theme in my scientific efforts. Most of my past work can be seen as an attempt to explore the structure and the consequences of economic behaviour. At least this is true if the term 'economic behaviour' is understood in a sufficiently wide sense to include applications of game theory to political science and evolutionary biology.

I do not want to convey the false impression that my research is single-mindedly organized around a grand question. I am easily attracted by the opportunity to shift my interests into unforeseen exciting new directions. The little coherence there is in my work is due to a desire to understand both fully rational and boundedly rational economic behaviour, especially in the context of game situations.

1. Early influences

In my high-school years I developed a strong interest in mathematics. My first exposure to economic theory was a history of economic thought by Heimann which I happened to find in a public library. Under the impression of this book, I shortly after considered studying economics, but in the end I did not change my original plans. In 1951, I began to study mathematics at the University of Frankfurt-on-Main.

As a student I did not work as hard as I should have done. I spent a lot of time listening to lectures and reading books in fields not directly connected to my studies. I had to take physics as subsidiary, a subject which did not interest me as much as economics and psychology. At this time the well-known *Gestalt* psychologist Edwin Rausch was teaching at Frankfurt. I went to his lectures and served as an experimental subject for a sufficient number of hours to be admitted to one of his proseminars. Later, it turned out that the time I spent on the study of experimental psychology was by no means wasted. Of course, the same is true of my early efforts to learn some economic theory. However, I also extended my curiosity to other fields such as astronomy and the history of science which at least up to then did not supply useful background knowledge for my scientific work.

Even before I began to study, I knew of the existence of an interesting new field called game theory about which I had read an article by Williams in *Fortune*. At the university, I read the book by von Neumann and Morgenstern (1944) in order to learn more about the subject. At first this also seemed to be an extracurricular effort, since nobody taught game theory at the Mathematics Department. To my surprise, somewhat later I discovered the announcement of a student seminar on game theory for economics students offered by Ewald Burger as a part of his teaching of mathematics for economists. I participated in this seminar and gladly followed the suggestion of Ewald Burger to write a master's thesis on the problem of defining a solution for general two-person games in normal form.

Burger later published an important textbook on game theory (1958) which has been translated into English (1963). It is worth looking at even today after so many years for its rigorous and concise proofs for the existence of equilibrium points in finite and infinite normal form games. He was an excellent teacher who was always willing to interact with his students. I owe much to his guidance.

2. Valuation of games

The work that I did under the supervision of Burger, first for my master's thesis and later for my PhD thesis, was an axiomatic value theory for extensive side payment games (1960, 1964). I first tried to base my theory heavily on a subgame truncation axiom which postulates that the value of a game remains unchanged if a subgame is cut off and replaced by its value as the pay-off vector at the decomposition point. I succeeded with this approach in the two-person case but soon discovered that for more than two players the subgame truncation property is incompatible with other plausible axioms. I obtained these results in my master's thesis. In my PhD thesis, finished in 1961, I presented an axiomatic value theory for n-person games. The English translation was published only four years later.

My work on value theory was in the spirit of the time. The problem of defining rationality in co-operative games was foremost in the minds of game theorists. In view of a popular belief that co-operation must take place wherever it is profitable, non-co-operative games seemed to be much less interesting.

My value theory is closely related to Harsanyi's general solution theory for co-operative games (1963). For the side payment case, to which my value is restricted, both theories come to the same conclusions, but from different points of departure. His work is based on a bargaining model. I avoided the interpretation of my value by a bargaining model, since I was disturbed by examples of three-person games, in which the threat pay-off of a player against the coalition of the two others is higher than his value.

3. The beginnings of experimental economics in Germany

After I had finished my master's thesis in 1957, I was hired by Heinz Sauermann who held a Chair of Economics at the University of Frankfurt-on-Main. He knew me as a participant of a colloquium organized by Burger, in which economists and mathematicians discussed applications of mathematics to economics. He must have been positively impressed by several talks I had given in this colloquium. He was one of the very few German economists who at this time already clearly saw the trend towards the use of more and more mathematics in economic theory. In spite of his lack of mathematical training, he was in favour of this development.

I worked under Heinz Sauermann for ten years in various assistant positions. At first it was my task to do research on 'Application of decision theory to the theory of the firm', a project financed by the Deutsche Forschungsgemeinschaft (the German counterpart of the National Science Foundation). I actually began to do something which fitted the name of the project. I tried to show that under certain assumptions on the economic environment, maximal long-run growth of the firm is achieved by the maximization of the expected logarithm of the firm's assets. However, I very soon abandoned this effort. H. A. Simon's pioneering work in his book *Models of Man* (1957) convinced me of the necessity to model economic behaviour as boundedly rational. Suddenly the application of normative decision theory to the theory of the firm seemed to be pointless.

Even before I was converted to Simon's views, I began to pursue the plan to perform oligopoly experiments. I had read a little book by Ricciardi (1957) which described the first computerized management game used by the American Management Association as a training device. It occurred to me that a much simpler business game could serve as an experimental research tool. No computer was necessary for this purpose. The idea to perform experiments was also due to the example of the pioneering work of Kalish et al. (1954), whose paper on the first characteristic function game experiments had impressed me very much.

At this time experiments in economics were almost unheard of. It was a standard textbook assertion that experiments are impossible in economics. I was lucky to work under the supervision of Heinz Sauermann, who had the courage and foresight to support my unusual plans and to get involved in experimental work himself. My first paper on oligopoly experiments (1959) was a joint work with him. He became a great supporter of experimental economics. He organized a number of successful conferences and edited a series of books, the *Contributions to Experimental Economics*. The first volume of this series (1967) collected experimental work done under his supervision at the University of Frankfurt in the time from 1959 to 1965.

Our paper of 1959 was very early, but it was by no means the first contribution to experimental economics. A bibliography compiled by Volker Häselbarth (1967) in the first volume of *Contributions to Experimental Economics* lists 20 publications before 1959. However, it can be said that our little group at Frankfurt was not preceded by any other group doing continuous research on experimental economics over many years. Experimental economics as a field of economic research did not emerge before the 1960s.

Our experimental work was not primarily directed at the testing of existing theories. We tried to build up interesting economic situations in the laboratory in order to gain insight into the structure of boundedly rational decision-making. The exploratory style of our research and our emphasis on the explanation of individual behaviour were not characteristic of the emerging field as a whole. In the USA, there was more interest in effects on the level of the system rather than the individual and the evaluation of experimental data was often guided by an unwavering confidence in the explanatory power of strong rationality assumptions. To some extent, this is still true even now.

4. A step on the thorny path towards a theory of bounded rationality

It is now clear to me that it will take many decades of painful experimental research until an empirically defendable general theory of bounded rationality emerges. At the beginning of my career, I was more optimistic in this respect. Together with my teacher Heinz Sauermann, I tried to develop a general theory of boundedly rational decision making of the firm. I had an idea how to model the search for alternatives and the final choice as a process guided by rules for the adaptation of aspiration levels on multiple goals. Finally, a complete theory emerged in long discussions with Heinz Sauermann. He was not only a professor of economics but also the owner of a textile wholesale business. His practical experience was the empirical background for our common work. We finally wrote a paper in German (1962) with the title 'Anspruchsanpassungstheorie der Unternehmung' (Aspiration-level adaptation theory of the firm).

This paper described a formal structure, but without any formulas. In this way, we hoped to reach a broad audience. Our theory never became known outside the German language area, but it seems to me that it still has something to offer as a point of departure for further work. At least we succeeded in constructing a coherent and plausible picture of non-optimizing decision making based on rules for the adaptation of aspiration levels on incomparable multiple goals. The question of optimization does not even arise in our theory, since it does not specify preferences among potential final choices, but only local priorities for the adaptation of aspiration levels. Bounded rationality is not conceived as an approximation to full rationality but as something different.

In the light of experimental evidence, I later came to the conclusion that boundedly rational behaviour is more complex than our theory of aspiration level adaptation suggests. We took goals, aspiration scales and adaptation rules as given, in the same way as preferences are taken as given by orthodox decision theory. I now think that, at least in some contexts, aspiration levels are constructed from more basic data. We need a theory on how this is done.

Our attempt to build a general theory of boundedly rational decision-making of the firm was, perhaps, premature. It was not more than a step on the long and thorny path towards an empirically sound theory of bounded rationality.

5. Causal diagrams

My second experimental paper on oligopoly (1967a) dealt with a model in which three firms had to choose capacities and production methods. Production was equal to capacity and the price was a linear function of total supply. The firms, represented by groups of three to five subjects, had to write protocols explaining the reasons for their decisions. Inspired by the influential paper on psycho-logic by Abelson and Rosenberg (1958) who had modelled political belief structures by signed graphs, I interpreted the protocols by graph structures which I called 'causal diagrams'. Formally, a causal diagram is a signed directed graph whose nodes correspond to variables and whose links represent positive or negative causal influences.

A causal diagram is 'balanced' if all causal chains from a decision parameter to a goal variable show an influence in the same direction. Subjects tend to achieve balance by the suppression of causal links implied by the rules of the game. This results in 'simplified' causal diagrams.

On the basis of the simplified causal diagrams, the firms determine the order in which capital widening and capital deepening investments are made. I called this order the 'strategic line'. Investment criteria determine when the next step on the strategic line is taken. The investment criteria have the character of aspiration levels on profit-related variables.

Similar qualitative belief structures were later explored by Axelrod (1976) under the name of 'cognitive maps'. It seems to be the case that many decisions are made in two stages. First, in a qualitative stage, it is decided which decision parameters should be changed in which direction. Second, in a quantitative stage, the amounts of change are determined. The qualitative decisions of the first stage are made by qualitative reasoning on the basis of qualitative information. Different principles, e.g. a conventional step size of change, are applied in the second stage.

6. Equal shares and coalition formation

In the early 1960s, a young psychologist, Klaus G. Schuster, joined our little research group at Frankfurt University. Together with him, I performed experiments on the five-person apex game, a special characteristic function game (1968). Listening to the face-to-face bargaining among the subjects, I had the impression that equal shares of coalition values have a great significance for the thinking of the players. Kalish et al. (1954) had already made a remark to that effect. However, since I was not able to substantiate my impression by a statistical analysis of the result of our experiments, we did not mention it in our paper.

I had the idea that players would tend to form a coalition with maximal equal share and that the agreed-upon pay-offs would be determined by aspiration levels derived from maximal equal shares of alternative coalitions. Within the region delineated in this way, I expected a numerically prominent pay-off division. Based on this principle, I was surprisingly successful in the prediction of classroom demonstration experiments on a specific seven-person game. For many years, I tried to generalize my tentative theory to all superadditive characteristic function games. For this purpose, I looked very hard at all published results of characteristic function experiments. I compiled a list of 211 plays. This data base was too small for a theory which specifies an explicit process of aspiration level formation. However, I found out that a simple theory called 'equal share analysis' fitted the available data quite well (1972). This theory does not assert that coalition values are split evenly. Unequal divisions are possible and may even be required. However, equal divisions of alternative coalition values bound the pay-off division within a coalition. Pay-off vectors must be in the 'equal division core', which means that no equal division of an alternative coalition value improves the pay-offs of all members.

Equal division is not only a fairness norm but also a natural point of departure for boundedly rational strategic thinking. A coalition member who is stronger than his partners has a reason to claim at least his equal share.

Equal share analysis was a better explanation of the data than its alternatives proposed by normative game theory, but it was not really satisfactory. As more data became available, I developed a new descriptive theory for zero-normalized superadditive three-person games in characteristic function form. This theory, called 'equal division pay-off bounds' (1983c, 1987), derives lower bounds for the players' aspiration levels based on simple computations involving various equal shares. The improved version of this theory (1987), in particular, has had a remarkable predictive success. A student of mine, Gerald Uhlich (1990), has proposed a generalization and modification called 'proportional pay-off bounds' which is applicable to three-person games with non-negative one-person values.

The descriptive theories developed by Uhlich and me are procedural in the sense that they specify the way in which the solution is determined. Moreover the theories are casuistic in the sense that many case distinctions based on simple criteria are made; simple principles are applied in every single case. Casuistic procedural structures seem to be more adequate for the description of boundedly rational coalition formation than solution concepts based on abstract general principles.

7. Presentation effects

Together with Claus C. Berg (1970), I performed oligopoly experiments in which we varied initial assets and profits in a way which did not change the relationship between strategic choices and final money pay-offs. Increased initial assets were compensated by increased fixed costs. This strategically inessential variation of the rules of the game had a marked influence on behaviour. One of the modes of observed co-operation was Pareto-optimal equal profits, a fairness concept which is not invariant to changes of fixed costs compensated by changes of initial assets. We called this a 'presentation effect'. In the terminology used now one would speak of 'framing' (Kahneman and Tversky, 1984). We were among the first to describe a phenomenon of this kind. Therefore I feel that I should be permitted to continue to speak of presentation effects. Another early description of a presentation effect can be found in a study of Pruitt (1970) who showed that the way in which a prisoner's dilemma game was presented to the subjects made a dramatic difference to the frequency of co-operation in repeated play.

Presentation effects are the most convincing evidence against utility maximization as an explanation of observed behaviour. In the case of the oligopoly experiment described above, the presentation effect becomes easily understandable if one thinks of behaviour as guided by aspiration levels. Zero profits are an obvious focus for the formation of a first aspiration level on pay-offs. If aspiration levels are formed in this way, equal profits are a natural mode of co-operation.

8. From oligopoly experiments to subgame perfectness

At the University of Frankfurt, I did experimental work on characteristic function bargaining (Selten and Schuster, 1968) and bilateral markets (Selten, 1970b), but most of my experimental research was concentrated on oligopoly. I was fascinated by the problems posed by the task of playing a business game. Usually the participants must reach their decision in a relatively short time. Obviously they cannot apply Bayesian decision theory or game theory. How should they approach their task? In this respect economic theory miserably fails to offer practical advice.

In my view it is of great importance to gain theoretical insight into the problem. With my work on oligopoly experiments, I tried to contribute to the accumulation of an empirical background for the development of a theory of boundedly rational decision making in business games. I think that it is worthwhile to work towards this goal even if we are still far away from it.

Even if the behaviour of experimental subjects cannot be expected to conform to normative game-theoretical solutions, it is useful to determine such solutions if this is possible. The way in which observed behaviour deviates from normative theory often conveys valuable insights into the structure of boundedly rational decision making.

One of the oligopoly experiments performed at Frankfurt (1967b) was based on a dynamic price variation model with demand inertia. When I constructed the model, I did not pay any attention to analytic tractability. The model has indivisibilities of production and investment and interest rates are different for positive and negative accounts. Only later it occurred to me that a game-theoretical analysis might be feasible for a simpler version of the model without these features. I determined an equilibrium solution by backward induction. Having done this, I discovered the presence of many other equilibrium points in pure strategies. However, the backward induction equilibrium was clearly the natural non-co-operative solution. In order to describe its distinguishing features in general game-theoretical terms, it was necessary to introduce the notion of subgame perfectness.

The paper was written in German (1965) and probably has not been read by many people even if it is quoted quite often for the definition of subgame perfectness. Experiments on the oligopoly model with demand inertia analysed in this paper were not performed until much later and not by me, but by my PhD student Claudia Keser (1992). Needless to say, the subgame perfect equilibrium solution did not prove to be a valid descriptive theory. I never had this irrational expectation.

9. My book on the multiproduct firm

When I began to work under the supervision of Heinz Sauermann, most German economists thought that mathematics had no future in economics. People like me, who entered economics by a side-door with a background in mathematics, were looked upon with suspicion. Therefore, I was first expected to acquire a master's degree and then a PhD in economics. I am grateful to Heinz Sauermann who made it possible for me to avoid these steps in my career. Several years after I had received my PhD in mathematics, he must have succeeded in convincing his colleagues that I should be permitted to write a habilitation thesis in economics without meeting the usual prior requirements. In Germany the habilitation, not the PhD, is the last formal barrier to eligibility for professorship. Heinz Sauermann asked me to write a habilitation thesis which would also be of interest to his colleagues teaching business administration. I decided to work on the multiproduct firm.

When I made this choice, I knew nothing about the literature on the subject-matter. What attracted me to the area was a strong interest in the theory of the firm combined with an inclination towards problems involving an unspecified integer number n in an essential way like the algebraic equation of nth degree which captured my imagination in my high-school days. I was not able to do anything about the nth degree equation, but I had successfully worked on the valuation of n-person games. I felt that it would be interesting to do something about the n-product firm.

Reading the literature, I was surprised by the unsuspected richness of the area. I learnt about the Edgeworth paradox of a monopoly price reducing excise tax and the phenomenon of the loss leader optimally sold below marginal cost. The classical German habilitation thesis is a monograph on a well-defined subject area. My book on the multiproduct firm (1970a) conformed to this pattern. I systematically represented the traditional theory of the multiproduct monopoly and extended it where I could. In addition to this I also developed a theory of Cournot–Nash equilibrium in multiproduct oligopoly. My special emphasis was on problems of aggregation. The problem of determining optimal prices for the multiproduct monopoly with linear demand and quadratic cost can be reduced to solving a system of three linear equations for three price indices, if the interrelationships of cost and demand are sufficiently unsystematic in a certain sense. A set of conditions on multiproduct oligopoly models permits an analysis in terms of an aggregated decision parameter called 'aggressivity'.

10. Bargaining under incomplete information

In 1965, I was invited to a workshop on game theory in Jerusalem. Only 17 participants attended this conference: game theory was still a very small field. The hottest topic of discussion was Harsanyi's new theory of games with incomplete information (Harsanyi, 1967–8). I supported Harsanyi's idea that it is natural to require a consistency condition on the players' beliefs which makes it possible to model incomplete information by imperfect information on a lottery performed by nature before the beginning of the game. The consistency condition was heavily criticized and the question arose of how to reduce incomplete to imperfect information without it. I contributed the idea to do this with the help of lotteries performed at the end. This was the beginning of a long co-operation with John C. Harsanyi.

Following his suggestion, the University of California, Berkeley, invited me for the academic year 1967–8 as a Visiting Full Professor at the business school. We worked on a theory of two-person bargaining under incomplete information (Harsanyi and Selten, 1972). It became more and more clear to us that a full understanding of rational co-operation requires non-co-operative modelling.

On the basis of our common paper, I later worked out a semi-numerical example (1975b). Moreover, I began a co-operation with Austin Hoggatt.

Together with several of his research assistants, we performed experiments on what was essentially a special case of this example (Hoggatt et al., 1978). Earlier than others Hoggatt had built up a computerized laboratory. He did this at a time when computers were very primitive compared to the technology available now. Writing the software for his laboratory required great skill and a tremendous amount of work. The results of our experiments were quite interesting. As the theory predicts, conflict usually occurred if both players were in a strong bargaining position, but, of course, observed behaviour was only boundedly rational.

11. Game models of international relations

In the late 1960s, I belonged to a group of game theorists hired by the research firm Mathematica to work on projects for the Arms Control and Disarmament Agency. We often met for several days at Washington DC or nearby. My common work with John Harsanyi on bargaining under incomplete information was also partly done in the framework of this group.

Herbert Scarf, who also was one of the members, proposed a model of nuclear deterrence which had the structure of a co-operative game whose stability could be explored with the help of core concepts. Together with Reinhard Tietz (1972), I developed a theory of 'security equilibria' for a modified version of this model. This theory involves aspects of bounded rationality and can be described as a hybrid of maximin and equilibrium theory for a game form akin to an extensive form. It is assumed that no player does anything which decreases his security level and that at every situation in the game a player's security level is the worst that can happen to him under this assumption.

With the help of numerical computations of security equilibria for special cases of the model, interesting results could be obtained which revealed reasons for stability and instability of nuclear deterrence systems. Neither the work of Scarf nor that of myself and Tietz was very well received in a presentation to the Arms Control and Disarmament Agency. Some of my difficulties were terminological. In the description of the pay-off functions I insisted on talking of a 'motive of revenge', even if in the audience there was strong resentment against the word 'revenge'. Somebody said that, maybe, the Soviet Union has a motive of revenge, but not the USA, since this is against Christian ethics – retaliation, yes, but not revenge. Stubbornly, I refused to talk about retaliation instead of revenge.

I had the impression that the practitioners did not really like to listen to anything applied. They felt they knew best how to solve their problems. They were much more pleased by qualitative interpretations of abstract game-theoretical results. Robert J. Aumann gave an excellent talk of this kind. He explained under what conditions it is advantageous to reveal or not to reveal information by strategic actions in repeated games with incomplete information.

I have been told that after Aumann's presentation a high official said to his entourage: 'We know these things, but we like to be reminded.'

My experiences with the Arms Control and Disarmament Agency did not discourage my inclination towards the theory of international relations. During one of my frequent visits to Berkeley, I met Amos Perlmutter, a political scientist, in the home of John Harsanyi. We decided to co-operate on the application of game theory to the theory of international relations. We developed a method called 'scenario bundle analysis' of constructing simple game models on the basis of expert judgements. Financial support by the Volkswagen Foundation enabled me to organize a small conference of area specialists on potential conflicts in the Persian Gulf. In this conference we applied our method. The participants had to work out various lists such as a list of actors, and for each actor lists of goals, fears and initial options. On the basis of qualitative information of this kind 'scenario bundles' were worked out. Scenario bundles are simple game structures, essentially games with perfect information, but with possibilities of coalition action.

Unfortunately, the report about the results of the conference (1977) was never formally published, mainly because our experts judged an Iranian revolution as extremely improbable. To my surprise, very soon this event happened and made our analysis obsolete. Nevertheless I think that our method is not without merit. It is useful for a chess player to analyse his situation before he makes his choice, even if the other player may easily respond with an unexpected move. I also think that our way of looking at the behaviour of governments as motivated by historically grown goals and fears rather than by an abstract objective such as power or national interest is the right approach to the modelling of concrete international conflict.

Several years ago, Peter Bernholz asked me to co-operate in a joint summer course on international relations at the Austrian College at Alpach. For this occasion I developed a parlour game with the intention to create an environment in which the traditional balance of power theory could work. The game is not without entertainment value. Up to 12 countries form changing alliances in a world of 30 provinces. A stationary equilibrium solution can be determined for a simplified version with only 6 provinces (1991c).

I am a supporter of the international language Esperanto and like most other Esperantists I would like to see the world peacefully united. Therefore I have a special interest in the theory of international relations. A better understanding of war and peace in an obsolete system of nation states may help to bring us nearer to a true world federation.

12. Oligopolistic general equilibrium

During frequent visits to Berkeley, I co-operated with John Harsanyi, Austin Hoggatt and Thomas Marschak. My joint work with Thomas Marschak resulted in a book (1974), *General Equilibrium with Price Making Firms*. In this book and in two articles (1977, 1978) we proposed a model of oligopolistic general equilibrium which took its point of departure from the German tradition of conjectural oligopoly theory. Expected responses to deviations from a status quo are described by 'reaction functions' and stability is defined by the absence of potential deviations which improve profits after the reaction of the competitors. We imposed a rationality criterion on the system of reaction functions which requires restabilizing reactions after a deviation. In this way we arrived at the concept of a 'convolution'. We chose this name for our solution concept because it was suggested to me by Robert J. Aumann in a dream.

Our theory describes oligopolistic competition by a specific convolution which combines aspects of kinked demand curve theory and limit pricing. The firms are assumed to have limited information about the whole economy. They form models on the basis of this information in a systematic way and stability is defined with respect to the reaction of the competitors in the model. It is possible to prove the existence of stable states for the whole economy.

Even if our book is often mentioned in the literature, it did not have as much influence on later work about general equilibrium with imperfect competition as we had hoped for. However, the problem area is still wide open.

13. Multistage game models of imperfect competition

In the early 1970s, I began to construct game models of imperfect competition, in which the competitors make different types of decisions sequentially in several stages. Later, multistage game models of this kind became more and more common in the new industrial economics literature.

My first multistage game model (1973) was an attempt to solve a basic problem of oligopoly theory. It is a widely held belief that the intensity of competition crucially depends on whether or not the number of competitors is small. The competition among few is believed to be qualitatively different from that among many, but where is the dividing line? I constructed several models in order to answer this question and finally worked out what seemed to be the most interesting one.

The model is a three-stage game. At the first stage, players decide whether they want to take part in cartel negotiations. At the second stage, quota cartels can be formed by unanimity bargaining and at the third stage production quantities are chosen. I analysed this game with the help of a selection theory which determined a unique 'perfect equilibrium set'. The solution shows that in this model four are few and six are many. From four to six, the probability of forming a cartel drops from 100 per cent to less than 2 per cent. This is due to a sudden increase in the incentive to become an outsider.

Since I wished to bring my theoretical thinking into closer contact with stylized empirical facts, I engaged in a systematic search for empirical regularities described in the industrial organization literature. I hoped to find an integrated explanation of as many of them as possible. The result of this effort was my paper, 'Oligopolistic size structure and profitability' (1983a), which presents a two-stage game model with entry and choice of technological level at the first stage and production quantity decisions at the second.

The model explains four empirical regularities: the typical skewed distribution of firm sizes in a market, the positive correlation between firm size and firm profitability, the positive correlation between concentration and market profitability and the negative one between marginal concentration and market profitability. Contrary to conventional explanations as direct causal relationships, the validity of the last three relationships in the model is due to the structure of the equilibrium and its response to changes of cost and demand parameters.

A third multistage game model (1984) permits the conclusion that under plausible conditions the prohibition of cartels increases the total profit of all firms in the economy. The effect is due to excessive entry into collusive markets.

Since I do not believe in the descriptive relevance of strong rationality assumptions, I prefer to think of game equilibrium in empirically oriented models as the result of adaptive dynamic processes. In principle, the justifiability of an interpretation in these terms should be explicitly discussed in every single case. Unfortunately, we do not yet have sufficiently well-developed theoretical tools for this purpose.

14. Perfectness

Soon after I had introduced subgame perfectness (1965), it became clear to me that more than that is needed in order to exclude equilibrium points which specify unreasonable behaviour at unreached information sets. I had some ideas how to solve this problem by a concept similar to sequential equilibrium (Kreps and Wilson, 1982b), but I saw no way to prove existence. I therefore abandoned this approach in favour of the trembling hand definition (1975a).

I regard perfectness as a necessary, but not sufficient, rationality requirement for equilibrium points in extensive games. Many stronger refinement concepts have been proposed in the literature. The book by van Damme (1987) provides an overview and a thorough discussion. In my view the stronger concepts go beyond the establishment of a minimal rationality criterion. Trembling hand perfectness is based on the idea of independent local mistakes. Nothing is assumed about the likely reasons for such mistakes. Stronger refinement concepts seem to imply the idea that some deviations can be excluded as less rational than others. Thus it is the essence of simple forward induction arguments that a player is unlikely to make a local mistake which results in playing a dominating strategy in the whole game.

If an equilibrium point is the solution of the game, then every deviation from it is non-rational. Therefore I see no justification for rationality arguments as a basis for excluding some deviations as unlikely. Of course, these short remarks cannot do full justice to the refinement discussion, but they may serve to indicate my point of view. Even if I admire the mathematical and conceptual ingenuity behind some of the stronger refinement concepts, I see no reason to think of perfectness as superseded by one of them.

15. Equilibrium selection

After we had finished our paper on bargaining under incomplete information (1972), John Harsanyi and I thought about extending this work from fixed threats to variable threats. However, we decided that we should first try to solve the general problem of selecting a unique equilibrium point for every non-co-operative game. We would then have a better basis for the treatment of the variable threat case of bargaining under incomplete information. We did not expect that it would take us 18 years to complete our work on equilibrium selection.

We discarded two fairly well-developed theories before we began to build our third one, the final version of which is presented in our book (1988). Each theory was based on a different concept of risk dominance of one equilibrium point over another. The first theory took its point of departure from what we called 'diagonal probabilities'. Suppose that all players deviate with the same probability p from one equilibrium point u to another equilibrium point v. Player i's diagonal probability for u is the maximum value of p such that player i's best reply to the resulting mixture is still his strategy at u. At first we considered a measure of risk dominance formed by the product of the diagonal probabilities at u divided by the product of the diagonal probabilities at v. Later we decided to modify the factors by exponents derived in a complex way as solutions of systems of linear equations. Finally we obtained a selection theory which seemed to have good properties but was practically unmanageable in the application to simple examples.

The risk dominance concept of our second theory and that of our final theory have a basic idea in common which leads to a prior probability distribution over the behaviour of the players. A player assumes that the others know whether u or v is the solution and that they will behave accordingly. He has a subjective probability p for u being the solution and maximizes his pay-off on the basis of this expectation. The subjective probability p is a random variable. In our final theory we always assume that p is uniformly distributed over the unit interval. In our second theory we also considered biased distributions obtained as follows. A public announcement of the solution by a game theorist who is known to be right with a probability r, transforms the initial subjective probability p to a new one by Bayes's rule. Risk dominance of u over v was measured by the size of the 'bias' r needed to bring the prior probability distribution into a 'formation' containing u, but not v. A formation is defined as a substructure closed with respect to best replies.

Our second theory was less unwieldy than our first one, but it still had the difficulty that in special cases risk dominance comparisons alone would not single out a unique equilibrium point. In such cases we had to rely on a very complicated tie-breaking procedure.

Harsanyi's invention of the tracing procedure (1975) made it possible to abolish the tie-breaking procedure and to reconsider the definition of risk dominance and its role in the process of equilibrium selection. Starting from an arbitrary mixed strategy combination interpreted as a prior distribution, the tracing procedure determines a unique equilibrium point.

In our first two theories, a cardinal measure was associated with risk dominance, but finally we decided to define risk dominance as an incomplete binary relation. In order to determine the risk dominance relationship between two equilibrium points, the tracing procedure is applied to the unbiased prior probability distribution.

The final version of our third theory proved to yield intuitively satisfactory results in all applications explored up to now, not only those included in our book but also others published elsewhere (Selten and Güth, 1982, 1991; Leopold, 1982, 1985; Selten and Leopold, 1983; Avenhaus, Güth and Huber, 1991; van Damme and Güth, 1991; Güth, 1991; Güth and Selten, 1991; Potters, van Winden and Mitzkewitz, 1991).

Each of our three theories was changed again and again as the result of heated discussions. Many ideas were proposed and many were discarded. It is only human that both of us wanted to exert as much influence as possible on the common product. However, both of us were willing to yield to stronger arguments at the end.

Probably some of the ideas which we discarded will reappear in later work on equilibrium selection. To some extent this has already happened. Güth and Kalkofen (1989) made use of diagonal probabilities, but their proposals are very different from our first theory.

It can be expected that the long discussion between John and me is only the prelude to a much longer public discussion with many participants. The problem of equilibrium selection by rational players is of great philosophical significance in spite of the limited descriptive relevance of rational game theory.

16. Non-co-operative characteristic function bargaining

Motivated by my experimental interests, I tried to find a non-co-operative procedure for characteristic function bargaining whose game-theoretic analysis yields conclusions akin to classical co-operative game theory. My first attempt in this direction was a simple infinite perfect information model (1981). At the beginning an 'initiator' is selected randomly. An initiator can propose a pay-off division for a coalition and name a member as the next 'responder'. If a responder rejects the proposal he becomes an initiator. If he accepts he has to name a next responder among those members other than the proposer who have not yet accepted the proposal. If there are no such players the coalition is formed and the game ends. An initiator can also shift the initiative to another player.

The game-theoretical analysis exhibited a close correspondence between stationary subgame perfect equilibria with certain additional properties and the stable demand vectors proposed by Albers (1975). For three-person quota games without the grand coalition or with an empty core this means that a two-person coalition with pay-offs according to the quotas must be formed. Experiments with the model do not confirm this prediction (Uhlich, 1990).

The model is an infinite game and strictly speaking only finite games can be played in the laboratory. Another difficulty with the model is posed by the presence of many additional non-stationary equilibrium points. For these reasons I constructed a finite model of characteristic function bargaining which yields quota agreements in the case of three-person quota games without the grand coalition as the only subgame perfect outcomes (1992). In this 'demand commitment model' players may temporarily commit themselves to pay-off demands or they may form coalitions with players who are already committed.

I am looking forward to experiments with this model. Here classical co-operative theory seems to have its best chance. Preliminary experiments with a similar but different model suggest that quota agreements may in fact have a chance to be learnt under favourable experimental conditions, even if up to now they were not typical as outcomes of characteristic function experiments.

17. Evolutionary game theory

The subject-matter of evolutionary game theory is applicable to biology. Luckily, I came into contact with this new field quite early, only about five years after the publication of the pioneering article by Maynard Smith and Price (1973). At that time I was attached to the Institute of Mathematical Economics at the University of Bielefeld. There, a unique institution, the Centre for Interdisciplinary Research, provides unusual opportunities for scientific contacts among people with very different backgrounds. Since I have always liked to look over the fence of my discipline, I made ample use of these opportunities. I became acquainted with many colleagues from other departments. One of these colleagues, the biologist Hubert Hendrichs, once approached me and told me that game theory is now applied to biology and that he would send me one of his students who knew more about this.

The student was Peter Hammerstein, a young mathematician who was working towards a PhD in biology. He explained to me the above-mentioned paper by Maynard Smith and Price (1973). This was the beginning of a long collaboration which is still continuing. In 1978, a conference on evolutionary game theory was held at the Centre for Interdisciplinary Research at Bielefeld. Hammerstein was one of the organizers of this excellent conference. John Maynard Smith and other leading researchers in the newly emerging field were present. Stimulated by exciting talks and lively discussions, an idea occurred to me about how to prove that evolutionarily stable strategies must be pure in two-player game models of animal conflicts with incomplete information in which the opponents always have different roles. Maynard Smith encouraged me to submit the paper (1980) to the *Journal of Theoretical Biology*.

The definition of an ESS (evolutionarily stable strategy) by Maynard Smith and Price is not well adapted to games in extensive forms. If this definition is applied literally, even very simple extensive game models with perfect information do not have any ESS. In order to overcome this difficulty, I developed a weaker concept of a limit ESS (1983b). For this purpose I made use of the same idea of small mistakes which underlies the definition of perfectness. However, in the new context I permitted such mistakes, but did not require them. Curiously enough, essentially the same apparatus which strengthens the notion of equilibrium to perfect equilibrium, weakens the notion of an ESS to that of a limit ESS.

Unfortunately, my paper (1983b) contains a serious mistake concerning decomposition into subgames and truncations. It is bad enough that I did not avoid an elementary error, but it is even worse that my intuition was wrong. I should have seen that an ESS does not decompose in the same way as a perfect equilibrium point. However, the main results of the paper remain valid. In a second paper (1988) I corrected the mistake and introduced a new kind of decomposition which can be used to simplify the analysis of extensive game models.

In the academic year 1987–8, I organized the research project 'Game Theory in the Behavioral Sciences' at the Centre for Interdisciplinary Studies at the University of Bielefeld. Among the participants who stayed there for extended periods up to one year were economists, biologists, mathematicians, political scientists, psychologists and a philosopher. The results of the project are collected in four volumes, *Game Equilibrium Models* (1991a).

The research year gave me the opportunity to pursue a number of co-operative efforts, among them a collaboration with Avi Shmida, a botanist at the Hebrew University of Jerusalem. We were interested in the question why flowers offer resources to pollinators, a problem which is deeper than one might think at first glance, and contributed a paper (1991) on the subject to volume I of *Game Equilibrium Models* which is devoted to evolution and game dynamics.

Recently Peter Hammerstein and I wrote a survey paper 'Evolutionary game theory' to be published in the *Handbook of Game Theory* (1992). We found ourselves forced to close some gaps in the literature. We were especially worried about the connection of evolutionary game theory and population genetics. The dynamics of population genetics on the basis of a fixed game pool does not necessarily work in the direction of fitness maximization. In order to justify the principle of fitness maximization one has to look at stability against mutations. For a standard model we proved that a monomorphic population state (i.e. a population state in which all individuals play the same pure or mixed strategy) is stable in this sense if and only if the population strategy is an ESS.

Evolutionary game theory is not normative but descriptive. Equilibrium is thought of as the result of a dynamic process. There is now a growing understanding among economists that in view of the limited rationality of economic agents a similar approach to equilibrium in economic game models should be taken. This makes it interesting to look at game learning processes. For a long time I was worried about the notorious instability of equilibrium points in mixed strategies with respect to learning processes. After much thinking about this problem, I discovered that an element of anticipation can be a stabilizing influence. I developed a theory of anticipatory learning in two-person games, published in volume I of *Game Equilibrium Models* (1991b).

18. Bounds of rationality

The problem of bounded rationality has occupied my mind much more than one would think if one looks at the moderate success of my efforts in this direction. One of the few basic insights I gained over the years is the necessity to distinguish between at least two different kinds of bounds of rationality, cognitive bounds and motivational bounds.

Cognitive bounds arise from the limits of the human capability to think and to compute. However, in some cases rationality is limited in a different way. This became clear to me when I discovered the chain store paradox (1978). This game has a clear backward induction solution which nevertheless is unacceptable even if all players are game theorists with common knowledge about this fact. I came to the conclusion that here, as in the finitely repeated prisoner's dilemma, we lack behavioural trust into abstract induction arguments.

I proposed a three-level theory of decision-making, in which decisions arise on up to three levels: the routine level, the level of imagination and the level of analysis. Routine decisions arise spontaneously without any thinking. The level of imagination derives decisions from selected scenarios, which are vividly imagined courses of future play of limited length. Backward induction arguments which span more than a few periods, require the abstract thinking performed on the level of analysis. If possible decisions arise on several levels, then a final decision has to select which of them becomes effective. This final decision is a spontaneous routine decision.

When I submitted this paper to a leading journal, the referee asked me to eliminate the three-level theory which he considered to be irrelevant. Since I thought that this part of my paper was very important I was not willing to do this and eventually, after several years of delay, I published the paper elsewhere.

The attempts to solve the chain store paradox in the framework of rational game theory (Kreps and Wilson, 1982a) evade the issue by addressing a modified game with elements of incomplete information instead of the original chain store game. A related backward induction paradox appears in the finitely repeated prisoner's dilemma, where one experimentally observes co-operation followed by an end-effect (Selten and Stoecker, 1986). The incomplete information explanation (Kreps et al., 1982) of this phenomenon is inadequate, since the typical pattern is learnt only after some experience and not established at once by rational analysis. Stoecker and I have explained our experimental data by a behavioural learning model.

The backward induction paradox is due to what I call motivational bounds of rationality. It is not a lack of cognitive power but rather a failure to behave according to one's rational insights that is important here. Many phenomena of everyday life can be understood as caused by motivational bounds of rationality. Somebody who is convinced that it would be best for him to stop smoking may nevertheless find himself unable to do this.

Recently, in my paper on anticipatory learning (1991b), I developed a related and, maybe, even more radical view of the nature of motivational bounds of rationality, the decision emergency hypothesis. This hypothesis maintains that conscious thinking does not have the task of making decisions. The rational mind is like an adviser to a king. The king is a subconscious hidden mechanism who makes the final decision. The king may or may not listen to the adviser. The adviser does not necessarily have to propose a decision. In some cases he may restrict himself to pointing out advantages and disadvantages of various alternatives. Decisions are not made, they emerge.

19. The strategy method

My greatest hope for progress towards an empirically based theory of bounded rationality is the strategy method: after having gained sufficient experience with a game situation, subjects are asked to write strategies in the form of computer programs. The strategies are matched in tournaments and the subjects can change their programs in the light of the result. This process is repeated several times.

I first described the strategy method in a paper (1967c) which presents an application to my experimental oligopoly game with demand inertia. At that time computer technology was not yet as developed as it is now. The participants wrote their strategies in flow-chart form and did this only once. A tournament of everybody against everybody was not within the bounds of our computing possibilities and therefore we had to be content with runs against one reference strategy written by ourselves.

The evaluation of the strategies provides valuable insights on the structure of boundedly rational behaviour. Of course, the planned behaviour produced by the strategies may be different from the spontaneous behaviour in experimental games, but one can expect that similar principles will be at work. Moreover both spontaneous and planned strategic behaviour are important in real life.

In a study by Mitzkewitz, Uhlich and myself (Selten, Mitzkewitz and Uhlich, 1988), the strategy method was applied to a 20-period supergame of a numerically specified asymmetric Cournot duopoly game. The final strategies after three rounds of programming exhibited a typical structure which suggests a new duopoly theory. Usually oligopoly theories, including game-theoretical approaches, assume optimization against definite expectations about the opponents' behaviour. Contrary to this, it is typical for the final strategies that no expectations are formed and nothing is optimized.

The subjects know by their game-playing experience that they must co-operate in order to earn higher profits. This leads to a typical approach to the problem of constructing a strategy. At first the question is asked where co-operation should take place. The answer is what we call an 'ideal point', a quantity pair derived by one of various fairness criteria, such as equal additional profits above Cournot profits or profits proportional to Cournot profits.

Now the question arises as how to induce the opponent to co-operate at one's own ideal point. The answer is what we call a 'measure-for-measure strategy', which responds to a move of the opponent nearer to one's own ideal point or further away from it by a similar move in the same direction. Thereby one hopes to induce the opponent to co-operate on one's own terms.

Typically, a measure-for-measure strategy is not used at the beginning and end of the game. In a short initial phase co-operativeness is signalled by a descending sequence of history-independent quantities. Near the end, co-operation breaks down.

Interestingly, a strategy was more successful in the final tournament, the more typical it was. This could be shown by a highly significant rank correlation of success with a measure of typicality based on 13 strategy characteristics.

The use of a measure-for-measure strategy in order to achieve co-operation at an ideal point is a reasonable approach to the finitely repeated Cournot duopoly. However, the procedural rationality of this approach is very different from the rationality of Bayesian game theory.

20. From obscurity to the mainstream

In the past I have often worked in obscure areas far away from the mainstream. This is partly due to a spirit of opposition which makes me favourably inclined towards radical departures from commonly held views and partly to an inability to be quick enough to compete successfully on the hot topics of the day. Since I am slow, I have to try to be early. However, it has happened to me several times that an esoteric specialty which attracted my attention later became a flourishing area of research. The number of game theorists was very small when I began to do work in this field. Research was concentrated on co-operative theory. It was the ruling opinion that co-operation would take place wherever it is profitable. In this situation I began to turn my attention to non-co-operative game theory.

During my stay at Berkeley in the academic year 1967–8, I gave a talk at Stanford on the demand inertia oligopoly paper (1965) in which I had introduced subgame perfectness. When people were leaving the room, I heard a young economist say to somebody else that 'this is only non-co-operative theory', a remark which captures the mood of the time.

Nowadays, non-co-operative game theory is used almost everywhere in economic theory. Game theory is no longer a specialty but rather a basic tool of economic theory.

When I began to do experimental work, experiments in economics were almost unheard of. For a long time laboratory research on economic behaviour was not taken seriously by most economists. However, the last decade has seen a tremendous rise of interest in experimental economics. The field has become respectable and it may even become fashionable.

In the many years since I became convinced of the necessity to do work on bounded rationality, I have had countless discussions with economists who firmly believe in utility maximization as an explanation of decision behaviour. The accumulating experimental evidence has made it easier and easier for me to make my point. Defendants of the orthodox view have become forced to take various positions of retreat, such as the idea that evolutionary forces must have produced an approximation to full rationality. In a fictitious dialogue on evolution, learning and economic behaviour (1990), I explained why I feel that attempts to save the rationalistic view of economic man by minor modifications have no chance of succeeding.

The interest in bounded rationality is increasing. We can hope for the emergence of a body of experimentally based descriptive theory. It is possible that the rationalistic approach has already reached its peak and that attention is going to shift to more realistic explanations of behaviour. It is tempting to speculate about the consequences of a trend in this direction. I expect that game models will continue to be very important in economics, but applied game theory will have to change its character. Rational game theory will remain of great philosophical significance, but a new kind of descriptive game theory will have to be developed for economic application.

Of course, my expectations for the future may be completely mistaken. The fact that I have been right before does not mean that I must be right again. However, I would not be surprised if, after some time, empirically based theorizing about boundedly rational behaviour becomes a substantial part of mainstream economics.

References

Abelson, R. P. and Rosenberg, M. J. (1958) Symbolic psycho-logic: a model of attitudinal cognition. *Behavioral Science*, 3, 1–13.
Albers, W. (1975) Zwei Lösungskonzepte für kooperative Mehrpersonenspiele, die auf Anspruchsniveaus der Spieler basieren, *OR-Verfahren*, XXI, Meisenheim, 1–13.
Avenhaus, R., Güth, W. and Huber, K. (1991) Implications of the defense efficiency hypothesis for the choice of military force structures. In R. Selten (ed) *Game Equilibrium Models*, Vol. IV. Berlin: Springer-Verlag, 256–318.
Axelrod, R. (1976) *Structure of Decision, The Cognitive Maps of Political Elites*. Princeton, NJ: Princeton University Press.
Burger, E. (1958) *Einführung in die Theorie der Spiele*. Berlin: de Gruyter.
Burger, E. (1963) *Introduction to the Theory of Games*. Hemel Hempstead: Prentice Hall.
van Damme, D. (1987) *Stability and Perfection of Nash Equilibria*. Berlin: Springer-Verlag.
van Damme, E. and Güth, W. (1991) Equilibrium selection in the Spence signaling game. In R. Selten (ed), *Game Equilibrium Models*, Vol. II. Berlin: Springer-Verlag, 283–88.
Güth, W. (1991) The stability of the Western defense alliance – a game theoretic analysis. In R. Selten (ed), *Game Equilibrium Models*, Vol. IV. Berlin: Springer-Verlag, 229–55.
Güth, W. and Kalkofen, B. (1989) *Unique Solutions for Strategic Games – Equilibrium Selection Based on Resistance Avoidance*. Lecture Notes in Economics and Mathematical Systems 328. Berlin: Springer-Verlag.
Güth, W. and Selten, R. (1991) Majority voting in the Condorcet paradox as a problem of equilibrium selection. In R. Selten (ed), *Game Equilibrium Models*, Vol. IV. Berlin: Springer-Verlag, 7–40.
Hammerstein, P. and Selten R. (1992) Evolutionary game theory. To be published in R. Aumann and S. Hart (eds), *Handbook of Game Theory*.
Harsanyi, J. C. (1963) A simplified bargaining model for the n-person cooperative game. *International Economic Review*, 4, 194–220.
Harsanyi, J. C. (1967–8) Games with incomplete information played by 'Bayesian' players. *Management Science*, 14, 157–82, 320–4, 486–502. Part I (Nov. 1967) The basic model. Part II (Jan. 1968) Bayesian equilibrium points.
Harsanyi, J. C. (1975) The tracing procedure: A Bayesian approach to defining a solution for n-person non-co-operative games. *International Journal of Games Theory*, 4, 61–94.
Harsanyi, J. C. and Selten, R. (1972) A generalized Nash solution for two-person bargaining games with incomplete information. *Management Science*, 18(5), P80–P106.
Harsanyi, J. C. and Selten, R. (1988) *A General Theory of Equilibrium Selection in Games*. Cambridge, Mass.: MIT Press.
Häselbarth, V. (1967) Literaturhinweise zur experimentellen Wirtschaftsforschung. In H. Sauermann (ed), *Beiträge zur experimentellen Wirtschaftsforschung*, Vol. 1. Tübingen: J. C. B. Mohr (Siebeck), 267–73.
Hoggatt, A., Selten, R., Crochett, D., Gill, S. and Moore, J. (1978) Bargaining experiments with incomplete information. In H. Sauermann (ed), *Beiträge zur experimentellen Wirtschaftsforschung*, Vol. 7. Tübingen: J. C. B. Mohr (Siebeck), 127–78.
Kahneman, D. and Tversky, A. (1984) Prospect theory: an analysis of decision under risk. *Econometrica*, 47, 263–91.
Kalish, G., Milnor, J. W., Nash, J. and Nering, E. D. (1954) Some experimental n-person games. In R. M. Thrall, C. H. Coombs and R. L. Davis (eds), *Decision Processes*. New York/London, 301–27.
Keser, C. (1992) Experimental duopoly markets with demand inertia. Unpublished PhD dissertation, Bonn.

Kreps, D., Milgrom, P., Roberts, J. and Wilson, R. (1982) Rational co-operation in the finitely repeated prisoner's dilemma supergame. *Journal of Economic Theory*, **27**, 245–52.
Kreps, D. M. and Wilson, R. (1982a) Reputation and imperfect information. *Journal of Economic Theory*, **27**, 253–79.
Kreps, D. M. and Wilson, R. (1982b) Sequential equilibria. *Econometrica*, **50**, 863–94.
Leopold, U. (1982) *Gleichgewichtsauswahl in einem Verhandlungsspiel mit Opportunitätskosten*. Bielefeld: Pfeffersche Buchhandlung.
Leopold, U. (1985) Equilibrium selection in a bargaining problem with transaction costs. *International Journal of Game Theory*, **14**, 151–72.
Maynard Smith, J. and Price, G. R. (1973) The logic of animal conflict. *Nature*, **246**, 15–18.
Marschak, Th. and Selten, R. (1974) *General Equilibrium with Price Making Firms*. Berlin: Springer-Verlag.
Marschak, Th. and Selten, R. (1977) Oligopolistic economies as games of limited information. *Zeitschrift für die gesamte Staatswissenschaft*, **133**, 385–410.
Marschak, Th. and Selten, R. (1978) Restabilizing responses, inertia supergames and oligopolistic equilibria. *Quarterly Journal of Economics*, **92**, 71–93.
von Neumann, J. and Morgenstern, O. (1944) *Theory of Games and Economic Behavior*. Princeton, NJ: Princeton University Press.
Potters, J., van Winden, F. and Mitzkewitz, M. (1991) Does concession always prevent pressure? In R. Selten (ed), *Game Equilibrium Models IV: social and political interaction*. Berlin: Springer-Verlag, 41–63.
Pruitt, P. G. (1970) Motivational processes in the decomposed prisoner's dilemma game. *Journal of Personality and Social Psychology*, **14**, 227–38.
Ricciardi, F. M. (1957) *Top Management Decision Simulation, The AMA Approach*. New York.
Sauermann, H. (ed) (1967, 1970, 1972, 1978) *Beiträge zur experimentellen Wirtschaftsforschung*, Vols 1, 2, 3 and 7, respectively. Tübingen: J. C. B. Mohr (Siebeck).
Sauermann, H. and Selten, R. (1959) Ein Oligopolexperiment. *Zeitschrift für die gesamte Staatswissenschaft*, **115**, 427–71.
Sauermann, H. and Selten, R. (1962) Anspruchsanpassungstheorie der Unternehmung. *Zeitschrift für die gesamte Staatswissenschaft*, **118**, 577–97.
Selten, R. (1960) Bewertung strategischer Spiele. *Zeitschrift für die gesamte Staatswissenschaft*, **116**, 221–81.
Selten, R. (1964) Valuation of n-person games. In *Advances in Game Theory*, Annals of Mathematics Studies no. 52, Princeton, NJ: Princeton University Press, 565–78.
Selten, R. (1965) Spieltheoretische Behandlung eines Oligopolmodells mit Nachfrageträgheit. *Zeitschrift für die gesamte Staatswissenschaft*, **121**, 301–24, 667–89.
Selten, R. (1967a) Investitionsverhalten im Oligopolexperiment. In H. Sauermann (ed), *Beiträge zur experimentellen Wirtschaftsforschung*, Vol. 1. Tübingen: J. C. B. Mohr (Siebeck), 60–102.
Selten, R. (1967b) Ein Oligopolexperiment mit Preisvariation und Investition. In H. Sauermann (ed), *Beiträge zur experimentellen Wirtschaftsforschung*, Vol. 1. Tübingen: J. C. B. Mohr (Siebeck), 103–35.
Selten, R. (1967c) Die Strategiemethode zur Erforschung des eingeschränkt rationalen Verhaltens im Rahmen eines Oligopolexperiments. In H. Sauermann (ed), *Beiträge zur experimentellen Wirtschaftsforschung*, Vol. 1. Tübingen: J. C. B. Mohr (Siebeck), 136–68.
Selten, R. (1970a) *Preispolitik der Mehrproduktenunternehmung in der statischen Theorie*. Berlin: Springer-Verlag.
Selten, R. (1970b) Ein Marktexperiment. In H. Sauermann (ed), *Beiträge zur experimentellen Wirtschaftsforschung*, Vol. 2. Tübingen: J. C. B. Mohr (Siebeck), 33–98.

Selten, R. (1972) Equal share analysis of characteristic function experiments. In H. Sauermann (ed), *Beiträge zur experimentellen Wirtschaftsforschung*, Vol. 3. Tübingen: J. C. B. Mohr (Siebeck), 130–65.

Selten, R. (1973) A simple model of imperfect competition where 4 are few and 6 are many. *International Journal of Game Theory*, 2(3), 141–201. (Reprinted in Selten, R. 1988: *Models of Strategic Rationality*, Dordrecht: Kluwer.)

Selten, R. (1975a) Reexamination of the perfectness concept for equilibrium points in extensive games. *International Journal of Game Theory*, 2(3), 25–55. (Reprinted in Selten, R. 1988: *Models of Strategic Rationality*, Dordrecht: Kluwer.)

Selten, R. (1975b) Bargaining under incomplete information – a numerical example. In O. Becker and R. Richter (eds), *Dynamische Wirtschaftsanalyse*. Tübingen: J. C. B. Mohr (Siebeck), 203–32. (Reprinted in Selten, R. 1988: *Models of Strategic Rationality*, Dordrecht: Kluwer.)

Selten, R. (1977) The scenario bundle method. In *Research Conference on Strategic Decision Analysis Focusing on the Persian Gulf*, 5–56.

Selten, R. (1978) The chain store paradox. *Theory and Decision*, 9, 127–59. (Reprinted in Selten, R. 1988: *Models of Strategic Rationality*, Dordrecht: Kluwer.)

Selten, R. (1980) A note on evolutionarily stable strategies in asymmetric animal contests. *Journal of Theoretical Biology*, 83, 147–61. (Reprinted in Selten, R. 1988: *Models of Strategic Rationality*, Dordrecht: Kluwer.)

Selten, R. (1981) A non-co-operative model of characteristic function bargaining. In V. Böhm and H. Nachtkamp (eds), *Essays in Game Theory and Mathematical Economics in Honor of Oskar Morgenstern*. Wissenschaftsverlag Bibliographisches Institut, 131–51. (Reprinted in Selten, R. 1988: *Models of Strategic Rationality*, Dordrecht: Kluwer.)

Selten, R. (1983a) A model of oligopolistic size structure and profitability. *European Economic Review*, 22, 33–57. (Reprinted in Selten, R. 1988: *Models of Strategic Rationality*, Dordrecht: Kluwer.)

Selten, R. (1983b) Evolutionary stability in extensive two-person games. *Mathematical Social Sciences*, 5, 269–363.

Selten, R. (1983c) Equal division payoff bounds for three-person characteristic function experiments. In R. Tietz (ed), *Aspiration Levels in Bargaining and Economic Decision Making*, Springer Lecture Notes in Economics and Mathematical Systems 213, 265–75. (Reprinted in Selten, R. 1988: *Models of Strategic Rationality*, Dordrecht: Kluwer.)

Selten, R. (1984) Are cartel laws bad for business? In H. Hauptmann, W. Krelle and K. C. Mosler (eds), *Operations Research and Economic Theory*. Berlin: Springer-Verlag, 86–117. (Reprinted in Selten, R. 1988: *Models of Strategic Rationality*, Dordrecht: Kluwer.)

Selten, R. (1987) Equity and coalition bargaining in experimental three-person games. In A. Roth (ed), *Laboratory Experimentation in Economics – Six Points of View*. Cambridge: Cambridge University Press, 42–98.

Selten, R. (1988) Evolutionary stability in extensive two-person games, correction and further development. *Mathematical Social Sciences* 16(3), 223–66.

Selten, R. (1990) Evolution, learning, and economic behavior. *Games and Economic Behavior*, 3, 3–24.

Selten, R. (ed) (1991a) *Game Equilibrium Models*, Vols I–IV. Berlin: Springer-Verlag.

Selten, R. (1991b) Anticipatory learning in two-person games. In R. Selten (ed), *Game Equilibrium Models*, Vol. I. Berlin: Springer-Verlag, 98–154.

Selten, R. (1991c) Balance of power in a parlor game. In R. Selten (ed), *Game Equilibrium Models*, Vol. IV. Berlin: Springer-Verlag, 150–209.

Selten, R. (1992) A demand commitment model of coalition bargaining. In R. Selten (ed),

Rational Interaction, Essays in Honor of John C. Harsanyi. Berlin: Springer-Verlag, 245–82.

Selten, R. and Berg, Claus C. (1970) Drei experimentelle Oligopolspielserien mit kontinuierlichem Zeitablauf. In H. Sauermann (ed), *Beiträge zur experimentellen Wirtschaftsforschung*, Vol. 2. Tübingen: J. C. B. Mohr (Siebeck), 162–221.

Selten, R. and Güth, W. (1982) Equilibrium point selection in a class of market entry games. In *Games, Economics, and Time Series Analysis*. Wurzburg: Physica-Verlag, 101–16.

Selten, R. and Güth, W. (1991) Original and fake – a bargaining game with incomplete information. In R. Selten (ed), *Game Equilibrium Models*, Vol. III. Berlin: Springer-Verlag, 186–229.

Selten, R. and Leopold U. (1983) Equilibrium point selection in a bargaining game with opportunity costs. *Economie Appliquée*, **XXXVI**(4), 611–48.

Selten, R., Mitzkewitz, M. and Uhlich, G. (1988) Duopoly strategies programmed by experienced players. Discussion Paper, Sonderforschungsbereich 303, No. B-106, University of Bonn.

Selten, R. and Schuster, Klaus G. (1968) Psychological variables and coalition forming behavior. In K. Borch and I. Mussin (eds), *Risk and Uncertainty*, London/Melbourne/Toronto/New York: St Martin's Press, 221–46.

Selten, R. and Shmida, A. (1991) Pollinator foraging and flower competition in a game equilibrium model. In R. Selten (ed), *Game Equilibrium Models*, Vol. I. Berlin: Springer-Verlag, 195–256.

Selten, R. and Stoecker, R. (1986) End behavior in sequences of finite prisoner's dilemma supergames. *Journal of Economic Behavior and Organization*, **I**, 47–70.

Selten, R. and Tietz, R. (1972) Security equilibria. In R. Rosecrance (ed), *The Future of the International Strategic System*, San Francisco/Scranton/London/Toronto: Chandler, 103–22.

Simon, H. A. (1957) *Models of Man*. New York: Wiley.

Uhlich, G. R. (1990) *Descriptive Theories of Bargaining*. Lecture Notes in Economics and Mathematical Systems, No. 341. Berlin: Springer-Verlag.

PART II

AXIOMATIC CHARACTERIZATIONS

[2]

VALUATION OF n-PERSON GAMES*

Reinhard Selten

INTRODUCTION

The problem of valuation in game theory can be loosely formulated in this way: What is the value of an n-person game for a player taking part in it? The von Neumann value [12], which is the natural consequence of a theory of rational behavior, is defined only for 2-person constant-sum games. Up to now there has been no satisfying theory of rational behavior for general n-person games on which a value concept could be based.

As far as the value problem is concerned, however, the difficult question of rational behavior can be avoided if another approach is taken, namely, axiomatic characterization of value functions. The first result in this direction, the characterization of the Shapley value [11], is not wholly satisfactory because it presupposes that general n-person games are adequately represented by their characteristic functions. This objection cannot be raised against the characterizations given here.

The main results of this paper are two characterization theorems. In §7 a theorem is proved characterizing the Shapley value as a value function for n-person constant-sum games. The value function characterized in §8 coincides with the Shapley value only for the region of n-person constant-sum games. This value function is the modified Shapley value introduced by John C. Harsanyi [2], [3]. Harsanyi has constructed bargaining models for which this value can be derived as an equilibrium solution with reference to a certain arbitration scheme.** Bargaining models and arbitration schemes offer

---

* This paper is essentially a translation of the author's doctoral dissertation at the Johann Wolfgang Goethe University, Frankfurt am Main, Germany. I am indebted to Professor Dr. E. Burger who gave valuable advice.

** Harsanyi's results can be considered as generalizations of the Nash theory of cooperative 2-person games; cf. [7].

an approach to the value problem basically different from the characterization by inner properties.

Most of the postulates used here for characterization have been intuitively justified already in [10]. The newly-introduced postulates (IXa) and (XI) are discussed in §2. As in [10], value functions are understood as valuations for finite cooperative games with transferable utility and unlimited side-payments; furthermore, it is implicitly assumed that the players want to maximize their expected payoff in the sense of the probability calculus. In [10] it is discussed at length why some of the postulates are reasonable only on these general assumptions.*

Although it is an immediate consequence of postulate (I) that the value depends on the normal form only, all the other postulates are formulated in terms of the extensive form. Probably it would have been possible to use the normal form exclusively, but the extensive form seemed to be more advantageous with respect to the interpretation of the postulates and the simplicity of the proofs. Postulate (I) can presumably be replaced by several simpler postulates concerning the extensive form. In this way, a system of postulates uniformly dealing with the extensive form could be obtained.

The characterization theorems proved in §7 and §8 can be considered generalizations of similar theorems on 2-person value functions already proved in [10], which are restated together with other results from [10] in §1. (There they appear as E 6 and E 7.) The 2-person theorems, however, are not contained in the general theorems as special cases, because fewer characterizing postulates are needed in the 2-person case.

The mutual independence of the characterizing postulates will not be discussed here. In [10], however, two theorems were proved which give a partial answer to this question for the characterization theorems in §7 and §8 (E 9 and E 10 in §1).

There are some open questions** with regard to the class of η-values described in §5. The η-values are value functions for general n-person games which satisfy all the postulates occurring in the characterization theorems of §7 and §8 with exception of postulate (X). The Shapley value and the modified Shapley value characterized in §8 are contained in this class

---

*   See the additional comment at the end of this paper.
** Cf. [10] p. 240.

VALUATION OF n-PERSON GAMES

as extreme special cases. Presumably it is possible to characterize the Shapley value, too, and even other η-values for the region of general n-person games, by using appropriate other postulates instead of postulate (X).

§ 1. FORMER RESULTS

In this paragraph some results of [10] are restated. As far as well-known game-theoretic concepts are concerned, we only explain our notation. Definitions in full detail are given in [10].

GAMES AND THEIR CONSTITUENTS. For games and their constituents, the following letters are used: Γ, and K, $\mathcal{P}$, $\mathcal{U}$, $\mathcal{A}$, p, h. Equal indices mean relationship to the same game. K with origin o is the game tree* of the game Γ. A vertex y comes "after" x, if x is between o and y, x comes "after" x. The set X(K) of vertices, other than endpoints, is partitioned by the player partition $\mathcal{P}$ into player sets P_i (i = 0, ..., n), and by the information partition $\mathcal{U}$ into information sets U. The move partition $\mathcal{A}$ partitions the alternative set A(K) into moves u.** (Alternatives are edges in K.) An alternative a is "at" x, if it connects x and a vertex after x. A move u is "at" U, and U "belongs" to u, if the alternatives of u are at vertices in U. The probability assignment p fixes the probabilities of the chance moves. The payoff function h defines a payoff vector $h(z) = (h_1(z), \ldots, h_n(z))$ for each z in the endpoint set Z(K). The symbol (x, b) stands for the "one-point game" with a single vertex x and payoff vector b. A player with $P_j = \emptyset$ and $h_j(z) = a$ for all $z \in Z(K)$ is called a "dummy with a".

GRAPHICAL REPRESENTATION. The graphical representation describes information sets by dashed lines and payoff vectors by column vectors. The symbol i to the left of a vertex x shows that x is in P_i.

STRATEGIES. Strategies of the player i are marked by the lower index i. A pure strategy π_i assigns moves $\pi_i(U)$ to information sets.

---
* We consider only games with finite trees.
** The use of the word "move" is not the same as in [5].

A mixed strategy fixes probabilities $q_i(\pi_i)$ for pure strategies. We identify a pure strategy π_i with the mixed strategy q_i which assigns 1 to π_i. A "combination" $\pi = (\pi_1, \ldots, \pi_n)$ of pure strategies contains a pure strategy for each of the players $1, \ldots, n$; if a component π_j in π is replaced by ζ_j the combination π/ζ_j results. The same notation is used for combinations $q = (q_1, \ldots, q_n)$ of mixed strategies. The strategy-set vector $\Pi = (\Pi_1, \ldots, \Pi_n)$ contains the sets Π_i of the pure strategies of a player i. $Q = (Q_1, \ldots, Q_n)$ is the corresponding strategy-set vector for mixed strategies.

EXPECTED PAYOFF. $H(\pi) = (H_1(\pi), \ldots, H_n(\pi))$ is the expected payoff for pure strategies. The maximum of $H_1(\pi) + \ldots + H_n(\pi)$ is called "maximum component sum $M(\Gamma)$". The game Γ is a "constant-sum game", if $H_1(\pi) + \ldots + H_n(\pi)$ is constant. $E(q) = (E_1(q), \ldots, E_n(q))$ is the expected payoff for mixed strategies. The "security level" $s(q_j)$ is the minimum of $E_j(q)$ for q_j in q.

THE NORMAL FORM. A strategy-set vector $\Pi = (\Pi_1, \ldots, \Pi_n)$ together with an expected payoff function H forms an n-person "normal form" $G = (\Pi, H)$. The constituents do not have to originate from a game. Two normal forms $G = (\Pi, H)$ and $G' = (\Pi', H')$ are called isomorphic if there are one-to-one payoff-preserving mappings $f_1, \ldots, f_n$ of the Π_i onto the Π_i'. $N(\Gamma)$ is the "normal form of Γ" with the constituents Π and H from Γ. To each normal form G there is a "related normal game" Γ, which can be described intuitively in this way: The players $i_1, \ldots, i_k$ ($i_1 < \ldots < i_k$) with more than one strategy in G successively select a strategy each, without being informed about previous moves of the others. The payoff is the expected payoff $H(\pi)$ corresponding to this selection.

OPERATIONS. The "sum" $\Gamma'' = \Gamma + \Gamma'$ is formed from Γ by substituting subgames Γ^z for the payoff vectors $h(z)$; apart from the payoff function h^z the Γ^z do not differ from Γ'; h^z is defined by $h^z(z') = h(z) + h'(z')$. The game $\lambda \cdot \Gamma$ results if all payoffs in Γ are multiplied by λ. "k dummies with 0 are added" by appending k 0-components to all payoff vectors. If a move u in Γ is "forbidden", all parts

VALUATION OF n-PERSON GAMES

of K only realizable through u are taken away; apart from this, the rules of Γ are transferred to the resulting game Γ'. Two players are "interchanged" by interchanging their player sets and their payoffs. "The payoff to player j is reduced" in Γ by substituting for the payoff function h a new payoff function h' with the following properties: (1) $h_j'(z) \leq h_j(z)$; (2) $h_i'(z) \geq h_i(z)$ for $i \neq j$; (3) for the resulting game Γ' we have $M(\Gamma') = M(\Gamma)$. From a 2-person game Γ, the "η-replenished" game $F_\eta(\Gamma)$ results by substituting for the payoff function h the following h_η:

$$h_{\eta,1}(z) = h_1(z) + \eta(M(\Gamma) - h_1(z) - h_2(z))$$
$$h_{\eta,2}(z) = h_2(z) + (1 - \eta)(M(\Gamma) - h_1(z) - h_2(z))$$
$$(0 \leq \eta \leq 1).$$

If the "indexed* mixed strategies $q_{j_1}^1, \ldots, q_{j_m}^m$ are added" to a normal form G, a normal form G' results which, besides the pure strategies of G, contains the $q_{j_r}^r$ (r = 1, ..., m) as pure strategies of the player j_r. The payoff for G' is given by $H'(\pi') = E(\pi')$.

LINEAR EQUIVALENCE. The normal forms G and G' are "linearly equivalent" if there is a chain of normal forms $G, G_1, \ldots, G_m, G'$, in which two neighboring links always satisfy one of the following conditions: (a) they are isomorphic, (b) one of them results from adding a mixed strategy to the other. Γ and Γ' are linearly equivalent if $N(\Gamma)$ and $N(\Gamma')$ are linearly equivalent. The notation $G \sim G'$ and $\Gamma \sim \Gamma'$ is used for linear equivalence.

VALUE FUNCTIONS. Value functions w assign values $w(\Gamma) = (w_1(\Gamma), \ldots, w_n(\Gamma))$ to games. The specifications "for general n-person games" and "for n-person constant-sum games" characterize the region on which w is defined. The "von Neumann value function" v is a value for 2-person constant-sum games; $v_i(\Gamma)$ in $v(\Gamma) = (v_1(\Gamma), v_2(\Gamma))$ is the maximum security level of player i. The "η-stable agreements" V_η $(0 \leq \eta \leq 1)$ are value functions for general 2-person games defined by $V_\eta(\Gamma) = v(F_\eta(\Gamma))$. The "2-person η-values" w_η $(0 \leq \eta \leq 1/2)$ are defined by $w_\eta(\Gamma) = \frac{1}{2} V_\eta(\Gamma) + \frac{1}{2} V_{1-\eta}(\Gamma)$.

---
* The same strategy may appear several times with different indices.

POSTULATES. Our postulates concern value functions. The postulates (IXa) and (XI), which did not occur in [10], are related to operations which will be described in §2. The postulates are listed below.

(I) LINEAR EQUIVALENCE. If $\Gamma \sim \Gamma'$ then $w(\Gamma) = w(\Gamma')$.

(II) SYMMETRY. If Γ' results from Γ by interchanging two of the players, then $w(\Gamma')$ is formed from $w(\Gamma)$ by interchanging the corresponding components.

(III) MAXIMUM JOINT PAYOFF. $w_1(\Gamma) + w_2(\Gamma) + \ldots + w_n(\Gamma) = M(\Gamma)$.

(IV) DUMMY. If player j in Γ is a dummy with a, then $w_j(\Gamma) = a$.

(V) STRATEGIC MONOTONICITY. If Γ' results from Γ by forbidding a move u of player j, then $w_j(\Gamma') \leq w_j(\Gamma)$.

(VI) HOMOGENITY. If $\Gamma' = \lambda \cdot \Gamma$ with $\lambda > 0$, then $w(\Gamma') = \lambda w(\Gamma)$.

(VII) ADDITIVITY. If $\Gamma'' = \Gamma + \Gamma'$ then $w(\Gamma'') = w(\Gamma) + w(\Gamma')$.

(VIIa) WEAK ADDITIVITY. If Γ' is a one-point game and $\Gamma'' = \Gamma + \Gamma'$, then $w(\Gamma'') = w(\Gamma) + w(\Gamma')$.

(VIII) TRUNCATION. If Γ' results from Γ by substituting for a subgame Γ_y its value $w(\Gamma_y)$, then $w(\Gamma') = w(\Gamma)$.

(IX) PAYOFF MONOTONICITY. If Γ' results from Γ by reducing the payoff of player j, then $w_j(\Gamma') \leq w_j(\Gamma)$.

(IXa) WEAK PAYOFF MONOTONICITY. If Γ' results from Γ by transferring payoff from player j to other players, then $w_j(\Gamma') \leq w_j(\Gamma)$.

(X) NORMALIZATION. $w_1(^n\Gamma^o) = w_2(^n\Gamma^o) = 1$. Here $^n\Gamma^o$ is the normalization game. It results from $^2\Gamma^o$ (Figure 1) by adding $n - 2$ dummies with 0.

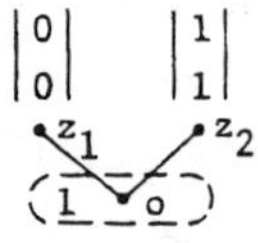

Figure 1: $^2\Gamma^o$

(XI) INFORMATION MONOTONICITY. If Γ' results from Γ by informing player k at U^1 about the move u of player j $(j \neq k)$, then $w_j(\Gamma') \leq w_j(\Gamma)$.

RESULTS. Results from [10] are listed below by numbers of the form E ... .

E 1: If G' results from G by adding the indexed mixed strategies $q_{j_1}^1, \ldots, q_{j_m}^m$, then $G' \sim G$ (§2, p. 247).

E 2: If the subgame Γ_y of a game Γ is replaced by a game Γ'_y with $\Gamma_y \sim \Gamma'_y$, a game Γ' results with $\Gamma \sim \Gamma'$ (§2, p. 248).

E 3: The von Neumann value function v satisfies postulates (I) - (IX) (§5, p. 256).

E 4: The η-stable agreements V_η satisfy postulates (I) and (III) - (IX) (§6, p. 261).

E 5: The 2-person η-values w_η satisfy postulates (I) - (VII) and (IX) (§6, p. 263).

E 6: There is one and only one value function for 2-person constant-sum games which satisfies postulates (I), (III), (IV), (V), and (IX), namely the von Neumann value function v (§5, p. 259).

E 7: There is one and only one value function for general 2-person games which satisfies postulates (I) - (V) and (VIII), namely $w_{1/2}$ (§7, p. 264).

E 8: There is one and only one value function for general 2-person games, which satisfies (I) - (V), (VII), and (X), namely $w_{1/2}$ (§7, p. 264).

E 9: (I), (III), (IV), (V) as postulates for value functions for 2-person constant-sum games are mutually independent (§8, p. 277).

E 10: (I) - (V), (VII), and (X) as postulates for value functions for general 2-person games are mutually independent (§8, p. 277).

E 11: There is no value function for general n-person games $(n \geq 3)$ satisfying postulates (I) — (V), and (VIII) (§9, p. 281).

§ 2. POSTULATES (IXa) AND (XI)

PAYOFF TRANSFER. Γ' results from Γ "by transferring payoff from player j to other players," if only the payoff functions h' and h'' differ in both n-person games and if, furthermore, the following conditions are satisfied:

(1) $h_j'(z) \leq h_j(z)$;

(2) $h_i'(z) \geq h_i(z)$ for $i \neq j$;

(3) $\sum_{i=1}^{n} h_i'(z) = \sum_{i=1}^{n} h_i(z)$.

If Γ' results from Γ by transferring payoff from player j to other players, then obviously Γ' results from Γ also by reducing the payoff of player j. Therefore postulate (IXa) already stated in §1 is weaker than postulate (IX).

As a postulate for value functions for general n-person games, postulate (IX) is too strong. If the payoff to player j is reduced, there may be an endpoint z where the payoff to another player k is increased to such an extent that both players together receive more than before; this could be an improvement of player j's position if he, together with player k, can enforce realization of z. Obviously this objection cannot be raised against (IXa). It will be shown in §5 that the value functions for general n-person games described here satisfy (IXa) but not (IX).

INFORMATION TRANSFER. Let U^2 be an information set of player j in an n-person game Γ and u a move at U^2. We define a vertex set $Y(u)$ "realizable with u". This set contains all vertices coming after vertices to which alternatives of u lead. An information set U of player k is called "partitionable by u", if it contains only vertices which come after vertices of U^2. Thus partitionability by u depends only on the information set belonging to u. Let U^1 be an information set of player k partitionable by u (we don't exclude j = k). U^1 can be partitioned into two sets $B = U^1 \cap Y(u)$ and $U^1 - B$. Each set is, if not empty, a set "into which u partitions U^1". A move "into which u partitions a move u^1 at U^{1}" contains all the alternatives of u^1 which are at vertices of a certain one of the sets into which u partitions U^1. We define the game Γ' which u results from Γ if "player k at U^1 is informed about the move u of player j". This game differs from Γ only with regard to the information partition $\mathcal{U}'$ and the move partition $\mathcal{Q}'$. $\mathcal{U}'$ differs from $\mathcal{U}$ by not containing U^1 and by containing the sets into which u partitions U^1, instead. Correspondingly, the move partition $\mathcal{Q}'$ does not contain the moves at U^1, but contains the moves into which u partitions the moves at U^1; otherwise $\mathcal{Q}'$ is not different from $\mathcal{Q}$. Speaking of player k at U^1 being informed about the move u of player j, we always presuppose that U^1 is partitionable by u.

The operation "information transfer" can be interpreted in the following way: The rules of the game are changed by an additional rule. Before the play begins, player j has to inform the "game director" whether his agent at U^2 will or will not select move u in case he will have to decide. After the beginning of the play, but before the decision of the agent at U^1, the game director gives this information to the agent at U^1. Thus the agent at U^1 always knows to which one of the two subsets B and $U^1 - B$ the play has advanced when and if he has to decide. If U^1 were not partitionable by u he would also always know that only one of two subsets of U^1 can be decisive, but the two subsets would have in common those vertices which do not follow vertices of U^2. Therefore we must always have partitionability of U^1 by u.

It is intuitively clear that the position of player j is not improved if another player is informed about one of his moves. Admittedly it could be advantageous for player j if player k receives information about u; in this case, however, he can also give this information to player k before the beginning of the play. Therefore postulate (XI) (information monotonicity) can be considered as reasonable with regard to our intuitive assumptions, because we deal only with fully cooperative games, in which the players can make arbitrary contracts before the beginning of the play which can be thought of as enforcible by law, once they are agreed upon.

THEOREM ON INFORMATION TRANSFER. Let Γ' be the game which results from Γ if player k at U^1 is informed about move u of player j. Then Γ' has the following properties: (a) $M(\Gamma') = M(\Gamma)$; (b) if $j = k$, then $\Gamma' \sim \Gamma$. (c) For any value function w, for general n-person games or for n-person constant-sum games, which satisfies (I), (III), and (V), we have

$$w_k(\Gamma') \geq w_k(\Gamma).$$

Furthermore, if $n = 2$, the value function w satisfies (XI).

COROLLARY. From (c), and E 3, E 4, E 5 it is clear that the 2-person value functions v, V_η, and w_η satisfy (XI).

PROOF. We can assume without loss of generality that both $B = U^1 \cap Y(u)$ and $U^1 - B$ are not empty, because otherwise Γ and Γ' are the same game and therefore our propositions are trivially true.

In Γ' all players with the exception of player k have the same strategies as in Γ. For every strategy π_k in Γ, player k has a strategy $\pi_k' = f(\pi_k)$ with $\pi_k'(U) = \pi_k(U)$ for the information sets with $U \neq U^1$ and $\pi_k'(B) = \pi_k'(U^1 - B) = \pi_k(U^1)$. For all combinations π in Γ we have

(1) $$H(\pi/\pi_k) = H'(\pi/f(\pi_k)).$$

The combinations π/π_k and $\pi/f(\pi_k)$ realize the same endpoints with the same probabilities, indeed, because for vertices x of player k we always have $\pi_k(x) = \pi_k'(x)$, if $\pi_k' = f(\pi_k)$. For every strategy π_k' in Γ' player k has two strategies $\pi_k^1 = g_1(\pi_k')$ and $\pi_k^2 = g_2(\pi_k')$ with $\pi_k^1(U) = \pi_k^2(U) = \pi_k'(U)$ for the sets U with $U \neq U^1$ and $\pi_k^1(U^1) = \pi_k'(B)$ and $\pi_k^2(U^1) = \pi_k'(U^1 - B)$. If $\pi_j'(U^2) = u$ is valid, a combination π' containing π_j' can lead only to plays which intersect with U^1 in B or not at all. For

$\pi_j'(U^2) \neq u$ we have the same situation with regard to $U^1 - B$. Therefore for combinations π' in Γ' we have

(2) $\qquad H(\pi'/\pi_j'/g_1(\pi_k')) = H'(\pi'/\pi_j'/\pi_k') \qquad$ for $\pi_j'(U^2) = u$

(3) $\qquad H(\pi'/\pi_j'/g_2(\pi_k')) = H'(\pi'/\pi_j'/\pi_k') \qquad$ for $\pi_j'(U^2) \neq u$.

If π is a combination with maximum component sum for $H(\pi)$, and if π_k is player k's component in π, then because of (1), the component sum for $H'(\pi/f(\pi_k))$ is $M(\Gamma)$, too. Therefore we have $M(\Gamma') \geq M(\Gamma)$. If π' is a combination with the maximum component sum for $H'(\pi')$ and if π_k' is player k's component in π', then because of (2) and (3) one of the two combinations $\pi'/g_1(\pi_k')$ and $\pi'/g_2(\pi_k')$ in Γ also results in the component sum $M(\Gamma')$. Consequently we have $M(\Gamma) \geq M(\Gamma')$. This proves (a). We now proceed to show (b). Thus we have $j = k$. Let $g(\pi_k') = g_1(\pi_k')$ for $\pi_k'(U^2) = u$ and $g(\pi_k') = g_2(\pi_k')$ for $\pi_k'(U^2) \neq u$. Because of (2) and (3) we have for all π'

(4) $\qquad H(\pi'/g(\pi_k')) = H'(\pi'/\pi_k')$.

It can also easily be seen that we always have $\pi_k = g(f(\pi_k))$. We extend the region, on which f is defined: For every π_k' which cannot be expressed by $\pi_k' = f(\pi_k)$, we construct an indexed mixed strategy q_k which is $g(\pi_k')$ completed by an index and then we define $\pi_k' = f(q_k)$. To $N(\Gamma)$ we add the set of all these q_k and get a normal form G^1. $N(\Gamma) \sim G^1$ follows from E 1 (cf. §1). Because of (4) we have

(5) $\qquad H'(\pi'/f(q_k)) = H(\pi'/g(f(q_k))) = H^1(\pi'/q_k);$

in addition, because of (1) we have

(6) $\qquad H'(\pi'/f(\pi_k)) = H(\pi'/\pi_k) = H^1(\pi'/\pi_k)$.

It can be seen from (5) and (6) that f together with the identical mappings of the strategy sets π_i with $i \neq k$ onto themselves gives the isomorphism of G^1 and $N(\Gamma')$. Because of $N(\Gamma) \sim G^1$ we have $\Gamma \sim \Gamma'$. This proves (b).

We now proceed to show (c). Here we assume $j \neq k$ and a value function w satisfying (I), (III), and (V). Let G^2 be the normal form which is different from $N(\Gamma')$ only in that player k has fewer strategies in G^2, that is, he has only those which can be expressed as $f(\pi_k)$; the payoff function is taken from $N(\Gamma')$ without change. Obviously we can get a game isomorphic to the normal game Γ^2 of G^2 by successively forbidding moves in the normal game of $N(\Gamma')$ where the forbidden moves correspond to the strategies π_k' which cannot be expressed as $\pi_k' = f(\pi_k)$. Because of (I) and (V) we have

$$w_k(\Gamma^2) \leq w_k(\Gamma').$$

It can be seen from (1) that f together with the identical mappings of the π_i with $i \neq k$ gives the isomorphism of $N(\Gamma)$ and G^2. Therefore we have

$$w_k(\Gamma') \geq w_k(\Gamma).$$

This is the first part of (c). Because of (a) and (III) we have $n = 2$

$$w_j(\Gamma') \leq w_j(\Gamma).$$

This proves the rest of the theorem.

§ 3. COALITION GAMES

COALITIONS. A coalition in an n-person game is a subset C of the set $N = (1, \ldots, n)$. The players i with $i \in C$ are the "members" of C. The coalition C' which results from C by "interchanging players i and j" contains one of the two players if and only if C contains the other; otherwise C' has the same members as C.

COALITION GAMES. We define for each coalition C in an n-person game Γ a coalition game $_C\Gamma$. Only with regard to the player partition and to the payoff function is this game different from Γ. $_C\Gamma$ is a 2-person game. The player-set $_CP_1$ of player 1 in $_C\Gamma$ is the union of all player sets of members of C in Γ. The player set $_CP_2$ is the union of all player sets of members of the complementary coalition $N - C$ in Γ. At all endpoints z, the payoff $_Ch_1(z)$ is the sum of the payoffs to the members of C in Γ. Similarly, $_Ch_2(z)$ is the sum of the payoffs to the members of $N - C$ in Γ. If we have $C = N$, this is to be understood as $_Ch_2(z) = 0$ everywhere; similarly we have $_Ch_1(z) = 0$ if $C \neq \emptyset$.

COALITION STRATEGIES AND COALITION FORMS. We use $|M|$ for the number of elements in a set M. Let $|C| = k$. A "coalition strategy" for C is a k-tuple π_C which contains a pure strategy for Γ for each member of C. By its components π_C assigns moves $\pi_C(U)$ to the information sets U in the player-set $_CP_1$, and therefore can be considered as a pure strategy of player 1 for $_C\Gamma$. Thus the coalition strategies for C are the pure strategies for player 1 in $_C\Gamma$ and the coalition strategies for $N - C$ are the pure strategies for player 2 in $_C\Gamma$.

In the same way we can define coalition strategies for normal forms, and for any normal form G we can form 2-person normal forms $_CG$, in which player 1 has the coalition strategies for C and player 2 has the coalition strategies for $N - C$. The payoff function $_CH$ is derived in the same way from H as $_Ch$ from h. Obviously, the "coalition form" $_CG$ of a normal form G and the normal form $N(_C\Gamma)$ of the coalition game $_C\Gamma$ derived from the normal game Γ related to G are isomorphic.

NOTATION. We always use the coalition as the left lower index for coalition games and coalition forms; otherwise the same indices are used as in in the original game or form.

THEOREM ON COALITION GAMES.

(a) If $\Gamma \sim \Gamma'$, then $_C\Gamma \sim _C\Gamma'$.

(b) $_{N-C}\Gamma$ results from $_C\Gamma$ if the players are interchanged.

(c) If Γ' and C' result from Γ and C respectively by interchanging players j and k, then $_C\Gamma$ and $_{C'}\Gamma'$ are the same game.

(d) $M(_C\Gamma) = M(\Gamma)$.

(e) If in Γ player j in C is a dummy with a, then $_C\Gamma = {}_{C-(j)}\Gamma + (x, (-a, 0))$.

(f) If Γ' results from Γ by forbidding the move u of player j, then $_C\Gamma'$ results from $_C\Gamma$ by forbidding u; in $_C\Gamma'$, u is a move of player 1 for $j \in C$ and u is a move of player 2 for $j \in N - C$.

(g) If Γ' results from Γ by transferring payoff from a member of C to other players, then $_C\Gamma'$ results from $_C\Gamma$ by transferring payoff from player 1 to other players.

(h) If $\Gamma' = \lambda\Gamma$, then $_C\Gamma' = \lambda\, _C\Gamma$.

(i) If $\Gamma'' = \Gamma + \Gamma'$, then $_C\Gamma'' = {}_C\Gamma + {}_C\Gamma'$.

(j) If Γ' results from Γ by informing player k at U^1 about the move u of player j and if $j \in C$ and $k \in N-C$, then $_C\Gamma'$ results from $_C\Gamma$ by informing player 2 at U^1 about the move u of player 1.

(k) If Γ' results from Γ by informing player k at U^1 about move u of player j and if $j \in C$ and $k \in C$, then we have $_C\Gamma' \sim {}_C\Gamma$.

PROOF. Statements (b) − (j) are immediate consequences of the definition of coalition game, and (k) follows from the theorem on information transfer in §1 (statement (b)). We now proceed to prove (a). As $\Gamma \sim \Gamma'$ and $N(\Gamma) \sim N(\Gamma')$ are equivalent, we have only to show that $_CG \sim {}_CG'$ is a consequence of $G \sim G'$. It is clear from the definition of linear equivalence that we have to consider only the special case in which G' is formed by adding a mixed strategy q_j of player j to G. Without loss of generality we can assume $j \in C$, because, if the theorem has been proved already for $j \in C$, we would have for $j \in N-C$ that $_{N-C}\Gamma \sim {}_{N-C}\Gamma'$ is a consequence of $\Gamma \sim \Gamma'$, then by interchanging both players in $_{N-C}\Gamma$ and $_{N-C}\Gamma'$ we do not disturb the linear equivalence, and we get $_C\Gamma \sim {}_C\Gamma'$ because of (b).

In $_CG'$ player 2 has the same strategies as in $_CG$; player 1 retains his strategies from $_CG$, too, but he has additional strategies π_C'

containing q_j. For each additional strategy π_C' we construct a related mixed strategy $q_C = f(\pi_C')$ for $_CG$; this is uniquely defined by $q_C(\pi_C'/\pi_j) = q_j(\pi_j)$. Obviously, $q_C = f(\pi_C')$ assigns the probability 0 to all coalition strategies which do not agree with π_C' about the behavior of the players $i \neq j$. We consider the normal form $_CG''$, which results from $_CG$ by adding the indexed set of all mixed strategies $q_C = f(\pi_C')$. We now define mappings which give the isomorphism of $_CG'$ and $_CG''$: the strategies of $_CG'$ which also occur in $_CG$ are mapped onto themselves, and the additional strategies of $_CG'$ are mapped by f onto the added strategies of $_CG''$. It can easily be seen that these mappings do not disturb the payoff function. Hence $_CG'$ and $_CG''$ are isomorphic. $_CG \sim {}_CG''$ follows from E 1. Therefore we have $_CG \sim {}_CG'$. This proves the theorem.

§ 4. CHARACTERISTICS

CHARACTERISTICS. An "n-person characteristic" is a function c which assigns real numbers c(C) to the coalitions $C \subseteq N$ and satisfies

(1) $\qquad c(C) + c(N - C) = c(N).$

A player j is a "dummy" in c, if for $j \in C$ we always have

(2) $\qquad c(C - (j)) = c(C) - c((j)).$

If all players are dummies, c is called "inessential". "Superadditivity" is defined in the usual way. The coalition function c' results from c by "interchanging players j and k", if for coalitions C and C', where C' results from C by interchanging players j and k, we always have $c'(C') = c(C)$. A characteristic c is "symmetric", if there is a coalition R called a "symmetry carrier" with the property that c depends only on $|C \cap R|$; clearly, because of $c(\emptyset) = 0$, all players outside R are dummies in c with $c((j)) = 0$. If c has N as symmetry carrier, c is "fully symmetric".

LEMMA ON SYMMETRY. A nontrivial symmetric characteristic has only one symmetry carrier R. (The "trivial" characteristic assigns 0 to all coalitions.)

PROOF. The proof is indirect. If c has several symmetry carriers, then there is at least one player j who belongs to one of the carriers R, but not to another, R'. Player j is a dummy in c and we have $c((j)) = 0$. As $c(C)$ depends only on $|C \cap R|$, it follows that we have $c(C) = 0$ for $|C| = 1$. If $c(C) = 0$ is true for $|C \cap R| = k < |R|$, then a coalition C with $|C \cap R| = k$, which does not contain j, can be extended by adding the dummy j; thus we get a coalition C_1 with $c(C_1) = 0$ and $|C_1 \cap R| = k + 1$; hence we have $c(C) = 0$ also for $|C \cap R| = k + 1$. By induction $c(C) = 0$ is always true. This proves the lemma.

COMBINATION THEOREM: All characteristics can be linearly combined from superadditive symmetric characteristics.

This theorem has already been proved by Isbell in [4], p. 393-394; Isbell has shown that all characteristics are linear combinations of the characteristic functions of direct majority games. Obviously these characteristic functions are superadditive symmetric characteristics.

η-CHARACTERISTICS. The "η-characteristic" $c_\eta (0 \leq \eta \leq 1/2)$ of an n-person game Γ is the n-person characteristic given by $c_\eta(C) = w_{\eta,1}(_C\Gamma)$.

We must justify this definition by showing that c_η really is a characteristic. From E 5 we know that w_η satisfies (I) — (IV). The theorem in §3 states under (b) that $_{N-C}\Gamma$ results from $_C\Gamma$ by interchanging both players, and under (d) we find $M(_C\Gamma) = M(\Gamma)$. By (I) and (b) we get $c_\eta(N-C) = w_{\eta,2}(_C\Gamma)$. Therefore, because of (III) and (d), we have $c_\eta(C) + c_\eta(N-C) = M(\Gamma)$; this is condition (1) from the definition of the characteristic. $c_\eta(\emptyset) = 0$ is an immediate consequence of (IV).

The symbol η is always used as lower index for η-characteristics; otherwise the same indices are used as in the original game.

CHARACTERISTIC FUNCTION. From our definitions it can easily be seen that the characteristic function introduced by von Neumann and Morgenstern is given by

VALUATION OF n-PERSON GAMES

$$V(C) = V_{0,1}(_C\Gamma) \qquad \text{for all} \quad C \subseteq N.$$

THEOREM ON CONSTANT-SUM GAMES. All the η-characteristics c_η ($0 \leq \eta \leq 1/2$) of an n-person constant-sum game Γ are the same, namely the characteristic function V of Γ.

PROOF. For 2-person constant-sum games the value functions V_η, W_η, and the von Neumann value function v coincide, because here η-replenishment does not change the expected payoff. From this we get the theorem as an immediate consequence.

THEOREM ON η-CHARACTERISTICS.

(a) If $\Gamma \sim \Gamma'$, then $c_\eta = c'_\eta$.

(b) If Γ' results from Γ by interchanging players j and k, then c'_η results from c_η by interchanging players j and k.

(c) $c_\eta(N) = M(\Gamma)$.

(d) If player j in Γ is a dummy with a, then he is a dummy in c_η, too, and we have $c((j)) = a$.

(e) If Γ' results from Γ by forbidding move u of player j, then for $j \in C$ we always have

$$c'_\eta(C) \leq c_\eta(C) + \tfrac{1}{2}(M(\Gamma') - M(\Gamma)).$$

(f) If

(1) $$\Gamma = \sum_{s=1}^{r} \lambda_s \Gamma^s \quad \text{where} \quad \lambda_s \geq 0 \quad \text{for} \quad s = 1, \ldots, r,$$

then we have

(2) $$c_\eta = \sum_{s=1}^{r} \lambda_s c_\eta^s.$$

(g) If Γ' results from Γ by transferring payoff from player j to other players, then for $j \in C$ we always have $c'_\eta(C) \leq c_\eta(C)$.

(h) If Γ' results from Γ by informing player k at U^1 about the move u of player j (j = k is not excluded), then for j ∈ C we always have $c'_\eta(C) \leq c_\eta(C)$.

PROOF. From §1, E 5, and §2 we know that the 2-person η-values satisfy (I) − (VII), (IX), and (XI). Considering this, we get (a) − (d) and (f) − (h) as immediate consequences of the theorem on coalition games in §3. We now proceed to prove (e).

$_C\Gamma'$ results from $_C\Gamma$ by forbidding move u of player 1 (statement (f) in the theorem of §3). The η-replenished game $F_\eta(_C\Gamma')$ results from $F_\eta(_C\Gamma) + (x, u)$ by forbidding move u, where u is the vector $(\eta, 1-\eta)$ multiplied by $M(\Gamma') - M(\Gamma)$ (cf. [10], §6, p. 260*). Using $M(_C\Gamma) = (M(\Gamma)$, (statement (d) in the theorem of §3) and considering the fact that v satisfies (V) and (VIIa), we conclude:

$$V_{\eta,1}(_C\Gamma') \leq V_{\eta,1}(_C\Gamma) + \eta(M(\Gamma') - M(\Gamma))$$

$$V_{1-\eta,1}(_C\Gamma') \leq V_{1-\eta,1}(_C\Gamma) + (1-\eta)(M(\Gamma') - M(\Gamma)).$$

By the definition of the 2-person η-values, this yields

$$c'_\eta(C) = w_{\eta,1}(_C\Gamma') \leq c_\eta(C) + \tfrac{1}{2}(M(\Gamma') - M(\Gamma)).$$

§ 5. THE n-PERSON η-VALUES

In this paragraph a class of value functions for general n-person games is described. These value functions are generalizations of the 2-person η-values (§1).

In the following, $\mathcal{C}_i$ stands for the set of all coalitions $C \subseteq N$ containing i; $\mathcal{C}$ is the set of all coalitions $C \subseteq N$.

n-PERSON η-VALUES. An η-value $(0 \leq \eta \leq 1/2)$ is a value function w_η for general n-person games; the components $w_{\eta,i}$ are given by

---

* There is an error in the third line of the theorem "Satz über das Verbot eines Zuges". The first letter should have been Γ'', not Γ'.

VALUATION OF n-PERSON GAMES

(A) $$w_{\eta,i}(\Gamma) = \sum_{C \in \mathcal{C}_i} \gamma_n(C)(c_\eta(C) - c_\eta(N-C))$$

where

$$\gamma_n(C) = (n - |C|)!\,(|C| - 1)!/n!$$

It can easily be seen that this definition for $n = 2$ agrees with the definition of w given in §1.

LEMMA ON γ_n. γ_n has the following properties:

(1) $$\sum_{C \in \mathcal{C}_i} \gamma_n(C) = 1$$

(2) $$\gamma_n(C) = \gamma_n((N - C) \cup (i)) \qquad \text{for all } C \in \mathcal{C}_i.$$

PROOF. It can easily be seen that (2) is true. There are exactly $\binom{n-1}{k-1}$ k-element coalitions in $\mathcal{C}_i$. Therefore we have

$$\sum_{C \in \mathcal{C}_i} \gamma_n(C) = \sum_{k=1}^{n} \binom{n-1}{k-1} \cdot \frac{(n-k)!\,(k-1)!}{n!} = 1.$$

This proves (1).

OTHER EXPRESSIONS FOR THE η-VALUES.

(B) $$w_{\eta,i}(\Gamma) = \sum_{C \in \mathcal{C}_i} \gamma_n(C)(c_\eta(C) - c_\eta(C - (i)))$$

(C) $$w_{\eta,i}(\Gamma) = -M(\Gamma) + 2 \sum_{C \in \mathcal{C}_i} \gamma_n(C) c_\eta(C)$$

for $i = 1, \ldots, n$.

PROOF. Expression (A) can be split into two sums:

$$\sum_{C \in \mathcal{C}_i} \gamma_n(C) c_\eta(C) - \sum_{C \in \mathcal{C}_i} \gamma_n(C) c_\eta(N - C).$$

If $(N-C') \cup (i)$ is substituted for C in the second sum, expression (B) is seen to result with the aid of (2). Also, because of $c_\eta(N) = M(\Gamma)$, we have

$$c_\eta(C) - c_\eta(N-C) = 2c_\eta(C) - M(\Gamma).$$

Using this, (A) is easily transformed into (C) with the aid of (1).

THEOREM ON THE n-PERSON η-VALUES. The n-person η-values satisfy postulates (I) − (VII), (IXa), and (XI).

PROOF. We use (a) − (h) from the theorem on η-characteristics (§4). (I) is satisfied because of (a). With the aid of (C), (b) gives (II). From (d) and property (1) of γ_n it can be seen with the aid of (B) that (IV) is satisfied. (V) follows from (e) and (C). Statement (f) gives (VI) and (VII) with the aid of (A). Postulates (IXa) and (XI) are given by (g) and (h) with the aid of (C), where property (a) from the theorem on information transfer (§2) is used with regard to (XI). Expression (A) is formed in the same way from c_η as the components of the Shapley value are formed from the characteristic function (cf. [11], p. 312, (13), or formula (S) in the next section). Therefore Shapley's axiom 2 ([11], p. 309), which states that the value-components sum up to the characteristic function value of N, together with (c) gives (III). (Shapley's axioms 1 and 3 could have been used in the same way in order to show that (II) and (VII) are satisfied.)

THE SHAPLEY VALUE. Let V be the characteristic function of Γ. The value function s with components

(S) $\quad s_i(\Gamma) = \sum_{C \in \mathcal{C}_i} \gamma_n(C)(V(C) - V(C - (i)))\quad$ for $i = 1, \ldots, n$

is called the "Shapley value" (cf. [11]).

THEOREM ON THE SHAPLEY VALUE. The Shapley value s and the η-value w_0 are the same value function.

VALUATION OF n-PERSON GAMES

PROOF. If the 2-person game Γ' results from Γ by interchanging both players, then $F_{1-\eta}(\Gamma')$ results from $F_\eta(\Gamma)$ by interchanging both players (cf. [10], p. 260). This, together with (b) from the theorem on coalition games (§3), gives

$$v_{1,1}(_C\Gamma) = v_{0,2}(_{N-C}\Gamma).$$

Hence, because (III) is satisfied by v_0 (E 4, §1), we have

$$v_{1,1}(_C\Gamma) = M(_C\Gamma) - v_{0,1}(_{N-C}\Gamma).$$

Therefore the definition of w_0 with the aid of $M(_C\Gamma) = M(\Gamma)$ and $V(C) = v_{0,1}(C)$ yields

$$c_0(C) = w_{0,1}(_C\Gamma) = \tfrac{1}{2}V(C) - \tfrac{1}{2}V(N - C) + M(\Gamma)/2.$$

Therefore the right side of (B) can be transformed into the following expression:

$$\tfrac{1}{2} \sum_{C \in \mathcal{C}_i} \gamma_n(C)(V(C) - V(C - (i))) + \tfrac{1}{2} \sum_{C \in \mathcal{C}_i} \gamma_n(C)(V((N-C) \cup (i)) - V(N-C)).$$

In the second sum the coefficient $\gamma_n(C)$ can be replaced by $\gamma_n((N-C) \cup (i))$ with the aid of property (2) of γ_n; obviously $\mathcal{C}_i$ is also the set of all coalitions $(N-C) \cup (i)$ with $C \in \mathcal{C}_i$. Therefore the second sum equals the first one. Consequently the expression equals the right side of equation (S). This proves the theorem.

THEOREM ON CONSTANT-SUM GAMES. For n-person constant-sum games all η-values are equal to the Shapley value.

PROOF. It follows from the theorem on constant-sum games in §4 that all w_η are equal for constant-sum games. Therefore the proposition is a consequence of the above theorem on the Shapley value.

FURTHER PROPERTIES OF THE η-VALUES. It follows from §1, E 11, that the n-person η-values do not satisfy (VIII) for $n > 2$. The behavior of the w_η with regard to (IX) and (X) will be described by two theorems.

THEOREM. Postulate (IX) is not satisfied by the n-person η-values for $n > 2$.

PROOF. In all games occurring in this proof the players $4, \ldots, n$ are dummies with 0. Only the 3-person case is graphically represented in Figures 2 and 3. For $n > 3$ all payoff vectors must be completed by adding vanishing components for the players $4, \ldots, n$. The game Γ^2 (Figure 3) results from the game Γ^1 (Figure 2) by reducing the payoff of player 1. We have $M(\Gamma^1) = M(\Gamma^2) = 1$. We investigate the η-characteristics c_η^1 and c_η^2 of Γ^1 and Γ^2 by comparing $c_\eta^1(C)$ and $c_\eta^2(C)$ for $C \in \mathfrak{C}_1$. We consider four cases: (1) $2 \in C$, $3 \in C$. $_C\Gamma^1$ and $_C\Gamma^2$ are linearly equivalent; hence we have $c_\eta^1(C) = c_\eta^2(C)$. (2) $2 \in C$, $3 \notin C$. Player 1 can secure 1 for himself in $F_\eta(_C\Gamma^1)$ and $F_\eta(_C\Gamma^2)$; player 2 always receives at least 0. Therefore we have $c_\eta^1(C) = c_\eta^2(C) = 1$. (3) $2 \notin C$, $3 \notin C$. In $F_\eta(_C\Gamma^1)$ and $F_\eta(_C\Gamma^2)$ player 1 can secure η for himself and player 2 always receives at least $1 - \eta$. Therefore we have $V_\eta(_C\Gamma^1) = V_\eta(_C\Gamma^2) = (\eta, 1 - \eta)$. Consequently we have $c_\eta^1(C) = c_\eta^2(C) = 1/2$. (4) $2 \notin C$, $3 \in C$. In $F_\eta(_C\Gamma^1)$ player 1 can secure η and player 2 always receives at least $1 - \eta$. As in (3) we have $c_\eta^1(C) = 1/2$. But player 1 in $F_\eta(_C\Gamma^2)$ can secure 1 for himself while player 2 always receives at least 0. Therefore we have $c_\eta^2(C) = 1$. Since in the cases (1), (2), (3) we have $c_\eta^1(C) = c_\eta^2(C)$, but $c_\eta^1(C) < c_\eta^2(C)$ holds in case (4), we see from the formula (C) for $w_{\eta,1}$ that $w_{\eta,1}(\Gamma^1) < w_{\eta,1}(\Gamma^2)$ is true. This contradicts (IX).

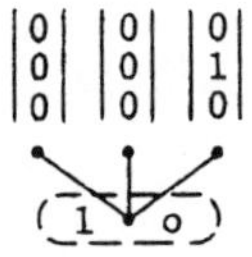

Figure 2: Γ^1

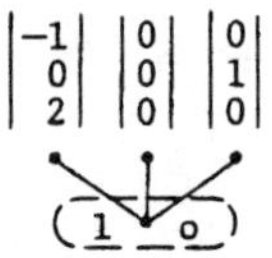

Figure 3: Γ^2

THEOREM. The n-person value function $w_{1/2}$ satisfies (X).

PROOF. Let $\mathfrak{C}_{1,2}$ be the set of all coalitions containing players 1 and 2. If C is not in $\mathfrak{C}_{1,2}$ but in $\mathfrak{C}_1$ or $\mathfrak{C}_2$, then we have

$^n_c\!o_{1/2}(C) - {^n_c}o_{1/2}(N\!-\!C) = 0$, because then the coalition games $_c{^n}\Gamma^o$ and $_{N-C}{^n}\Gamma^o$ of the normalization game from (X) and the 2-person normalization game $^2\Gamma^o$ with value (1, 1) are the same game. Therefore, applying formula (A) to $w_{1/2,i}(^n\Gamma^o)$ for $i = 1, 2$, we may change the regions of summation from $\mathfrak{C}_1$ and $\mathfrak{C}_2$ to $\mathfrak{C}_{1,2}$. This shows that both value-components are equal; they are equal to 1 since we have $M(^n\Gamma^o) = 2$, and (III) and (IV) are satisfied by $w_{1/2}$. This proves the theorem.

COROLLARY. $w_{1/2}$ is the only n-person η-value satisfying (X).

This is an immediate consequence of the characterization theorem which will be proved in §8.

§ 6. LEMMATA

In this paragraph eight lemmata are proved which prepare for the characterization theorems in §7 and §8.

LEMMA 1. For every n-person constant-sum game Γ there is an n-person constant-sum game Γ' with the following properties: (1) Γ' is the normal game of a normal form G'; (2) each coalition $C \in \mathfrak{C}_1$ has an optimal pure strategy $\bar{\pi}_{C'}$ for Γ'; (3) if w is a value function for n-person constant-sum games which satisfies (I) and (XI), we have $w_1(\Gamma') \leq w_1(\Gamma)$; (4) the characteristic functions V and V' of Γ and Γ' are equal to each other.

PROOF. Let Γ^1 be the normal game of Γ. We construct Γ^2: The origin o of Γ^2 belongs to the player set of player 1. There is a one-to-one correspondence between the alternatives a_C at o and the coalitions $C \in \mathfrak{C}_1$. The alternative a_C leads to a vertex x_C. Each vertex x_C forms a one-element information set U_C of player 1. The alternative $a(\pi_C)$ leads from x_C to a vertex $x(\pi_C)$. There is a one-to-one correspondence between the pure coalition strategies π_C in Γ and the alternatives $a(\pi_C)$. Each vertex $x(\pi_C)$ is origin of a subtree of Γ^2 isomorphic to the tree K^1 of Γ^1. The vertices in these subtrees which correspond to vertices in the same information set U_i^1 of player i in Γ^1, together form an information set U_i^2 of player i ($i = 1, \ldots, n$). The alternatives on vertices from U_i^2, which correspond to alternatives in the same move u_i^1, together form a move u_i^2 belonging to U_i^2. In this way exactly one information set U_i^2 or move u_i^2 respectively corresponds to each information set U_i^1 and to each move u_i^1 of Γ^1. We write $U_i^1 \to U_i^2$ or $u_i^1 \to u_i^2$, respectively, if this correspondence holds. We assign the same payoffs to the endpoints of the subtree on the $x(\pi_C)$ as to the corresponding endpoints of Γ^1. This completes the construction of Γ^2.

We now proceed to prove the linear equivalence of Γ and Γ^2. It is sufficient to show $\Gamma^1 \sim \Gamma^2$. We write $\pi_i^2 \to \pi_i^1$ if $\pi_i^1(U^1) \to \pi_i^2(U^2)$ is true for $U_i^1 \to U_i^2$, and we write $\pi^2 \to \pi^1$ if the components of both combinations in this way correspond to each other; we use the same notation for coalition strategies. For $i = 2, \ldots, n$ this correspondence evidently is one-to-one, but to each π_1^1 correspond several π_1^2. For all these π_1^2 we have $H^2(\pi^2/\pi_1^2) = H^1(\pi^1/\pi_1^1)$ if $\pi^2 \to \pi^1$, since the payoff is only influenced by the decisions after $x(\pi_C)$. Therefore it is clear that a normal form isomorphic to $N(\Gamma^2)$ can be formed by adding a set of indexed mixed strategies to $N(\Gamma^1)$. Therefore we have $\Gamma^1 \sim \Gamma^2$ (§1, E 1).

Let U_0 be the information set containing the origin of Γ^2. The game Γ^2 is changed into Γ^3 by transferring information about all moves u at U_0 in the following way: All players are informed about u successively at all information sets partitionable by u.

The game Γ^3 is changed into Γ^4 by transferring information about all moves u at the U_C in the following way: All members of C successively are informed about u at all their information sets partitionable by u. The information sets U^4 of a coalition $C \in \mathcal{C}_1$ in Γ^4 evidently have the following property: If U^4 contains vertices coming after $x(\pi_C)$, then U^4 contains only such vertices. Therefore in Γ^4 we can define special strategies for all members of C called "C-strategies": The alternatives assigned to vertices after any $x(\pi_C)$ by a C-strategy are indirectly

VALUATION OF n-PERSON GAMES

indicated by π_C through the isomorphism of the subtree at $x(\pi_C)$ and K^1; the decisions correspond to those of the strategy taking the position of π_C in Γ^1. Furthermore, a C-strategy of player 1 assigns a_C to the origin.

For each π_C we arbitrarily select one of player 1's C-strategies assigning $a(\pi_C)$ to x_C; this strategy is called "the π_C-strategy" of player 1. Let q_C be a mixed coalition strategy for $C \in \mathcal{C}_1$ (i.e., a mixed strategy of player 1 in $_C\Gamma$). We construct a mixed strategy q_1^4 "related" to q_C in the following way: if π_1^4 is the π_C-strategy of player 1, then we define $q_1^4(\pi_1^4) = q_C(\pi_C)$; consequently, if π_1^4 is not a π_C-strategy, we have $q_1^4(\pi_1^4) = 0$. Let $\overline{Q}_1^4$ be a set of indexed mixed strategies which for each $C \in \mathcal{C}_1$ contains exactly one mixed strategy $\overline{q}_1^4$ related to a fixed optimal strategy $\overline{q}_C$ of player 1 in $_C\Gamma$; let only these strategies be in $\overline{Q}_1^4$. We add $\overline{Q}_1^4$ to $N(\Gamma^4)$ and thereby obtain the normal form G' with the normal game Γ'. We have $\Gamma' \sim \Gamma^4$ (§1, E 1).

We have to show that Γ' has properties (1) – (4). Obviously (1) is true. Γ^4 results from Γ^2 by successively transferring information about moves of player 1. Hence, if w satisfies (I) and (XI), we have $w_1(\Gamma^4) \leq w_1(\Gamma^2)$. Here, beside (XI), we use (I) and (b) from the theorem on information transfer in §1. Because of $\Gamma \sim \Gamma^2$ and $\Gamma' \sim \Gamma^4$ it follows with the aid of (I) that Γ' has the property (3). For each $C \in \mathcal{C}_1$ let $\overline{q}_C^4$ be a mixed coalition strategy which is used if player 1 uses the strategy $\overline{q}_1^4 \in \overline{Q}_1^4$ related to $\overline{q}_C$ and all other members of C use arbitrary but fixed C-strategies. A strategy π_{N-C}^4 is in the "π_{N-C}-class" $B(\pi_{N-C})$ if in the subgame at x_C the decisions prescribed by π_{N-C}^4 are indirectly indicated by π_{N-C} in the following way: Let π_{N-C}^1 correspond to π_{N-C} in Γ^1 and let π_{N-C}^2 correspond to π_{N-C}^1 by $\pi_{N-C}^2 \to \pi_{N-C}^1$; in the subgame at x_C the decisions of strategies in $B(\pi_{N-C})$ coincide with those of π_{N-C}^2. This definition makes use of the fact that in the subgame at x_C every information set U^4 of a member of N–C is a subset of a unique information set U^2 for Γ^2. It can be seen easily that the payoff which results, if a strategy π_{N-C}^4 is played against $\overline{q}_C^4$, depends only on the π_{N-C}-class of π_{N-C}^4, and, moreover, this payoff is equal to the payoff which results if the corresponding π_{N-C} is used against $\overline{q}_C$. Therefore $\overline{q}_C^4$ must have the same security level in $_C\Gamma^4$ as $\overline{q}_C$ in $_C\Gamma$. Hence we have $v^4(C) \geq v(C)$ for $C \in \mathcal{C}_1$. On the other hand, as Γ^4 results from Γ^2 by transferring information about moves of player 1, with the aid of $V = c_\eta$ (theorem on constant-sum games, §4) it follows from (h) in the theorem on η-characteristics (§4) that we have $v^4(C) \leq v^2(C)$ for $C \in \mathcal{C}_1$. But V and v^2 coincide because of $\Gamma \sim \Gamma^2$ (§4, (a) in the theorem on η-characteristics). Hence we have $v^4(C) = V(C)$ for $C \in \mathcal{C}_1$. Since V is a characteristic, V and v^4 coincide. Therefore $\bar{q}_C^{-4}$ has the security level $v^4(C)$ in $_C\Gamma^4$. It follows that $\bar{q}_C^{-4}$ is optimal. V' and v^4 coincide because of $\Gamma' \sim \Gamma^4$. Therefore Γ' has the property (4). Obviously the pure coalition strategy $\bar{\pi}_C{}'$ corresponding to $\bar{q}_C^{-4}$ in Γ' is optimal, too. This proves (2).

SUPERFLUOUS ALTERNATIVES. A "superfluous" alternative a_0 at a vertex x of player j is characterized by the following properties: (1) x belongs to a one-element information set; (2) a_0 leads to an endpoint z_0; (3) there is another alternative a_1 at x which leads to an endpoint z_1 with $h_j(z_1) \geq h_j(z_0)$ and $h_i(z_1) \leq h_i(z_0)$ for $i \neq j$; besides we have

$$\sum_{i=1}^n h_i(z_0) = \sum_{i=1}^n h_i(z_1)$$

A move is called "superfluous", if it contains a superfluous alternative.

LEMMA 2. Let w be a value function for general n-person games or for n-person constant-sum games which satisfies (I), (III), (IV), (V), (VII), and (IXa). Let u_0 be a superfluous move in an n-person game Γ. For the game Γ' which results from Γ by forbidding u_0, we have $w(\Gamma') = w(\Gamma)$.

PROOF. We consider the game Γ^k (Figure 4; $k = 1, \ldots, n$); the dots stand for vanishing payoff components). By forbidding the move from o to y to player k in Γ^k, a game with value $(0, \ldots, 0)$ results; this is because of (III) and (IV). Therefore because of (V) we have $w_k(\Gamma^k) \geq 0$. Thus $w_j(\Gamma^k) \geq 0$ holds for $j = k$. By forbidding in Γ^k the move from y to

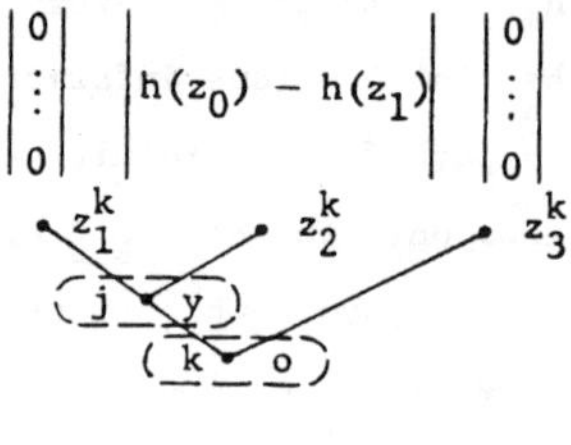

Figure 4: Γ^k

z_2^k a game is formed which is linearly equivalent to a one-point game with vanishing payoffs. Therefore, because of (I), (IV), and (V), we have $w_j(\Gamma^k) \geq 0$ also for $j \neq k$. For $i \neq j$ and $i \neq k$ we have $h_i(z_0) - h_j(z_1) \geq 0$; therefore by substituting 0 for all payoffs at z_2^k in Γ^k we form a game which results from Γ^k by transferring payoff from player i to other players. Since player i in this game is a dummy with 0, it follows with the aid of (IV) and (IXa) that we have $w_i(\Gamma^k) \geq 0$. Hence, because of (III) and $M(\Gamma^k) = 0$, the value of Γ^k is $(0, \ldots, 0)$. We form the games $\Gamma^{n+k} = \Gamma' + \Gamma^k$. Let u_1 be the move of Γ^{n+k} at the endpoint z_1 (from property (3) of a_0) which corresponds to the move from o to z_3^k in Γ^k. By forbidding in Γ^{n+k} first u_1 and then all the moves at the endpoints $z \neq z_1$ which correspond to the move from o to y in Γ^k, a game Γ^{2n+k} results. The subgame of Γ^{2n+k} at the endpoints $z \neq z_1$ and at the vertex x_1 coming immediately before z_1 are linearly equivalent to the subgames of Γ at the same points. It follows with the aid of §1, E 2, that Γ^{2n+k} and Γ are linearly equivalent. Because of (I) we have $w(\Gamma^{2n+k}) = w(\Gamma)$. Because of (V) we have $w_k(\Gamma^{n+k}) \geq w_k(\Gamma^{2n+k})$. Postulate (VII) gives $w(\Gamma^{n+k}) = w(\Gamma')$. With the aid of $M(\Gamma) = M(\Gamma')$ it follows by (III) that $w(\Gamma) = w(\Gamma')$ is true. This proves the lemma.

THE EQUALIZING GAME. For $n = 2, 3, \ldots$ we now construct an n-person game ${}^n\Gamma$, the "equalizing game". The intuitive motive for this name will be made clear by later definitions. ${}^n\Gamma$ is a constant-sum game. The origin o is a vertex of player 1. There are $n - 1$ alternatives at o leading to the endpoints $z_2, \ldots, z_n$.

The payoff is given by

$${}^n h_i(z_k) = \begin{cases} 0 & \text{for } i = k \text{ or } i = 1 \\ 1 & \text{for } i \neq k \text{ and } i \neq 1 \end{cases}$$

for $k = 2, \ldots, n$. Since at each endpoint z_k at most one of the members of a nonempty coalition $C \neq \mathcal{C}_1$ receives 0, the common payoff of C is always at least $|C| - 1$. Player 1 can secure $|C| - 1$ for any coalition $C \in \mathcal{C}_1 - (N)$ by realizing a z_k with $k \in N-C$. Therefore the characteristic function $^n v$ of $^n \Gamma$ can be described as follows: (1) $^n v(\emptyset) = 0$; (2) $^n v(N) = n - 2$; (3) $^n v(C) = |C| - 1$ for $0 < |C| < n$. Evidently $^n v$ is fully symmetric.

> LEMMA 3. If a value function w for n-person constant-sum games satisfies (I) – (V), (VII), and (IXa), then we have $w_i(a \cdot {}^n\Gamma) = a(n - 2)/n$ for $a \geq 0$ and $i = 1, \ldots, n$.

PROOF. Let Γ^1 be the game $a \cdot {}^n\Gamma$. We construct an n-person game Γ^2: the trees of Γ^1 and Γ^2 are isomorphic; the endpoints $z_2, \ldots, z_n$ of Γ^1 correspond to the endpoints $z_2^2, \ldots, z_n^2$ of Γ^2. In Γ^2 all players, with exception of players 1 and 2, are dummies with 0; we have $h_1^2(z_k) = +a$ and $h_2^2(z_k) = -a$ for $k \neq 2$ and $h_1^2(z_2) = h_2^2(z_2) = 0$; player 2 decides at the origin. Then $w_i(\Gamma^2) = 0$ holds for $i = 3, \ldots, n$ because of (IV). By transferring payoff from player 1 to other players, Γ^2 can be changed into a game in which all players receive 0 everywhere. Because of (III) and (IV) the value of this game is $(0, \ldots, 0)$. Hence by (IXa) we have $w_1(\Gamma^2) \geq 0$. By forbidding in Γ^2 all moves not leading to z_2^2 a game is formed the value of which is $(0, \ldots, 0)$ because of (III) and (IV). Therefore by (V) we have $w_2(\Gamma^2) \geq 0$. With the aid of (III) it follows from what has already been said that $(0, \ldots, 0)$ is the value of Γ^2.

Next we form $\Gamma^3 = \Gamma^2 + \Gamma^1$. Because of (VII) we have $w(\Gamma^3) = w(\Gamma^1)$. By forbidding for $k = 2, \ldots, n$ all moves at z_k^2 which do not correspond to the move leading to z_k in Γ^1, the game Γ^4 results from Γ^3. By substituting linearly equivalent one-point games for the subgames at the z_k^2, the game Γ^4 is changed into Γ^5. Clearly $\Gamma^4 \sim \Gamma^5$ (§1, E 2). By interchanging the players 1 and 2, a game Γ^6 results from Γ^1. Obviously Γ^5 and Γ^6 are isomorphic. Thus by (I) and (II) it follows that $w_1(\Gamma^5) = w_2(\Gamma^1)$. By (I) and (V) we have $w_1(\Gamma^5) \leq w_1(\Gamma^3)$. Since $w(\Gamma^3) = w(\Gamma^1)$, we see that $w_2(\Gamma^1) \leq w_1(\Gamma^1)$. It will be shown that $w_1(\Gamma^1) \leq w_2(\Gamma^1)$ holds, too. From $w_1(\Gamma^1) = w_2(\Gamma^1)$ it can be seen by (II) that $w_1(\Gamma^1) = w_i(\Gamma^1)$ for $i = 2, \ldots, n$; indeed, Γ^1 is not changed if two of the players $2, \ldots, n$ are interchanged. Since $M(\Gamma^1) = a(n-2)$ the proposition of the theorem is implied by this. Therefore, only $w_1(\Gamma^1) \leq w_2(\Gamma^1)$ remains to be shown.

By substituting 0 for all payoffs we change Γ^2 into Γ^7, which by (III) and (IV) has (0, ..., 0) as value. We form $\Gamma^8 = \Gamma^7 + \Gamma^1$. We construct a game Γ^9 which differs from Γ^8 only by a different payoff function: At the endpoint z_k^2 which corresponds to z_k in Γ^1 we substitute $h^6(z_k)$ for $h^1(z_k)$; we do this for $k = 2, ..., n$; the payoff at all other endpoints of Γ^9 is $h^1(z_2) = h^6(z_2)$. These changes nowhere increase the payoff of player 2, since player 2 in Γ^6 receives 0 everywhere. The vector $h^6(z_k)$ differs from $h^1(z_k)$ only with regard to the first two components and to the advantage of player 1. Similarly $h^1(z_2)$ is at least as good as $h^1(z_k)$ for all players with exception of player 2. Hence Γ^9 results from Γ^8 by transferring payoff from player 2 to other players. Therefore (IXa) implies $w_2(\Gamma^9) \leq w_2(\Gamma^8)$. By (VII) we have $w(\Gamma^8) = w(\Gamma^1)$. Clearly, $w_2(\Gamma^9) \leq w_2(\Gamma^1)$. It can easily be seen that in Γ^9 all the moves at z_k^2, $k = 3, ..., n$, which do not lead to the endpoint corresponding to z_k of Γ^1 are superfluous in the sense of Lemma 2. A game Γ^{10} results from Γ^9 by forbidding all these superfluous moves. By Lemma 2 we have $w(\Gamma^{10}) = w(\Gamma^9)$. Hence $w_2(\Gamma^{10}) \leq w_2(\Gamma^1)$. In Γ^{10} we can substitute linearly equivalent subgames for the subgames at the z_k^2, $k = 2, ..., n$. A game Γ^{11} results. By §1, E 2, we have $\Gamma^{11} \sim \Gamma^{10}$. It follows from (I) that $w(\Gamma^{11}) = w(\Gamma^{10})$. Hence $w_2(\Gamma^{11}) \leq w_2(\Gamma^1)$. Evidently Γ^{11} and Γ^6 are isomorphic. This yields $w_2(\Gamma^6) \leq w_2(\Gamma^1)$. By (II) we have $w_2(\Gamma^6) = w_1(\Gamma^1)$. Hence $w_1(\Gamma^1) \leq w_2(\Gamma^1)$. This proves the lemma.

EQUALIZING. Since in $\Gamma = a \cdot {}^n\Gamma$ only player 1 has more than one pure strategy, the expected payoff for mixed strategies can be described as a function $E(q_1) = (E_1(q_1), ..., E_n(q_1))$ of player 1's mixed strategy q_1 alone. Let C be a nonempty coalition not containing player 1. The mixed strategy q_1 "equalizes" the vector $b = (b_1, ..., b_n)$ with regard to C, if the following conditions are satisfied:

1) $b_i + E_i(q_1)$ has the same value for all $i \in C$

2) $E_i(q_1) = a$ for all $i \in N-C - (1)$.

Because of 2) the behavior prescribed by q_1 is optimal for N-C.

LEMMA 4. For every vector $b = (b_1, \ldots, b_n)$ there is a number $a_0 \geq 0$ such that for $a \geq a_0$ there is a mixed strategy q_1 in $\Gamma = a \cdot {}^n\Gamma$ which equalizes b with regard to a nonempty coalition $C \subseteq N - (1)$.

PROOF. The vectors $E(q_1) = (E_1(q_1), \ldots, E_n(q_1))$ can be considered as points in an n-dimensional "payoff space". $E(q_1)$ satisfies 2) if and only if q_1 assigns positive probabilities only to those $|C|$ pure strategies leading to the payoff 0 for one of C's members. All these $E(q_1)$ form a $(|C| - 1)$-dimensional simplex S in the payoff space. The center of gravity of S is $m = (m_1, \ldots, m_n)$, where $m_1 = 0$ and $m_i = a$ for $i \in N-C - (1)$; for $i \in C$ we have $m_i = a(|C| - 1)/|C|$. Let b_C be the sum of all b_i with $i \in C$. Evidently q_1 satisfies 1) from the definition of "equalizing", if we have

(1) $$b_i + E_i(q_1) = m_i + b_C/|C| \qquad \text{for } i \in C.$$

By substituting $b_C/|C|$ for all b_i with $i \in C$, the vector b is changed into a vector $\bar{b}$. (1) is implied by

(2) $$b + E(q_1) = m + \bar{b}.$$

(2) in turn is equivalent to

(3) $$E(q_1) = m + \bar{b} - b.$$

(3) implies not only 1) but also 2), because $\bar{b}_i = b_i$ holds for $i \in N-C$. Let r be the radius of the biggest $(|C| - 1)$-dimensional sphere which has m as center and is wholly contained in S. Clearly, if $(\bar{b} - b)^2 < r^2$, then a q_1 can be found which satisfies (3). Since r proportionally depends on a, the inequality for r can be satisfied by choosing a_0 sufficiently great. This proves the lemma.

THE NORMAL GAME OF A SYMMETRIC SUPERADDITIVE CHARACTERISTIC: For every symmetric superadditive characteristic c we construct a normal form

G which has c as characteristic function. If c is the trivial characteristic with $c(C) = 0$ for all $C \in \mathcal{C}$, then G is the normal form of the one-point game with vanishing payoffs. If c is different from this trivial characteristic, then c has a unique symmetry carrier R (cf. the lemma in §4). G is constructed as follows: Each member of $N - R$ receives 0 everywhere and has exactly one pure strategy. The strategies of a player $i \in R$ are the coalitions $C \in \mathcal{C}_i$ contained in R. A combination π of pure strategies "realizes" C, if it contains C as component for all members of C. Let $C_1, \ldots, C_s$ be all coalitions realized by π; define

$$u(\pi) = \frac{1}{|R|} (c(R) - \sum_{j=1}^{s} c(C_j)).$$

If player $i \in R$ is in the realized coalition C_j, then

$$H_i(\pi) = c(C_j)/|C_j| + u(\pi);$$

if player $i \in R$ is in none of the realized coalitions, then

$$H_i(\pi) = u(\pi).$$

This completes the construction of G.

Since u is nonnegative, because of the superadditivity of c, the common payoff $c(C)$ is enforced for C if all members of $C \cap R$ select $C \cap R$ as their strategy. Therefore c is indeed the characteristic of G.

We call the normal game Γ of G the "normal game of c". Let w be a value function satisfying (I) – (IV). We prove that w assigns the Shapley value $s(\Gamma)$ to Γ. If in Γ two members of R are interchanged, a linearly equivalent game results. Therefore by (I) and (II) all members of R have the same value. Since all players in $N - R$ are dummies with 0, their value is 0 by (IV). It follows by (III) that $w_i(\Gamma) = c(R)/|R|$ for $i \in R$. Thus the value of Γ is fixed by (I) – (IV). Since the Shapley value satisfies these postulates, we have $w(\Gamma) = s(\Gamma)$.

The results of this section are stated in the following Lemma 5.

LEMMA 5. For every symmetric superadditive characteristic c there is a normal game Γ which has c as characteristic. Every value function satisfying (I) – (IV) assigns the Shapley value $s(\Gamma)$ to Γ.

THE LEVELLING GAME. For $n = 2, 3, \ldots$ we construct an n-person game ${}^n\Gamma'$, the "levelling game": n alternatives $a_1, \ldots, a_n$ lead from the origin o to the endpoints $z_1, \ldots, z_n$. Player 1 decides at o. At the endpoint z_1 all players receive 0. At z_k ($k = 2, \ldots, n$) the players 1 and k receive -1, while all other players receive 0.

LEMMA 6. Let w be a value function for general n-person games; if w satisfies (I) – (V), (VII), and (X), then $w_i(a \cdot {}^n\Gamma') = 0$ for $a \geq 0$ and $i = 1, \ldots, n$.

PROOF. A game isomorphic to $\Gamma^1 = a \cdot {}^n\Gamma'$ results from Γ^1 by interchanging any two members of $N - (1)$. Therefore by (I) and (II) all members of $N - (1)$ have the same value. Thus because of (III) it is sufficient to show that $w_1(\Gamma^1) = 0$. By forbidding all moves not leading to z_1, a game with value $(0, \ldots, 0)$ results from Γ^1; this is because of (III) and (IV). Hence by (V) we have $w_1(\Gamma^1) \geq 0$. In the following way a value function w^2 for general 2-person games is induced by w: if Γ' is the n-person game resulting from the 2-person game Γ by adding $n - 2$ dummies with 0, then $w_i^2(\Gamma) = w_i(\Gamma')$ for $i = 1, 2$. It is easily seen that (I) – (V), (VII), and (X) carry over from w to w^2. It has been proved in [10], §7, that, with regard to 2-person value functions, (I) – (V), (VII), and (X) imply the following property (VIa).

(VIa) $$w(\lambda^2\Gamma^o) = \lambda w({}^2\Gamma^o) \quad \text{for } \lambda > 0.$$

Hence w^2 satisfies (VIa). (E 8, §1, could be used instead of (VIa), but we don't need this stronger result.) Since $a \cdot {}^n\Gamma^o$ results from $a \cdot {}^2\Gamma^o$ by adding $n - 2$ dummies with 0, by (IV) we have $w(a \cdot {}^n\Gamma^o) = aw({}^n\Gamma^o)$. By adding the one-point game $(x, (-a, -a, 0, \ldots, 0))$ to $a \cdot {}^n\Gamma^o$, a game Γ_1 results. Since by (IV) the value of a one-point game is its payoff vector, the value of Γ_1 is $(0, \ldots, 0)$ by (VII) and (X). By interchanging players 1 and i, a game Γ_i results from Γ_1. We form

$$\Gamma^2 = \sum_{i=2}^{n} \Gamma_i .$$

By (II) we have $w(\Gamma_i) = (0, \ldots, 0)$. Hence $w(\Gamma^2) = (0, \ldots, 0)$, by (VII). In Γ^2 we designate n "special" strategies of player 1: If x is one of the vertices corresponding to the origin of Γ_k, for $k \neq i$ the move corresponding to the move of Γ_k which leads to the payoff $(0, \ldots, 0)$ is assigned to x by the ith special strategy; for $k = i$ the opposite decision is prescribed $(i = 2, \ldots, n)$. Only moves corresponding to moves leading to $(0, \ldots, 0)$ are selected by the first special strategy. A game Γ^3 results from the normal game of Γ^2 by forbidding all moves not corresponding to special strategies. Since the kth special strategy leads to the same payoff as the move to z_k in Γ^1, the games Γ^1 and Γ^3 are isomorphic. Thus by (V) we have $w_1(\Gamma^1) \leq w_1(\Gamma^2)$. Hence $w_1(\Gamma^1) \leq 0$. We have already shown that $w_1(\Gamma^1) \geq 0$; hence $w_1(\Gamma^1) = 0$. This proves the lemma.

LEVELLING. Since in $\Gamma = a \cdot {}^n\Gamma'$ only player 1 has more than one pure strategy, the expected payoff for mixed strategies can be described as a function $E(q_1) = (E_1(q_1), \ldots, E_n(q_1))$ of player 1's mixed strategy q_1 alone. A strategy q_1 is "levelling against j", if it assigns positive probabilities only to moves leading to z_1 and to z_j $(j = 2, \ldots, n)$; if moreover,

$$\sum_{i=1}^{n} b_i + E_i(q_1) = g,$$

then q_1 "levels the vector $b = (b_1, \ldots, b_n)$ to g". For $C \in \mathfrak{C}_1$ the payoff in $F_{1/2}(_C\Gamma)$ is $(0, 0)$ at the endpoints z_k with $k = 1$ or $k \in N-C$, and $(-a, +a)$ at the other endpoints. Therefore the moves to the endpoints z_1 and z_j constitute 1/2-optimal* coalition strategies for C if $j \notin C$. Hence, q_1 is 1/2-optimal for each coalition $C \in \mathfrak{C}_1$ not containing j, if q_1 is levelling against j.

---
* A strategy is called "η-optimal" for Γ, if it is optimal for $F_\eta(\Gamma)$.

LEMMA 7. For every vector $b = (b_1, \ldots, b_n)$ and to every $g \leq b_1 + \ldots + b_n$ there is an $a_0 \geq 0$, so that for $a \geq a_0$ mixed strategies q_1^j ($j = 2, \ldots, n$) exist in $\Gamma = a \cdot {}^n\Gamma'$ which level b against j to g.

PROOF. It is easily seen that it is sufficient to take

$$a_0 = \frac{1}{2}(-g + \sum_{i=1}^{n} b_i)$$

LEMMA 8. The Shapley value s is the only value function for n-person constant-sum games which satisfies (I) — (IV), (V), and (VII), and which furthermore assigns $s(\Gamma)$ to all Γ with fully symmetric characteristic functions.

PROOF. Let Γ be an arbitrary n-person constant-sum game. Its characteristic function V can be linearly combined from symmetric superadditive characteristics (§4, combination theorem). Therefore we can assume that

(1) $$V = \sum_{k=1}^{m} \lambda_k V_k,$$

where the λ_k are real numbers and the V_k are symmetric superadditive characteristics. In the following Equation (2) the negative terms of (1) are collected on the left side:

(2) $$V + \sum_{k=1}^{m} \lambda_k' V_k = \sum_{k=1}^{m} \lambda_k'' V_k, \quad (\lambda_k' \geq 0, \lambda_k'' \geq 0, \quad k = 1, \ldots, m).$$

The $\lambda_k' V_k$ are symmetric and superadditive, since these properties are not lost by nonnegative multiplication. Therefore a normal game Γ_k exists for each $\lambda_k' V_k$. We form

(3) $$\Gamma^1 = \Gamma + \sum_{k=1}^{m} \Gamma_k.$$

It follows from Lemma 5 with the aid of (VII) that

$$
(4) \qquad w(\Gamma^1) = w(\Gamma) + \sum_{k=1}^{m} s(\Gamma_k).
$$

On the other hand we have

$$
(5) \qquad s(\Gamma^1) = s(\Gamma) + \sum_{k=1}^{m} s(\Gamma_k).
$$

Comparison of (4) and (5) shows that $w(\Gamma^1) = s(\Gamma^1)$ implies $w(\Gamma) = s(\Gamma)$. Therefore it is sufficient to prove $w(\Gamma^1) = s(\Gamma^1)$. Let Γ'_k be the normal game of $\lambda''_k V_k$, $k = 1, \ldots, m$. We form

$$
(6) \qquad \Gamma''_1 = \sum_{k=1}^{m} \Gamma'_k.
$$

There are exactly $n!$ games $\Gamma''_1, \ldots, \Gamma''_{n!}$ which can result if, starting from Γ''_1, several interchanges of players take place successively. There is one such game for every permutation of the players $1, \ldots, n$. It follows from Lemma 5 with the aid of (VII) that $w(\Gamma''_1) = s(\Gamma''_1)$. Since w and s both satisfy (II), this yields

$$
(7) \qquad w(\Gamma''_r) = s(\Gamma''_r) \qquad \text{for } r = 1, \ldots, n!.
$$

We form

$$
(8) \qquad \Gamma^2 = \Gamma^1 + \sum_{r=2}^{n!} \Gamma''_r.
$$

Since Γ''_1 and Γ^1 have the same characteristic function, because of (2), (3) and (6), the characteristic function of Γ^2 is given by

$$
(9) \qquad V^2 = \sum_{r=1}^{n!} V''_r.
$$

Evidently V^2 is fully symmetric. Hence $w(\Gamma^2) = s(\Gamma^2)$. It follows from (8) by (VII) and Lemma 5 that

$$
(10) \qquad s(\Gamma^2) = w(\Gamma^1) + \sum_{r=2}^{n!} s(\Gamma''_r).
$$

Since (10) holds for $w = s$, we must have $w(\Gamma^1) = s(\Gamma^1)$. This proves the lemma.

§ 7. CHARACTERIZATION OF THE SHAPLEY VALUE FOR n-PERSON CONSTANT-SUM GAMES

THEOREM. There is one and only one value function for n-person constant-sum games which satisfies (I) − (V), (VII), (IXa), and (XI), namely the Shapley value s.

PROOF It has already been proved in §5 that s satisfies the postulates of the theorem. Therefore we only have to show that $w(\Gamma) = s(\Gamma)$ holds for any w satisfying these postulates. Because of Lemma 8, we can assume that the characteristic function V of Γ is fully symmetric. Since s and w both satisfy (III) it is sufficient to show

(1) $\qquad s_i(\Gamma) \leq w_i(\Gamma) \qquad$ for $i = 1, \ldots, n$.

Since (II) is satisfied by s and w, the inequalities (1) are implied by

(2) $\qquad s_1(\Gamma) \leq w_1(\Gamma)$,

because (2) can be applied to the game which results from Γ by interchanging players 1 and i. Let Γ' be related to Γ as in Lemma 1. It follows from (3) in Lemma 1 that $w_1(\Gamma') \leq w_1(\Gamma)$, and because of (4) in Lemma 1, we have $s_1(\Gamma') = s_1(\Gamma)$. Therefore (2) is implied by

(3) $\qquad s_1(\Gamma') \leq w_1(\Gamma')$.

Because of Lemma 4, the parameter a in $\Gamma^1 = a \cdot {}^n \Gamma$ can be selected sufficiently great to ensure the existence of a set Q_1^1 containing exactly one indexed strategy q_1^1 for every pair $h'(z')$, C, where $h'(z')$ is a payoff vector of Γ' and C is a nonempty coalition not in $\mathcal{C}_1$, and q_1^1 equalizes $h'(z')$ with regard to C. By adding such a set Q_1^1 to $N(\Gamma^1)$, a normal form G^2 results. Let Γ^2 be the normal game of G^2. Since s and w satisfy the postulates of Lemma 3, we have $s(\Gamma^1) = w(\Gamma^1)$. Clearly, $\Gamma^1 \sim \Gamma^2$

VALUATION OF n-PERSON GAMES

by §1, E 1. Since (I) is satisfied by s and w, it follows that

(4) $$s(\Gamma^2) = w(\Gamma)^2.$$

We now construct an n-person normal form G^3. In this construction a parameter $b \geq 0$ appears which will be fixed later. For $k = 1, \ldots, n$, the strategies of player k in G^3 are the numbers $j \in N - (k)$. If a player $k \neq 1$ chooses 1 as his strategy, he recieves 0; his payoff is $-b$ if he selects another strategy. Player 1 has the following payoff:

(5) $$H_1^3(\pi^3) = -\sum_{i=2}^{n} H_i^3(\pi^3).$$

Let Γ^3 be the normal game of G^3. Evidently the characteristic function v^3 of Γ^3 assigns 0 to all coalitions. By transferring payoff from player 1 to other players, we can change Γ^3 into a game Γ^4 where all players receive 0 everywhere. Since Γ^4 is linearly equivalent to a one-point game with vanishing payoffs, the value of Γ^4 is $(0, \ldots, 0)$ by (I) and (IV). Thus by (IXa) we have $w_1(\Gamma^3) \geq 0$. By forbidding all the moves of player $j \neq 1$ in Γ^3 which do not correspond to the strategy 1 in G^3, a game Γ^5 results; player j is a dummy with 0 in the normal game of Γ^5. Hence $w_j(\Gamma^3) \geq 0$ because of (I), (IV), and (V). By (III) we have

(6) $$w(\Gamma^3) = s(\Gamma^3) = (0, \ldots, 0).$$

We form $\Gamma^6 = \Gamma^3 + \Gamma' + \Gamma^2$. The application of (VII) with regard to s_1 and w_1 shows with the aid of (4) and (6) that the difference between $w_1(\Gamma^6)$ and $w_1(\Gamma')$ is the same as that between $s_1(\Gamma^6)$ and $s_1(\Gamma')$. Therefore (3) is implied by

(7) $$s_1(\Gamma^6) \leq w_1(\Gamma^6).$$

Since (7) implies (3) which implies (2) which implies (1), only (7) remains to be shown.

The proof of (7) will use the following sequence of games: Γ^6, Γ^8,

Γ^9, Γ^{10}, Γ^{11}, Γ^{12}. The construction of this sequence will ensure that w_1 is never increased by the transition from one of these games to the next. The last game Γ^{12} will have strong symmetry properties which will be sufficient to prove $w(\Gamma^{12}) = s(\Gamma^6)$. This result will imply (7).

We now construct Γ^8: After every endpoint z^7 of $\Gamma^7 = \Gamma^3 + \Gamma'$ we change Γ^6 in the following way: Let z^3 be the endpoint leading to z^7, and let π^3 be the strategy combination corresponding to z^3 in G^3. In the future $C_k(z^3)$ will always denote the coalition formed by the player having k as strategy in π^3. Let z^7 correspond to z' in Γ'. If z^7 comes after a z^3 with $C_1(z^3) \neq \emptyset$, we forbid at z^7 all those moves of player 1 which do not correspond to the equalizing strategy $q_1^1 \in Q_1^1$ for $h'(z')$ with regard to $C_1(z^3)$. If, however, z^7 comes after a z^3 with $C_1(z^3) = \emptyset$, we forbid at z^7 all moves with exception of one which can be selected arbitrarily. This completes the construction of Γ^8. We have $w_1(\Gamma^8) \leq w_1(\Gamma^6)$ because of (V). By substituting linearly equivalent one-point games for the subgames at the endpoints of Γ^7, a game Γ^9 results from Γ^8. Because of §1, E 2, we have $\Gamma^8 \sim \Gamma^9$. It follows by (I) that $w_1(\Gamma^9) = w_1(\Gamma^8)$.

We now construct Γ^{10} by changing the payoff function of Γ^9. Later it will be shown that Γ^{10} results from Γ^9 by transferring payoff from player 1 to other players. We define h^{10} by the following description of the payoff in the subgame $\Gamma_{z^3}^{10}$ after any endpoint z^3 of Γ^3. The subgame has at most $|N - C_1(z^3)| + 1$ different payoff vectors. One of those is the vector $u(z^3)$ with components given by

(8) $\quad u_i(z^3) = \dfrac{1}{|C_k(z^3)|} v^6(C_k(z^3)) + \dfrac{1}{n}v^6(N) - \dfrac{1}{n}\sum_{r=1}^{n} v^6(C_r(z^3))$, $\quad i = 1, \ldots, n$,

where $C_k(z^3)$ is that coalition among the $C_1(z^3), \ldots, C_n(z^3)$, which contains i. This means that k is player i's component in the π^3 corresponding to z^3. v^6 is the characteristic function of Γ^6. For each player $j \notin C_1(z^3)$, there is another payoff vector $u^j(z^3)$ which may appear. The components are given by

(9) $\quad u_i^j(z^3) = \begin{cases} u_i(z^3) & \text{for } i \in N - C_1(z^3) \text{ and } i \neq j \\ u_j(z^3) - \bar{g}|C_1(z^3)| + \sum_{s \in C_1(z^3)} u_s(z^3) & \text{for } i = j \\ \bar{g} & \text{for } i \in C_1(z^3) \end{cases}$

where the parameter $\bar{g}$ must be chosen in accordance with three inequalities (10), (11), and (12), but may be fixed arbitrarily otherwise. (11) and (12) ensure that $u^j(z^3)$ is worse than $u(z^3)$ for player j, but not for the other players.

(10) $$\bar{g} > \max_{z^9 \in Z(K^9)} \max_{i \in N-(1)} h_i^9(z^9)$$

(11) $$u_j^j(z^3) < u_j(z^3)$$

(12) $$u_i^j(z^3) > u_i(z^3) \quad \text{for } i \in C_1(z^3).$$

v^3 assigns 0 to all coalitions, independently of b. Therefore v^6 does not depend on b either. The inequalities (10), (11), and (12) only require that $\bar{g}$ be sufficiently great; $\bar{g}$ can be selected independently of b.

Let $\zeta(z^3)$ be a pure coalition strategy for $N-C_1(z^3)$ in the subgame $\Gamma_{z^3}^{10}$ which corresponds to an optimal pure coalition strategy in Γ'. Because of (2) in Lemma 1 such a strategy exists in Γ'. Let $Z(z^3)$ be the set of all endpoints of $\Gamma_{z^3}^{10}$ which can be realized, if $\zeta(z^3)$ is used. If an endpoint z^9 of the subgame is not in $Z(z^3)$, then the way from z^3 to z^9 contains at least one alternative which cannot be realized, if $\zeta(z^3)$ is used. Among those players who have a move containing such an alternative, let player j be the lowest in the natural order; we call player j "first deviator" for z^9. Evidently the first deviator is always a member of $N-C_1(z^3)$. Let $Z_j(z^3)$ be the set of all endpoints z^9, for which player j is the first deviator. h^{10} is given by

(13) $$h^{10}(z^9) = \begin{cases} u(z^3) & \text{for } z^9 \in Z(z^3) \\ u^j(z^3) & \text{for } z^9 \in Z_j(z^3). \end{cases}$$

This completes the construction of Γ^{10}.

Γ^{10} results from Γ^9 by transferring payoff from player 1 to other players, provided that the following conditions (14) – (18) are satisfied:

(14) $\quad u_1(z^3) \leq h_1^9(z^9)$ $\Big\}$ for $z^9 \in Z(z^3)$
(15) $\quad u_i(z^3) \geq h_i^9(z^9)$ for $i = 2, \ldots, n$

(16) $\quad u_1^j(z^3) \leq h_1^9(z^9)$ $\Big\}$ for $z^9 \in Z_j(z^3)$,
(17) $\quad u_i^j(z^3) \geq h_i^9(z^9)$ for $i = 2, \ldots, n$ $\quad j \in N - C_1(z^3)$

(18) $\quad \sum_{i=1}^{n} h_i^{10}(z^9) = \sum_{i=1}^{n} h_i^9(z^9) = v^6(N)$.

We shall show that (14) — (18) can be ensured by an appropriate selection of the parameter b in G^3; it is only required that b be sufficiently great.

We consider the subgames $\Gamma_{z^3}^6$ of Γ^6 at the endpoints z^3 of Γ^3. Evidently the game $\Gamma^{z^3} = \Gamma' + \Gamma^2 + (x, h^3(z^3))$ is isomorphic to $\Gamma_{z^3}^6$. A strategy which is optimal for $\Gamma' + \Gamma^2$ is optimal for Γ^{z^3}, too; the addition of a one-point game does not destroy the optimality. Since the members of a coalition $C \in \mathcal{C}_1$ always have a nonnegative common payoff in Γ^3, the characteristic function v^{z^3} of Γ^{z^3} satisfies $v^{z^3}(C) \geq v^6(C)$ for $C \in \mathcal{C}_1$. The strategy $\varsigma(z^3)$ corresponds to an optimal strategy for $N-C_1(z^3)$ in Γ'. When Γ^8 was constructed, only one move at every endpoint z^7 of $\Gamma^3 + \Gamma'$ was not forbidden. For $C_1(z^3) \neq \emptyset$ this move corresponds to a strategy q_1^1, which equalizes $h'(z')$ with regard to $C_1(z^3)$; for $C_1(z^3) = \emptyset$ an arbitrary move was left over. In both cases, the only move remaining at z^7 in Γ^8 corresponds to a coalition strategy in Γ^2 which is optimal for $N-C_1(z^3)$. Therefore $\varsigma(z^3)$, together with the only possible decisions at z^7, forms an optimal strategy for the subgame $\Gamma_{z^3}^8$ in Γ^8. It follows that $\varsigma(z^3)$ is optimal for the subgame $\Gamma_{z^3}^9$ in Γ^9. Clearly, $\varsigma(z^3)$ secures $v^{z^3}(N-C_1(z^3))$ for $N-C_1(z^3)$ in $\Gamma_{z^3}^9$. Therefore the common payoff for $N-C_1(z^3)$ is at least $v^6(N-C_1(z^3))$ at endpoints $z^9 \in Z(z^3)$.

We consider the special case $N-C_1(z^3) = (1)$: Condition (14) is satisfied, since $u_1(z^3) = v^6((1))$. In (16) we must have $j = 1$. Because of (10) the components of $u^1(z^3)$ are better for the members of $N - (1)$ than their best payoffs in Γ^9. Therefore player 1's component is worse than his least payoff in Γ^9; this implies (16).

If player 1 is not the only member of $N-C_1(z^3)$, we have

(19) $\quad h_1^9(z^9) \geq h_1'(z') + b$

where z' is the endpoint of Γ' which corresponds to z^9. (19) implies (14) and (16), if b is sufficiently great. Thus we have seen that (14) and (16) can be ensured.

We now proceed to show the same for (15) and (17). If b is increased, $h_i^9(z^9)$ is decreased for $i \in N-C_1(z^3) - (1)$. Therefore we can ensure (15) and (17) for these players. In the following we can assume $C_1(z^3) \neq \emptyset$, because otherwise nothing remains to be shown about (15) and (17). Let z^7 be an endpoint of $\Gamma^7 = \Gamma^3 + \Gamma'$ which comes after z^3. Since all members of $C_1(z^3)$ receive 0 at z^3 in Γ^3, the vector $h^7(z^7)$ is equalized with regard to $C_1(z^3)$ by the strategy q_1^1 which corresponds to the only move left over at z^7 in Γ^8. Therefore in $\Gamma_{z^3}^9$ the common payoff of $C_1(z^3)$ is always divided equally among the members of $C_1(z^3)$. Since the common payoff of $N-C_1(z^3)$ is at least $v^6(N-C_1(z^3))$ for $z^9 \in Z(z^3)$, the common payoff of $C_1(z^3)$ is at most $v^6(C_1(z^3))$. Since v^6 is a characteristic function, it follows with the aid of (8) that

(20) $h_i^9(z^9) \leq v^6(C_1(z^3))/|C_1(z^3)| \leq u_i(z^3)$ for $i \in C_1(z^3)$ and $z^9 \in Z(z^3)$.

(20) implies (15). Condition (17) for $i \in C_1(z^3)$ and (18) are immediate consequences of (8) and (9).

Now, having shown the possibility of doing so, we fix b in a way which ensures that Γ^{10} results from Γ^9 by transferring payoff from player 1 to other players. It follows by (IXa) that $w_1(\Gamma^{10}) \leq w_1(\Gamma^9)$.

For every endpoint z^3 of Γ^3 we construct a game $\Gamma_{z^3}^{11}$. Let $S(z^3)$ be the set of all members of $N-C_1(z^3)$, which have more than one strategy for the subgame $\Gamma_{z^3}^{10}$ at z^3 in Γ^{10}. If $S(z^3)$ is empty, $\Gamma_{z^3}^{11}$ is the one-point game $(x_1, u(z^3))$. In the following we assume $S(z^3) \neq \emptyset$. Let $i_1, \ldots, i_r$ with $i_1 < i_2 < \ldots < i_r$ be the members of $S(z^3)$. In $\Gamma_{z^3}^{11}$ the player i_k has exactly one vertex x_k in his player set, $k = 1, \ldots, r$. The vertex x_1 is the origin of $\Gamma_{z^3}^{11}$. The alternative a_k connects x_k and x_{k+1}, $k = 1, \ldots, r-1$. The alternative a_r at x_r leads to the endpoint z_0. There are only two alternatives at each x_k; the second one, a_k', leads to the endpoint z_k, $k = 1, \ldots, r$. The payoff is $u(z^3)$ at z_0 and $u^{i_k}(z^3)$ at z_k. This completes the construction of $\Gamma_{z^3}^{11}$.

Now we proceed to show that $\Gamma_{z^3}^{10} \sim \Gamma_{z^3}^{11}$. Since Γ^{10} and

$\Gamma^7 = \Gamma^3 + \Gamma'$ differ only with regard to the payoff function and, since Γ' is a normal game (cf. (1) in Lemma 1, §6), the subgame $\Gamma_{z^3}^{10}$ can be considered to be the normal game of a normal form $G_{z^3}^{10}$. Therefore each player in $\Gamma_{z^3}^{10}$ has to decide only once, and the order of these decisions is given by the natural order of the players. Clearly, there must be a first deviator if $\zeta(z^3)$ is not used; he is in $S(z^3)$ since the other members of $N-C_1(z^3)$ have only one strategy which must be a component of $\zeta(z^3)$. Among those members of $S(z^3)$ who do not use their components of $\zeta(z^3)$, the first deviator is the lowest in the natural order. The payoff in $\Gamma_{z^3}^{10}$ is $u^j(z^3)$, if player j is the first deviator and it is $u(z^3)$ if there is no first deviator. The payoff in $\Gamma_{z^3}^{11}$ depends in the same way on the first deviator, if player i_k is called "first deviator" for z_k, $k = 1, \ldots, r$. It does not matter which one of his strategies a member of $C_1(z^3)$ is using in $\Gamma_{z^3}^{10}$, and it does not matter which one of the possibly many strategies deviating from $\zeta(z^3)$ a player of $S(z^3)$ is using. From what has already been said, it is easily seen that $N(\Gamma_{z^3}^{11})$ can be changed into a normal form isomorphic to $G_{z^3}^{10}$ by adding a set Q of indexed mixed strategies which contains sufficiently many copies of the strategies for the players of $C_1(z^3)$ and sufficiently many copies of the strategies which prescribe alternatives a_k'. Therefore we have $\Gamma_{z^3}^{10} \sim \Gamma_{z^3}^{11}$ by E 1, §1. By substituting the $\Gamma_{z^3}^{11}$ for the corresponding $\Gamma_{z^3}^{10}$, a game Γ^{11} results from Γ^{10}. We have $\Gamma^{11} \sim \Gamma^{10}$ by E 2, §1. Hence $w_1(\Gamma^{11}) = w_1(\Gamma^{10})$ because of (I).

By substituting $u(z^3)$ for $h^3(z^3)$ everywhere, a game Γ^{12} results from Γ^3. This game Γ^{12} will also result from Γ^{11} by a number of successive changes which will be made alternately in the following ways: (a) superfluous moves are forbidden; (b) linearly equivalent one-point games are substituted for subgames. The changes will take place successively within all subgames $\Gamma_{z^3}^{11}$ with $S(z^3) \neq \emptyset$. Since the required postulates are satisfied by w, it will be clear by Lemma 2 and by E 2, §1, that $w_1(\Gamma^{12}) = w_1(\Gamma^{11})$.

Let Γ_k be the game which results from $\Gamma_{z^3}^{11}$ by substituting the one-point game $(x_k, u(z^3))$ for the subgame at x_k. Let Γ_{r+1} be the game $\Gamma_{z^3}^{11}$. The changes mentioned above are the following: (a) the move containing a_k' is forbidden in Γ_{k+1}; (b) the resulting game Γ_{k+1}' is changed into Γ_k by substituting the one-point game $(x_k, u(z^3))$ for the subgame at x_k. This is done for $k = r, \ldots, 1$. It can be seen with the aid of (11) and (12) that the forbidden moves are indeed superfluous. Since Γ_1 is the one-point game $(x_1, u(z^3))$, the final result is indeed Γ^{12}.

Let Γ^{13} be the game which results from Γ^{12} by interchanging players j and k. The strategies of player i in G^3 are the numbers from $N - (i)$. Thereby an enumeration of player i's strategies for Γ^3 is given. We use this enumeration also for Γ^{13}, but with the following slight change: k is substituted for j and j is substituted for k. Thus the numbers of $N - (i)$ correspond to the strategies of player i in Γ^{13}, too. Every strategy combination for Γ^{12} or Γ^{13} partitions N into coalitions C_r of players whose strategies have the same number r. Since Γ', Γ^2, and Γ^3 have fully symmetric characteristic functions, v^6 is fully symmetric, too. Therefore the payoffs in Γ^{12} and Γ^{13} depend in the same way on the coalitions C_r. It follows that the normal forms of Γ^{12} and Γ^{13} are isomorphic. Consequently, we have $w(\Gamma^{12}) = w(\Gamma^{13})$ because of (I). It follows with the aid of (II) that $w_j(\Gamma^{12}) = w_k(\Gamma^{12})$. Since j and k can be chosen arbitrarily, all players must have the same value in Γ^{12}. Since v^6 is fully symmetric, it follows with the aid of (III) that

(21) $\qquad w_i(\Gamma^{12}) = M(\Gamma^{12})/n = v^6(N)/n = s_i(\Gamma^6), \qquad i = 1, \ldots, n.$

Γ^{12} is the last game in the sequence $\Gamma^6, \Gamma^8, \Gamma^9, \Gamma^{10}, \Gamma^{11}, \Gamma^{12}$. We have seen that the value of player 1 is not increased by the transition from one of these games to the next. Therefore we have (7). This proves the theorem.

§ 8. CHARACTERIZATION OF THE VALUE FUNCTION $w_{1/2}$ FOR GENERAL n-PERSON GAMES

THEOREM. There is one and only one value function for general n-person games which satisfies (I) — (V), (VII), (IXa), (X), and (XI), that is $w_{1/2}$.

PROOF. It has been already proved in §5 that $w_{1/2}$ satisfies the postulates of the theorem. Let w be a value function for general n-person games which satisfies these postulates. We must have $w(\Gamma) = s(\Gamma)$ for n-person constant-sum games, since otherwise we would have a contradiction to the characterization theorem of §7. We know from §5 that $w_{1/2}$ and s coincide for constant-sum games. Therefore we have $w(\Gamma) = w_{1/2}(\Gamma)$ for constant-sum games. It remains to be shown that the same is true for non-constant-sum games. Since $w_{1/2}$ and w both satisfy (III), it is sufficient to prove

(1) $$w_{1/2,i}(\Gamma) \leq w_i(\Gamma) \qquad \text{for } i = 1, \ldots, n.$$

Since (II) is satisfied by $w_{1/2}$ and w, the inequalities (1) are implied by

(2) $$w_{1/2,1}(\Gamma) \leq w_1(\Gamma)$$

because (2) can be applied to the game which results from Γ by interchanging players 1 and i. Thus only (2) remains to be shown.

Let Γ' be the normal game of Γ. Because of (I) we have

(3) $$w(\Gamma') = w(\Gamma).$$

Define

(4) $$g = \min_{z' \in Z(K')} \sum_{i=1}^{n} h'_i(z').$$

Because of Lemma 7, the parameter a in $\Gamma^1 = a \cdot {}^n\Gamma'$ can be selected sufficiently great to ensure the existence of a set Q_1^1 containing exactly one indexed mixed strategy q_1^1 for every pair $h'(z')$, j, where $h'(z')$ is a payoff vector of Γ' and j is a member of $N - (1)$, and q_1^1 levels $h'(z')$ against j to g. By adding such a set Q_1^1 to $N(\Gamma^1)$, a normal form G^2 results. Let Γ^2 be the normal game of G^2. Since $w_{1/2}$ and w satisfy the postulates of Lemma 6, we have $w(\Gamma^1) = w_{1/2}(\Gamma^1) = (0, \ldots, 0)$. Clearly $\Gamma^1 \sim \Gamma^2$ by E 1, §1. Since (I) is satisfied by w and $w_{1/2}$, it follows that $w(\Gamma^2) = w_{1/2}(\Gamma^2) = (0, \ldots, 0)$. We form $\Gamma^3 = \Gamma' + \Gamma^2$. Because of (VII), we have $w(\Gamma^3) = w(\Gamma')$.

Let z_0' be an endpoint of Γ' which satisfies

(5) $$\sum_{i=1}^{n} h_i'(z_0') = M(\Gamma').$$

In Γ^2 player 1 can enforce the payoff vector $(0, \ldots, 0)$ by a certain move u_0^2. Let u_0^3 be the move at z_0' in Γ^3 which corresponds to u_0^2. After every endpoint z' of Γ' we change Γ^3 in the following way: We forbid all the moves different from u_0^3 which do not correspond to a strategy q_1^1 levelling $h'(z')$ against a $j \in N - (1)$ to g. The result is Γ^4. Because of (5) the move u_0^3 leads to an endpoint z_0^4 of Γ^4 with

(6) $$\sum_{i=1}^{n} h_i^4(z_0^4) = M(\Gamma').$$

Since all the other moves at the z' in Γ^4 correspond to strategies which level $h(z')$ to g, we have

(7) $$\sum_{i=1}^{n} h_i^4(z^4) = g$$

for all endpoints $z^4 \neq z_0^4$. Because of (V) we have $w_1(\Gamma^4) \leq w_1(\Gamma^3)$. It follows with the aid of $w(\Gamma^3) = w(\Gamma')$ and (3) that

(8) $$w_1(\Gamma^4) \leq w_1(\Gamma).$$

By forbidding u_0^3, a game Γ^5 results from Γ^4. It follows by (5) that Γ^5 is a constant-sum game with $M(\Gamma^5) = g$. We now proceed to show that the characteristic function v^5 of Γ^5 and the 1/2-characteristic $c_{1/2}$ are related as follows

(9) $\quad v^5(C) = c_{1/2}(C) - \frac{1}{2}(M(\Gamma) - g) \qquad$ for all $C \in \mathcal{C} - (N) - (\emptyset)$.

It is sufficient to prove (9) for $C \in \mathcal{C}_1 - (N)$; this can be seen in the following way: If C is not in $\mathcal{C}_1 - (N)$, then $N-C$ is in $\mathcal{C}_1 - (N)$, and (9) can be applied to $N-C$. Since v^5 and $c_{1/2}$ are characteristic, we can substitute $g - v^5(C)$ for $v^5(N-C)$ and $M(\Gamma) - c_{1/2}(C)$ for $c_{1/2}(N-C)$. This yields an equation which is equivalent to (9) applied to C.

Let C be a coalition in $\mathcal{C}_1 - (N)$. We consider the coalition game $_C\Gamma^5$. There are two kinds of moves at every z': (a) moves corresponding to a q_1^1 which levels against a $j \in C$; (b) moves corresponding to a q_1^1 which levels against a $j \in N-C$. The moves of the kind (a) are worse for C and better for $N-C$; therefore they can be forbidden as superfluous in $_C\Gamma^5$. The result is a game $_C\Gamma^6$. Since the postulates of Lemma 2 are satisfied by the von Neumann value v, we have

(10) $$v_1(_C\Gamma^6) = v_1(_C\Gamma^5) = v^5(C).$$

We change the coalition game $_C\Gamma'$ into a game $_C\Gamma^7$ by substituting the following payoff function $_Ch^7$ for $_Ch'$:

(11) $$_Ch_i^7(z') = {_Ch_i}'(z') - \tfrac{1}{2}({_Ch_1}'(z') + {_Ch_2}'(z') - g), \quad i = 1, 2.$$

In $_C\Gamma^6$ all moves at a z' are of the kind (b); the corresponding strategies q_1^1 level to g by taking away equal amounts from the payoff of player 1 and from the payoff of a player $j \in N-C$. Therefore we have $_Ch^6(z^6) = {_Ch^7}(z')$ for all endpoints z^6 after z' in $_C\Gamma^6$. It follows that $_C\Gamma^6$ can be changed into $_C\Gamma^7$ by substituting all the subgames at the z' by linearly equivalent one-point games. Clearly, $_C\Gamma^7 \sim {_C\Gamma^6}$ by §1, E 2. Thus by (I) and (10) we have

(12) $$v_1(_C\Gamma^7) = v_1(_C\Gamma^6) = v^5(C).$$

Define

(13) $$_C\Gamma^8 = {_C\Gamma^7} + (x, \tfrac{1}{2}(M(\Gamma) - g), \tfrac{1}{2}(M(\Gamma) - g))).$$

The von Neumann value of a one-point game is its payoff vector. It follows by (VII) and (12) that

(14) $$v_1(_C\Gamma^8) = v^5(C) + \tfrac{1}{2}(M(\Gamma) - g).$$

Because of (11) we have

(15) $$_c h_1^7(z') - {_c h_1}'(z') = {_c h_2}^7(z') - {_c h_2}'(z').$$

It follows with the aid of $M(\Gamma) = M(\Gamma')$ that $_c\Gamma^8$ is the 1/2-replenished coalition game $F_{1/2}(_c\Gamma')$. Therefore we have

(16) $$v_1(_c\Gamma^8) = w_{1/2}(_c\Gamma') = c_{1/2}(C).$$

With the aid of (14), this yields (9).

Since v^5 is the 1/2-characteristic of Γ^5 (theorem on constant-sum games, §4), we can use formula (A) from §5 in order to describe $w_{1/2}(\Gamma^5)$ as follows:

(17) $$w_{1/2,i}(\Gamma^5) = \gamma_n(N)v^5(N) + \sum_{C \in \mathcal{C}_i-(N)} \gamma_n(C)(v^5(C) - v^5(N-C)), \quad i = 1, \ldots, n.$$

Because of (9) we have

(18) $$v^5(C) - v^5(N-C) = c_{1/2}(C) - c_{1/2}(N-C)$$

for all C which can occur in one of the sets $\mathcal{C}_i - (N)$. With the aid of $v^5(N) = g$ and $\gamma_n(N) = 1/n$ it follows that

(19) $$w_{1/2,i}(\Gamma^5) = g/n + \sum_{C \in \mathcal{C}_i-(N)} \gamma_n(C)(c_{1/2}(C) - c_{1/2}(N-C)), \quad i = 1, \ldots, n.$$

Let $d = (d_1, \ldots, d_n)$ be the vector with the components

(20) $$d_i = (M(\Gamma) - g)/n.$$

It follows that

(21) $$d_i + \frac{g}{n} = M(\Gamma)/n = \gamma_n(N)c_{1/2}(N).$$

Therefore, the right side of (19) can be changed into the right side of formula (A) applied to Γ, simply by adding d_i. Consequently we have

(22) $$w_{1/2}(\Gamma^5) + d = w_{1/2}(\Gamma).$$

We now construct Γ^9: Player i has exactly one vertex x_i in his player set ($i = 1, \ldots, n$). x_1 is the origin of Γ^9. There are two alternatives a_k and a_k' at each vertex x_i ($i = 1, \ldots, n$). For $i = 1, \ldots, n-1$, the alternative a_k connects x_i with x_{i+1}; but a_n leads to the endpoint z_0^9. The alternative a_k' leads to the endpoint z_k' ($k = 1, \ldots, n$). The payoff at all endpoints is 0 for all players.

By substituting a one-point game with vanishing payoffs for the subgame at x_k a game Γ_k^9 results from Γ^9. The players $k, \ldots, n$ in Γ^9 are dummies with 0.

By substituting d for the payoff vector at z_0^9, a game Γ^{10} results from Γ^9.

With the aid of (I) – (IV) it can be seen easily that

(23) $w_i(\Gamma^9) = w_{1/2,i}(\Gamma^9) = 0$

(24) $w_i(\Gamma_k^9) = w_{1/2,i}(\Gamma_k^9) = 0 \quad$ for $k = 1, \ldots, n \quad\Big\}\quad$ for $i = 1, \ldots, n$

(25) $w_i(\Gamma^{10}) = w_{1/2,i}(\Gamma^{10}) = d_i$

We form $\Gamma^{11} = \Gamma^9 + \Gamma^4$. By (VII) we have $w(\Gamma^{11}) = w(\Gamma^4)$. By (6) we have $M(\Gamma^4) = M(\Gamma')$. Clearly, $M(\Gamma') = M(\Gamma)$ by $\Gamma \sim \Gamma'$. Therefore we have $M(\Gamma^4) = M(\Gamma)$. This yields $M(\Gamma^{11}) = M(\Gamma)$.

We form $\Gamma^{12} = \Gamma^5 + (x, d)$. Since $M(\Gamma^5) = g$, we have $M(\Gamma^{12}) = M(\Gamma)$. We change Γ^{11} into Γ^{13} by substituting Γ^{12} for the subgame at z_0^9. Evidently we have $M(\Gamma^{13}) = M(\Gamma)$. We forbid the move containing a_k in Γ^{13} and then substitute the linearly equivalent game Γ^4 for the subgame at x_k; thus Γ^{13} is changed into $\Gamma_k^{13} = \Gamma_k^9 + \Gamma^4$. By (VII) and (24) we have

(26) $$w(\Gamma_k^{13}) = w(\Gamma^4).$$

By (I) and (V) we have

(27) $$w_k(\Gamma_k^{13}) \leq w_k(\Gamma^{13}).$$

With the aid of (26) this yields

(28) $$w_k(\Gamma^4) \le w_k(\Gamma^{13}).$$

Since $M(\Gamma^4)$ and $M(\Gamma^{13})$ coincide with $M(\Gamma)$, it follows by (III) that

(29) $$w(\Gamma^4) = w(\Gamma^{13}).$$

Hence, by (8) we have

(30) $$w_1(\Gamma^{13}) \le w_1(\Gamma).$$

We change Γ^{13} into a game Γ^{14} by forbidding all those moves in the subgames at the endpoints $z_1^9, \ldots, z_n^9$ of Γ^9 which correspond to the move u_0^3 in Γ^4. By (V) we have $w_1(\Gamma^{14}) \le w_1(\Gamma^{13})$. It follows by (30) that

(31) $$w_1(\Gamma^{14}) \le w_1(\Gamma).$$

The subgames of Γ^{14} at the $z_1^9, \ldots, z_n^9$ all are isomorphic to Γ^5; the subgame at z_0^9 is isomorphic to $\Gamma^{12} = \Gamma^5 + (x, d)$. This shows that Γ^{14} is isomorphic to the game $\Gamma^{10} + \Gamma^5$. It follows by (VII) that

(32) $$w(\Gamma^{14}) = w(\Gamma^{10}) + w(\Gamma^5).$$

Since Γ^5 is a constant-sum game, we have $w(\Gamma^5) = w_{1/2}(\Gamma^5)$. It follows from (32) with the aid of (22) and (25) that

(33) $$w_1(\Gamma^{14}) = d_1 + w_{1/2,1}(\Gamma^5) = w_{1/2,1}(\Gamma).$$

(33) and (31) imply (2). We have already seen that it is sufficient to show (2). This proves the theorem.

ADDED IN PROOF

In reference to the discussion in the Introduction, there is one important interpretative problem, which was not mentioned in [10] because then I was not aware of it. This is the question of commitment power, which has been raised by Schelling [9]. The monotonicity postulates (V), (IX), (IXa) and (XI) presuppose that nothing can be gained by some forms of commitment. This is reasonable if every player can commit himself to whatever he wants before the beginning of the play, because then the commitment possibilities should be reflected by the value. Therefore we should add the assumption of "full commitment power" to the general assumptions mentioned in the fourth paragraph of the Introduction.

BIBLIOGRAPHY

[1] BURGER, E., Einführung in die Theorie der Spiele, Berlin 1959.

[2] HARSANYI, J. C., "A bargaining model for the cooperative n-person game," Contributions to the Theory of Games, Vol. IV, Princeton 1959, pp. 325-355.

[3] HARSANYI, J. C., "A simplified bargaining model for the cooperative n-person game," (mimeographed), Canberra, about 1960.

[4] ISBELL, J. R., "Absolute games," Contributions to the Theory of Games, Vol. IV, Princeton 1959, pp. 357-396.

[5] KUHN, H. W., "Extensive games and the problem of information," Contributions to the Theory of Games, Vol. II, Princeton 1953, pp. 193-216.

[6] LUCE, R. D., and RAIFFA, H., Games and Decisions, New York, 1957.

[7] NASH, J., "Two-person cooperative games," Econometrica, Vol. 21 (1953), pp. 128-140.

[8] RAIFFA, H., "Arbitration schemes for generalized two-person games," Contributions to the Theory of Games, Vol. II, Princeton 1953, pp. 361-387.

[9] SCHELLING, T. C., The Strategy of Conflict, Cambridge, Massachusetts, 1960.

[10] SELTEN, R., "Bewertung strategischer Spiele," Zeitschrift für die gesamte Staatswissenschaft, 116. Band, 2. Heft (1960), pp. 221-282.

[11] SHAPLEY, L. S., "A value for n-person games," Contributions to the Theory of Games, Vol. II, Princeton 1953, pp. 307-317.

[12] VON NEUMANN, J., "Zur Theorie der Gesellschaftsspiele," Mathematische Annalen, 100 (1928), pp. 295-320.

[13] VON NEUMANN, J. and MORGENSTERN, O., Theory of Games and Economic Behavior, Princeton, 1944.

Johann Wolfgang Goethe-University
Frankfurt am Main, Germany

Reinhard Selten

PROPERTIES OF A MEASURE OF PREDICTIVE SUCCESS

Reinhard SELTEN

Department of Economics, University of Bonn, Adenauerallee 24-42, D-5300 Bonn 1, F.R.G.

Communicated by K.H. Kim
Received 26 September 1990
Revised 4 December 1990

> Area theories for the prediction of experimental results delineate regions of predicted outcomes within the set of all possible outcomes. The difference measure of predictive success for area theories introduced by Selten and Krischker (1983) is the difference between hit rate and area. The hit rate is the relative frequency of successful predictions and the area is the relative size of the predicted region within the set of all possible outcomes. It is argued that other measures proposed in the literature are unreasonable with respect to the implied structure of unimprovable theories. Two axiom systems for measures based on hit rate and area characterize the difference measure. The first characterization is ordinal and the second one is cardinal.

Key words: Measure; predictive success.

1. Introduction

The measure of predictive success investigated in this paper has been introduced as an instrument for the comparison of area theories for characteristic function experiments (Selten and Krischker, 1983). An *area theory* is a theory that predicts a subset of all possible outcomes.

The *hit rate* of a theory is the relative frequency of correct predictions. The hit rate is a measure of accuracy, but accuracy alone cannot be the aim of an area theory. No area theory can be more accurate than the trivial one, which simply predicts the set of all possible outcomes. This theory never fails to predict correctly, but it is useless in view of its complete lack of precision.

The precision of an area theory is related to the size of its set of predicted outcomes. The smaller this set is, the more precise is the theory. A measure of size, e.g. euclidic volume, may serve as the basis of a definition of *relative size* of the set of predicted outcomes. In applications it is not difficult to define a natural size measure, but it must be admitted that the choice of the right size measure is not always a trivial problem. In this paper we shall assume that an agreed-upon size measure is given, which permits us to assign a relative size between 0 and 1 to sets of predicted outcomes. The relative size of the empty set is 0 and the relative size of all outcomes is 1. The relative size of the subset predicted by a theory is called the *area* of this theory. (In the case of an infinite set of possible outcomes theories with non-measurable predicted subsets are excluded from consideration.)

The measure of predictive success developed by Selten and Krischker (1983) can be described as follows:

$$m = r - a, \tag{1}$$

where m = measure of predictive success, r = hit rate, the relative frequency of correct predictions, and a = the area, the relative size of the predicted subset compared with the set of all possible outcomes.

In discussions following verbal presentations of empirical results this measure of predictive success is sometimes criticized as arbitrary. The idea that m should be a function of r and a does not meet any opposition. Objections are raised against the functional form.

An obvious alternative to $m = r - a$ is $m = r/a$. Consider a theory A with a hit rate of 0.9 and an area of 0.1 and a theory B with a hit rate of 0.01 and an area of 0.0001. We obtain $r/a = 9$ for theory A and $r/a = 100$ for theory B. According to $m = r/a$ theory B is much more successful in spite of its low accuracy. It seems to be obvious to the author that this is an undesirable result. The measure r/a overemphasizes precision in comparison with accuracy, especially if the area is very small. The measure $r - a$ seems to be more reasonable. We obtain $r - a = 0.8$ for theory A and $r - a = 0.0099$ for theory B.

Since $m = r - a$ and $m = r/a$ are the simplest functional forms, one may take the point of view that considerations of simplicity together with examples like the one given above decide the issue. This paper tries to strengthen the case for $r - a$ as the preferred measure of predictive success for area theories by a deeper discussion of its properties.

The search for better area theories guided by a measure of predictive success aims at the maximization of this measure. Therefore it is of interest to ask: Which sets of outcomes maximize the expectation of the measure of predictive success for a given probability distribution? The answer reveals the implied structure of a theory which is unimprovable with respect to the measure. In Section 2 this problem will be discussed. Three measures of predictive success will be examined, $r - a$ and two other ones, namely r/a and $(r - a)/(1 - a)$, which also has been used in the literature (Forman and Laing, 1983).

In Section 3 axioms will be introduced that impose plausible requirements on the functional form of a measure of predictive success which depends only on the hit rate r and the area a. In Section 4 a theorem will be proved which shows that a subset of these axioms characterizes $m = r - a$ up to increasing monotonic transformations. Another subset characterizes $m = r - a$ up to positive linear transformations. This is shown in Section 5. The key axiom in the first characterization is Axiom 4, which requires that the comparison between two theories depends only on gains and losses in accuracy evaluated by hit rate differences and gains and losses in precision evaluated by area differences. The second characterization does not make use of

Axiom 4. Here the key axiom is Axiom 6, which requires 'sample aggregability' in the sense that the measures of predictive success of two samples with n_1 and n_2 observations, respectively, yield the measure of predictive success for the combined sample by taking the weighted average with weights $n_1/(n_1+n_2)$ and $n_2/(n_1+n_2)$, respectively.

2. The implied structure of unimprovable theories

It is possible to distinguish at least three types of theories for an experimental situation with an outcome space P. A *point theory* predicts a single element $p \in P$, to be interpreted as the predicted central tendency of observed outcomes. An *area theory* specifies a subset S of predicted outcomes. A *distribution theory* determines a probability distribution over P and predicts that observed outcomes will be independent random drawings according to this distribution.

Distribution theories are more informative than point theories and area theories. However, in many situations the nature of the theoretical arguments does not lead to distribution theories, but rather to point theories or area theories. Moreover, a meaningful comparison between empirical observations and distribution theories often requires more data than are available.

It is clear that a point theory can be looked upon as a less demanding substitute for a distribution theory. The difficult task of predicting the distribution is replaced by the easier attempt to predict its central tendency. Point theories also need measures of predictive success. Measures based on mean squared deviations or on mean absolute deviations can be used. In both cases one may ask: Which point prediction is *unimprovable* in the sense that it maximizes expected predictive success in the sense of the measure used? If predictive success is judged by the smallness of mean squared deviation, then the unimprovable point prediction is the mean; in the case of mean absolute deviations it is the median or its multidimensional generalization. The measure of predictive success used determines the statistical definition of central tendency, which describes the implied structure of an *unimprovable point theory*.

Our brief discussion of point theories has shown that the implied structure of an unimprovable theory reveals the aim of a search for better theories guided by a measure of predictive success. In what follows the same idea will be applied to area theories.

We shall consider three measures of predictive success, depending on the hit rate r and the area a:

$$m_1 = r - a, \tag{2}$$

the *ratio measure*:

$$m_2 = r/a, \tag{3}$$

and the measure

$$m_3 = \frac{r-a}{1-a}. \qquad (4)$$

Instead of (4) we can also write

$$m_3 = 1 - \frac{1-r}{1-a}. \qquad (5)$$

Whereas r/a can be looked upon as the density of observations *within* the predicted area, the ratio $(1-r)/(1-a)$ may be interpreted as the density of observations *outside* the predicted area. Therefore we call m_3 the *outside ratio measure*. Instead of rewarding a high density within the predicted set, the outside ratio measure aims at a low density outside the predicted set.

For the sake of simplicity we compare the three measures under the assumption of a finite outcome space $S = \{1, \ldots, N\}$ with equal area weights $1/N$ for each possible outcome. For $i = 1, \ldots, N$ let p_i be the probability of outcome i. For any subset T of S let $|T|$ be the number of elements in T. The area $a(T)$ of a subset T of S is the proportion of outcomes in T:

$$a(T) = \frac{|T|}{N}. \qquad (6)$$

The *expected hit rate* $r(T)$ is the probability for the event that the outcome is in T:

$$r(T) = \sum_{i \in N} p_i. \qquad (7)$$

2.1. Unimprovability with respect to the difference measure

The difference measure $m_1(T)$ for a set T of expected outcomes is as follows:

$$m_1(T) = r(T) - a(T). \qquad (8)$$

A subset T of S is *unimprovable* with respect to m_1 if it maximizes $m_1(M)$ over all subsets $M \subseteq S$, or in other words if it satisfies the following condition:

$$m_1(T) = \max_{M \subseteq S} m_1(M). \qquad (9)$$

Assertion 1. *$T \subseteq S$ is unimprovable with respect to m_1 if and only if the following two conditions, (i) and (ii), are satisfied:*
  (i) *Every $i \in S$ with $p_i > 1/N$ belongs to T.*
  (ii) *Every $i \in S$ with $p_i < 1/N$ does not belong to T.*

Proof. Assume that T is unimprovable. Let $i \in S$ be a point with $p_i > 1/N$. Suppose $i \notin T$. Then we have

$$m_1(T \cup \{i\}) = m_1(T) + p_i - \frac{1}{N} > m_1(T). \qquad (10)$$

Similarly, if T contains a point i with $p_i < 1/N$, the following is true:

$$m_1(T \setminus \{i\}) = m_1(T) - p_i + \frac{1}{N} > m_1(T). \tag{11}$$

This shows that (i) and (ii) must hold. It is also true that T must be unimprovable if (i) and (ii) hold; any change of T which takes out points i with $p_i > 1/N$ or puts in points i with $p_i < 1/N$ diminishes the measure m_1.

Remark. If $p_i \neq 1/N$ holds for $i = 1, \ldots, N$, then there is a unique unimprovable set T. Otherwise there are multiple unimprovable subsets T of S which differ only with respect to points i with $p_i = 1/N$.

Comment. In the case of a finite outcome space with equal area weights for all points, an unimprovable theory with respect to m_1 includes all points with more than average probability and excludes all points with less than average probability. A search for better area theories guided by m_1 aims at the inclusion of points with above-average probability and the exclusion of points with below-average probability. This is a reasonable goal for the construction of area theories.

Remark. Suppose that the area weights of the points in S are not equal. Let $a_1, \ldots, a_N$ be the area weights of the points $i = 1, \ldots, N$. It is easy to generalize the definition of unimprovability and Assertion 1 to this case. In (i) and (ii) the probability per area p_i/a_i takes the place of p_i.

2.2. Unimprovability with respect to the ratio measure

We continue to consider the case of a finite outcome space $S = \{1, \ldots, N\}$ with equal area weights for all outcomes. For every non-empty subset T of S the ratio measure $m_2(T)$ is as follows:

$$m_2(T) = \frac{r(T)}{a(T)}. \tag{12}$$

A non-empty subset T of S is *unimprovable* with respect to m_2, if it maximizes $m_2(M)$ over all non-empty subsets of S:

$$m_2(T) = \max_{\emptyset \subset M \subseteq S} m_2(M). \tag{13}$$

Obviously the case $T = \emptyset$ has to be excluded in order to avoid division by zero. We shall use the notation $p_{\max}$ for the maximum of the probabilities $p_1, \ldots, p_N$ of outcomes in S.

Assertion 2. *A non-empty subset T of S is unimprovable with respect to m_2 if and only if the following condition* (iii) *is satisfied:*
  (iii) *For every $i \in T$ we have $p_i = p_{\max}$.*

Proof. If (iii) is satisfied, then $m_2(T)$ assumes the value $p_{\max}$. It can be seen without difficulty that $m_2(T)$ is lower if (iii) is not satisfied.

Remark. If p_i assumes its maximum at a single point j, then $T = \{j\}$ is the uniquely determined non-empty subset of S which is unimprovable with respect to m_2.

Comment. For the case of a finite outcome space with equal area weights for all outcomes, unimprovability with respect to $m_2 = r/a$ has the consequence that only points with maximal probability are predicted. Apart from special cases, an unimprovable theory with respect to m_2 predicts a single point, the mode of the distribution. One may say that r/s aims at a point theory. For sufficiently large N an unimprovable theory with respect to m_2 will typically have a very small area and a very small hit rate, even if other theories with a high hit rate and a moderate area are available. It seems to be clear that the structure of an unimprovable theory implied by m_2 cannot be accepted as reasonable. In the construction of area theories one should not be willing to make extreme sacrifices of accuracy in favor of gains of precision. Nevertheless, this is exactly what m_2 recommends to do in the case of a finite outcome space with equal area weight for all outcomes, if there is only one point with maximal probabibility. In this situation, which is by no means special, the use of m_2 aims at the maximization of precision without any regard to costs in terms of accuracy.

2.3. Unimprovability with respect to the outside ratio measure

For every proper subset T of the outcome space $S = \{1, \ldots, N\}$ the outside ratio measure $m_3(T)$ is as follows:

$$m_3(T) = \frac{r(T) - a(T)}{1 - a(T)}. \tag{14}$$

A proper subset T of S is *unimprovable* with respect to m_3 if it maximizes $m_3(M)$ over all proper subsets of S:

$$m_3(T) = \max_{M \subset S} m_3(M). \tag{15}$$

Obviously, the case $T = S$ must be excluded in order to avoid division by zero. We shall use the notation $p_{\min}$ for the minimum of the probabilities $p_1, \ldots, p_N$.

Assertion 3. *A proper subset T of S is unimprovable with respect to m_3 if and only if the following condition (iv) is satisfied:*
  (iv) *The set T contains all $i \in S$ with $p_i > p_{\min}$.*

Proof. In view of (5) it is clear that maximization of $m_3(T)$ is equivalent to the minimization of the outside ratio $(1 - r(T))/(1 - a(T))$. Obviously this ratio is at least as great as $p_{\min}$ and it is equal to $p_{\min}$ if (iv) is satisfied.

Remark. If p_i assumes its minimum at a single point j, then $T = S \setminus \{j\}$ is the uniquely determined proper subset of S which is unimprovable with respect to m_3.

Comment. In the case of a finite outcome space with equal area weights for all outcomes, unimprovability with respect to m_3 has the consequence that all points with more than minimum probability are predicted. Apart from special cases, an unimprovable theory with respect to m_3 predicts all points with the exception of a single one. Whereas m_2 aims at extreme precision regardless of the cost in accuracy, the measure m_3 aims at extreme accuracy regardless of the cost in precision. The implied structure of unimprovable theories with respect to m_3 is unacceptable, since it cannot be the aim of area theory construction to end up with an unimprovable theory that predicts almost everything. The minimization of the outside ratio $(1-r)/(1-a)$ is as unreasonable as the maximization of the 'inside ratio' r/a. The measures m_2 and m_3 drive theory construction to opposite extremes. A reasonable measure of predictive success should favor a compromise between accuracy and precision. Among the three measures examined in this section, only m_1 has this property. As far as the author knows, no other measures of predictive success, depending on hit rate and area only, have been used in the literature.

3. The axioms

In this section plausible axioms will be introduced which express the desirable properties of measures of predictive success for area theories. The object to be axiomatized is a function m which assigns a real number $m(r,a)$ to every pair (r,a) with $r \in [0,1]$ and $a \in [0,1]$. The function m is interpreted as a *measure of predictive success*. A theory with the hit rate r and the area a is judged to be the more successful the greater is $m(r,a)$. We shall refer to (r,a) as the *hit rate–area combination* of the theory under consideration.

It is convenient to introduce the following *improvement indicator function* $\Delta(r_1, a_1, r_2, a_2)$ defined for r_1, a_1, r_2, a_2 in $[0,1]$ for a given measure m of predictive success:

$$\Delta(r_1, a_1, r_2, a_2) = \begin{cases} +1 & \text{for } m(r_1, a_1) > m(r_2, a_2), \\ 0 & \text{for } m(r_1, a_1) = m(r_2, a_2), \\ -1 & \text{for } m(r_1, a_1) < m(r_2, a_2). \end{cases} \quad (16)$$

The improvement indicator function summarizes the implication of the measure $m(r,a)$ for ordinal comparisons between the predictive success of two theories T_1 and T_2 with different hit rate-area combinations (r_1, a_1) and (r_2, a_2), respectively.

In the following six axioms will be introduced, which impose plausible requirements on a measure of predictive success $m(r,a)$ and the improvement indicator function Δ arising from $m(r,a)$.

Axiom 1 (monotonicity with respect to r). *For every $a \in [0, 1]$ we have*

$$m(r_1, a) > m(r_2, a) \quad \text{for } 0 \le r_2 < r_1 \le 1. \tag{17}$$

Interpretation. If the areas of two theories are equal, the theory with the greater hit rate should be considered to be more successful.

Axiom 2 (monotonicity with respect to a). *For every $r \in [0, 1]$ we have*

$$m(r, a_1) > m(r, a_2) \quad \text{for } 0 \le a_1 < a_2 \le 1. \tag{18}$$

Interpretation. If the hit rates of two theories are equal, the theory with the smaller area should be considered to be more successful. The two monotonicity axioms, Axioms 1 and 2, are obvious requirements for any measure of predictive success depending only on r and a.

Axiom 3 (continuity). *$m(r, a)$ is continuous everywhere on $[0, 1] \times [0, 1]$.*

Interpretation. It is reasonable to require that a measure of predictive success depends continuously on the hit rate r and the area a, if it is a function of these two variables. Roughly speaking, continuity means that small changes in (r, a) correspond to small changes in $m(r, a)$.

Axiom 4 (cost-benefit evaluations). *A function $\delta(r_1 - r_2, a_1 - a_2)$ defined on $[-1, +1] \times [-1, +1]$ exists such that for $r_1, a_1, r_2, a_2 \in [0, 1]$ the following is true:*

$$\Delta(r_1, a_1, r_2, a_2) = \delta(r_1 - r_2, a_1 - a_2). \tag{19}$$

Interpretation. The axiom requires that a comparison between two theories T_1 and T_2 with hit rate-area combinations (r_1, a_1) and (r_2, a_2) depends only on the differences $r_1 - r_2$ and $a_1 - a_2$. Assume that both differences are positive. Suppose that T_1 is a new theory to be compared with an older theory T_2. The new theory is less precise in the sense that its area a_1 is greater than a_2. On the other hand, T_1 is more accurate in the sense that its hit rate r_1 is greater than r_2. Does the *gain in accuracy* $r_1 - r_2$ more than compensate for the *loss in precision* $a_1 - a_2$? In order to answer this question it should be sufficient to compare the gain in accuracy $r_1 - r_2$ with the loss of precision $a_1 - a_2$. Nothing else should matter. In this sense Axiom 4 requires that theory comparisons are determined by cost-benefit evaluations. The task of comparing two theories is reduced to the task of comparing gains and losses on accuracy and precision.

Admittedly Axiom 4 imposes a strong restriction on the measure of predictive success. However, it is reasonable to insist on theory comparison by cost-benefit evaluations. In this way one makes sure that theory comparisons are impartial with respect to the part of the (r, a)-space in which they take place. Why should comparisons between gains and losses of accuracy and precision depend on whether both theories have high or low hit rates or on whether both theories have big or small areas? It seems to be more adequate to take the point of view that, regardless of the theory to be improved, the same change in hit rate and area is evaluated in the same way.

Axiom 5 (equivalence of trivial theories). *We have*

$$m(0,0) = m(1,1). \tag{20}$$

Interpretation. Assume that the set of all possible outcomes is the interval $[0, 1]$. Let T_1 be a theory which predicts a single point, say 0.3, which never occurs as the outcome of an experiment. If the area is measured in the usual way, the hit rate–area combination for T_1 is $(0, 0)$. Obviously T_1 is a useless theory. It is extremely precise but completely inaccurate. Now consider a theory T_2 which predicts the whole outcome space $[0, 1]$ and therefore never fails to predict accurately. T_2 has the hit rate–area combination $(1, 1)$. Obviously T_2, too, is a useless theory. T_2 is completely accurate but without any precision. According to Axiom 5 both theories are judged to be equally useless. But it does not matter whether a theory is completely precise and inaccurate or completely accurate without any precision. It seems to be reasonable to assign the same measure of predictive success to the hit rate–area combinations $(0, 0)$ and $(1, 1)$.

Axiom 6 (sample aggregability). *Let (r_1, a_1) and (r_2, a_2) be two hit rate–area combinations. For every α with $0 \leq \alpha \leq 1$ we have*:

$$m(\alpha r_1 + (1-\alpha)r_2, \alpha a_1 + (1-\alpha)a_2) = \alpha m(r_1, a_1) + (1-\alpha)m(r_2, a_2). \tag{21}$$

Interpretation. Consider a theory for a class of experiments with varying parameters. We may, for example, think of a theory for two-person games in characteristic function form varying with respect to the values assigned to the two one-person coalitions and the two-person coalition. The theory may predict outcomes sets of different relative size for different parameter combinations. In order to test the predictive success of the theory it is necessary to perform experiments over a broad range of parameter combinations. How should the measure $m(r, a)$ be applied to such heterogeneous samples?

The hit rate is the relative frequency of correct predictions. No problems arise in this respect in connection with a heterogeneous sample. However, the question arises: How should the area of a heterogeneous sample be defined? In applications the convention is used that the area of a heterogeneous sample is the arithmetic mean of all areas, computed on the basis of the parameter combinations underlying the observations; every observation has the same weight in this arithmetic mean. We shall now argue that it is natural to use this convention.

Consider the case of just two parameter combinations, 1 and 2. Assume that n

observations have been made, n_1 for parameter combination 1 and n_2 for parameter combination 2. Let (r_1, a_1) and (r_2, a_2) be the hit rate-area combinations for the subsamples of all observations on parameter combination 1 and 2, respectively. Define $\alpha = n_1/n$. Then the sample of all n observations has the hit rate $r = \alpha r_1 + (1-\alpha) r_2$. The definition of the hit rate implies a rule of aggregation which permits the computation of r as a weighted arithmetic mean of r_1 and r_2 with weights $\alpha = n_1/n$ and $1 - \alpha = n_2/n$. It is natural to use the same rule for the aggregation of areas. This leads to the convention for the computation of areas which has been described above.

Axiom 6 requires that the same convention which must be used for the aggregation of hit rates, and therefore is naturally applied to areas, can also be directly applied to measures of predictive success.

If one wants to construct a measure of predictive success which depends only on hit rate and area, one implicitly accepts the linearity properties of hit rates. This suggests the use of a sample area definition and a measure of predictive success with the same linearity properties.

Strictly speaking the interpretation of Axiom 6 given above works for rational α only. Axiom 6 also contains a continuity assertion which extends the validity of (21) to all α with $0 \leq \alpha \leq 1$.

The need for the construction of a measure of predictive success arises from the inadequacy of the hit rate. This suggests the construction of a 'corrected hit rate' with the same linearity properties as the uncorrected hit rate. In this sense a measure of predictive success must satisfy Axiom 6 if it permits an interpretation as a corrected hit rate.

4. Ordinal characterization

We continue to use the definitions and notations of Section 3. The following theorem shows that the measure $m = r - a$ is characterized by Axioms 1-5, as far as its ordinal properties are concerned.

Theorem 1. *The measure of predictive success,*

$$m(r, a) = r - a, \tag{22}$$

satisfies Axioms 1-5. Let $\bar{m}(r, a)$ be another measure of predictive success which satisfies Axioms 1-5. Then the improvement indicator function Δ of $m(r, a) = r - a$ is also the improvement indicator function of $\bar{m}(r, a)$. Moreover a continuous monotonically increasing function φ defined on $[-1, +1]$ exists, such that

$$\bar{m}(r, a) = \varphi(\bar{m}(r, a)) \tag{23}$$

holds for all $r, a \in [0, 1]$.

Proof. For each of the five axioms it can be seen immediately that $m(r,a) = r - a$ satisfies these axioms. Assume that $\bar{m}(r,a)$ is a different measure of predictive success which satisfies Axioms 1–5.

We first show that

$$\bar{m}\left(\frac{1}{n}, \frac{1}{n}\right) = \bar{m}(0,0) \tag{24}$$

holds for every positive integer. Consider a fixed n and assume

$$\bar{m}\left(\frac{1}{n}, \frac{1}{n}\right) > \bar{m}(0,0). \tag{25}$$

It follows by Axiom 4 that we must have

$$\bar{m}\left(\frac{k+1}{n}, \frac{k+1}{n}\right) > \bar{m}\left(\frac{k}{n}, \frac{k}{n}\right) \quad \text{for all } k = 0, \ldots, n-1. \tag{26}$$

Gains and losses of accuracy and precision are the same in the comparison of both sides of (26) as in the case of (25). This has the consequence that $m(1,1)$ must be greater than $m(0,0)$, contrary to Axiom 5. An analogous argument excludes the possibility

$$\bar{m}\left(\frac{1}{n}, \frac{1}{n}\right) < \bar{m}(0,0). \tag{27}$$

This inequality implies (26) with '<' instead of '>'. It follows that (24) holds. Axiom 3 has the consequence that we have

$$\bar{m}(x,x) = \bar{m}(0,0) \quad \text{for } 0 \leq x \leq 1. \tag{28}$$

Let $\bar{\delta}$ be the function connected with $\bar{m}(r,a)$ corresponding to δ in (19). Together with Axiom 4, equation (28) permits the conclusion:

$$\bar{\delta}(x,x) = 0. \tag{29}$$

With the help of Axioms 1 and 2 we obtain:

$$\bar{\delta}(r_1 - r_2, a_1 - a_2) = \begin{cases} 1 & \text{for } r_1 - a_1 > r_2 - a_2, \\ 0 & \text{for } r_1 - a_1 = r_2 - a_2, \\ -1 & \text{for } r_1 - a_1 < r_2 - a_2. \end{cases} \tag{30}$$

This shows that the improvement indicator function $\bar{\Delta}$ of $\bar{m}(r,a)$ agrees with the improvement indicator function Δ of $m(r,a)$. We must have

$$\bar{m}(r_1, a_1) > \bar{m}(r_2, a_2) \quad \text{for } r_1 - a_1 > r_2 - a_2 \tag{31}$$

and

$$\bar{m}(r_1, a_1) = \bar{m}(r_2, a_2) \quad \text{for } r_1 - a_1 = r_2 - a_2. \tag{32}$$

Equation (32) shows that $\bar{m}(r,a)$ is a function of $r - a$. Inequality (31) implies that $\bar{m}(r,a)$ is monotonically increasing in $r - a$. In view of Axiom 3 the measure $\bar{m}(r,a)$

must be a continuous function of $m(r,a) = r - a$. This completes the proof of the theorem.

5. Cardinal characterization

We continue to use the definitions and notations of Section 3. The following Theorem 2 characterizes the measure $m = r - a$ up to positive linear tranformations. Theorem 1 makes use of Axioms 1-5. Theorem 2 is based on Axioms 1, 2, 5 and 6. The continuity axiom, Axiom 3, and cost-benefit evaluation, Axiom 4, are replaced by the sample aggregability axiom, Axiom 6.

Theorem 2. *The measure of predictive success,*

$$m(r,a) = r - a, \tag{33}$$

satisfies Axioms 1, 2, 5 and 6. Let $\bar{m}(r,a)$ be another measure of predictive success which satisfies Axioms 1, 2, 5 and 6. Then a positive number γ and a real number β exist, such that

$$\bar{m}(r,a) = \beta + \gamma m(r,a) \tag{34}$$

holds for all $r, a \in [0, 1]$.

Proof. It can be seen immediately that $m(r,a) = r - a$ satisfies Axioms 1, 2, 5 and 6. Let $\bar{m}(r,a)$ be a different measure of predictive success which satisfies Axioms 1, 2, 5 and 6. Let (r, a) be a hit rate-area combination with

$$r + a \leq 1. \tag{35}$$

With the help of Axiom 6 we can derive the following equations:

$$\bar{m}(r,a) = (r+a)\bar{m}\left(\frac{r}{r+a}, \frac{a}{r+a}\right) + (1 - r - a)\bar{m}(0,0), \tag{36}$$

$$\bar{m}\left(\frac{r}{r+a}, \frac{a}{r+a}\right) = \frac{r}{r+a}\bar{m}(1,0) + \frac{a}{r+a}\bar{m}(0,1). \tag{37}$$

Equations (36) and (37) yield:

$$\bar{m}(r,a) = r\bar{m}(1,0) + a\bar{m}(0,1) + (1 - r - a)\bar{m}(0,0) \tag{38}$$

for $r + a \leq 1$. In view of Axiom 6 we have

$$\bar{m}(\tfrac{1}{2}, \tfrac{1}{2}) = \tfrac{1}{2}\bar{m}(0,0) + \tfrac{1}{2}\bar{m}(1,1). \tag{39}$$

This, together with Axiom 5, yields:

$$\bar{m}(\tfrac{1}{2}, \tfrac{1}{2}) = \bar{m}(0,0). \tag{40}$$

With the help of (38) we obtain:

$$\bar{m}(\tfrac{1}{2}, \tfrac{1}{2}) = \tfrac{1}{2}\bar{m}(1,0) + \tfrac{1}{2}\bar{m}(0,1). \tag{41}$$

This yields:

$$\bar{m}(0,0) = \frac{\bar{m}(1,0) + \bar{m}(0,1)}{2}. \tag{42}$$

With the help of (34), equation (30) can be rewritten as follows:

$$\bar{m}(r,a) = \frac{\bar{m}(1,0) - \bar{m}(0,1)}{2}(r-a) + \bar{m}(0,0). \tag{43}$$

Up to now the validity of this equation has been shown for $r+a \le 1$ only. In what follows we shall prove that (43) holds for $r+a>1$, too. Let (r,a) be a hit rate-area combination with

$$r+a>1. \tag{36}$$

In view of Axiom 6 we have

$$\bar{m}(r,a) = (2-r-a)\bar{m}\left(\frac{1-a}{2-r-a}, \frac{1-r}{2-r-a}\right) + (r+a-1)\bar{m}(1,1). \tag{45}$$

With the help of (38) we obtain:

$$(2-r-a)\bar{m}\left(\frac{1-a}{2-r-a}, \frac{1-r}{2-r-a}\right) = (1-a)\bar{m}(1,0) + (1-r)\bar{m}(0,1). \tag{46}$$

In view of Axiom 5 equations (45) and (46) yield:

$$\bar{m}(r,a) = r(\bar{m}(0,0) - \bar{m}(0,1)) + a(\bar{m}(0,0) - \bar{m}(1,0)) + \bar{m}(0,0). \tag{47}$$

With the help of (42) it can be seen that (47) is equivalent to (43). Therefore (43) holds for every hit rate-area combination (r,a). Axioms 1 and 2 permit the conclusion that $\bar{m}(1,0)$ is greater than $\bar{m}(0,1)$. It follows by (43) that $\bar{m}(r,a)$ is a positive linear tranformation of $r-a$. This completes the proof of the theorem.

Remark. The proof of the theorem shows that Axioms 1 and 2 could be replaced by the much weaker requirement that $m(1,0)$ is greater than $m(0,1)$. Equation (35) is a consequence of Axioms 5 and 6 alone.

6. Discussion

This paper is devoted to the interpretation and justification of a very simple measure of predictive success for area theories: the difference between hit rate and area. Section 2 has raised a problem of interpretation. What is an area theory trying to achieve? The answer to this question depends on the evaluation of predictive success. The use of a specific measure of predictive success gives a direction to the search for better area theories and determines the structure of an ideal unimprovable area theory. If the search is guided by the hit rate minus area measure, then an ideal area theory should have at least average probability per area unit everywhere in its set of predicted outcomes, and outside of this set probability per area unit should be at most average. This is a reasonable goal for the development of an area theory.

For every proposed measure of predictive success for area theories it is important to look at the implied structure of an ideal unimprovable area theory. The examination of the ratio measure r/a and the outside ratio measure $(r-a)/(1-a)$ has shown that both measures fail to yield a plausible implied structure of an unimprovable theory. This has been shown in the framework of a finite outcome space with equal area weights for all outcomes. Apart from special cases, unimprovability with respect to the ratio measure implies the prediction of a single outcome, namely the most probable one. The ratio measure overemphasizes precision and does not give enough weight to accuracy. Apart from special cases, unimprovability with respect to the outside ratio measure implies the prediction of all outcomes except one, namely the least probable one. The outside ratio measure overemphasizes accuracy and does not give enough weight to precision.

Neither the prediction of a single outcome nor the prediction of almost all outcomes is a reasonable aim in the construction of area theories. The ratio measure and the outside ratio measure are both unacceptable in view of their implications for the structure of an unimprovable theory. Contrary to this the difference measure $r-a$ implies a reasonable structure of an unimprovable theory.

The comparsion of the three measures used in the literature does not yet single out $r-a$ among all possible measures of predictive success depending on hit rate and area only. Therefore plausible axioms have been introduced which express desirable properties of such measures of predictive success. One set of axioms yields an ordinal characterization of the difference measure $r-a$ and another leads to a cardinal characterization.

The ordinal characterization uniquely determines the difference measure up to continuous monotonically increasing transformations. Admittedly Axiom 4, the key axiom in this characterization, is very strong. It requires that ordinal comparisons between two theories depend always in the same way on hit rate differences and area differences.

The ordinal axiomization uniquely determines the difference measure up to a positive linear transformation. The key axiom in this characterization is Axiom 6 which imposes a linearity property on the measure. It has been argued that, in view of the necessity to evaluate the predictive success of heterogeneous samples, it is natural to construct a measure with the same linearity properties as the hit rate. Even if together with the other axioms, Axiom 6 is much stronger than Axiom 4, the arguments in favor of Axiom 6 seem to be more compelling than those in favor of Axiom 4.

It has been shown that the functional form of the difference measure is by no

It has been shown that the functional form of the difference measure is by no means arbitrary. Other functional forms which seem to be plausible at first glance have unacceptable implications for the structure of an unimprovable theory. Moreover, two axiomatic characterizations of the difference measure lend further support to the point of view that the difference measure is the best method for the evaluation of the predictive success of area theories by means of hit rates and areas.

In this paper attention has been restricted to measures of predictive success depending on hit rate and area only. More complicated measures can be constructed based on additional infomation, such as the distance of unpredicted observations to predicted outcomes. However, it is doubtful whether such additional information should be taken into account.

Compared with point theories and distribution theories, area theories have the advantage that for every observed outcome it is clear whether the prediction was correct or not. The strict distinction between success and failure is blurred by a different treatment of near misses and far deviations. This is undesirable from the point of view that a measure of predictive success should provide a clear guidance for the improvement of area theories. If there are many observations outside the predicted set, but near to it, the theory should be changed in the appropriate direction. Since area theorists are free to choose their sets of predicted outcomes, they should not be permitted to use the nearness of observed deviations as an excuse. A hit is a hit and a miss is a miss.

References

R. Forman and J.D. Laing, Game-theoretic expectations, interest groups, and salient majorities in committees, in: R. Tietz, ed., Aspiration Levels in Bargaining and Economic Decision Making (Springer, Berlin and Heidelberg, 1983) 321–336.

R. Selten and S. Krischker, Comparison of two theories for characteristic function experiments, in: R. Tietz, ed., Aspiration Levels in Bargaining and Economic Decision Making (Springer, Berlin and Heidelberg, 1983) 259–264.

An axiomatic approach to consumers' welfare

Reinhard Selten

Department of Economics, University of Bonn, 5300 Bonn 1, Germany

Eyal Winter*†

Department of Economics, Hebrew University of Jerusalem, Jerusalem 91905, Israel

Communicated by K.H. Kim
Received 3 June 1993
Revised 6 July 1993

Abstract

This paper addresses the issue of consumers' welfare using an axiomatic approach, and avoiding the assumption of utility maximization. We take here the demand function as a primitive, and construct a measure for the consumers' welfare from a change of prices. This is done by imposing a set of axioms on this measure which *uniquely* determines it.

Key words: Welfare; Axiomatic approach; Demand function; Measure

1. Introduction

What are the effects of a price change on consumers' welfare? Is it possible to answer this question based on an aggregate demand function alone? Obviously it depends on the definition of consumers' welfare whether this is the case or not. How should we define consumers' welfare in order to be able to form a meaningful judgement on its change as a consequence of policy measures influencing the price system? This is the problem that this paper addresses.

The classical approach to the problem posed here is the notion of consumers' surplus introduced by Dupuit (1884). Dupuit was inspired by the special case of the demand function of a single commodity originating from the demand of many individuals each of whom has a reservation price up to which he is willing to buy exactly one unit and not more. For Dupuit, consumers' surplus was the total amount saved by the buyers compared with the reservation price they are willing

---

* Corresponding author.
† This research was supported by the Deutscheforschungsgemeinschafts through the SFB 303.

to pay. The difficulty that arises with the concept of consumers' surplus in a more general context is well known. Serious criticisms have been raised against the concept, e.g., by Samuelson. There also have been attempts to defend it in a modified form (Hicks, 1941). However, the discussion was always based on the assumption that aggregate demand can be derived from consistent preferences of a representative consumer. There is no reason to suppose that aggregate demand functions permit the construction of underlying preferences even if one is willing to assume that individual households are utility maximizers.

In this paper we do not want to proceed from the highly dubious assumption of utility maximization. It is our aim to provide a basis of welfare judgement without any references to consistent preferences. For us the aggregate demand will be a primitive concept taken at face value. Our approach will be axiomatic. The experimental evidence against the assumption of utility maximization is by now very strong. Nevertheless, it is a matter of great practical importance to be able to form judgements of consumers' welfare in a systematic way even if utility theory cannot serve as a point of departure. We therefore take the point of view that we should not ask whether the measurement of consumers' surplus is possible, but rather how should consumer's surplus be defined in such a way that it becomes measurable on the basis of aggregate demand functions.

We shall axiomatize a money equivalent of the gain or loss in consumers' welfare caused by the change in the price system. For the one-commodity case our concept of 'consumers' gain' coincides with the change of consumers' surplus. It is not clear how the classical concept of consumers' surplus should be generalized to the case of many commodities. Several solutions to this problem have been proposed (see Silberberg, 1972). Our axiomatization yields a result which, as far as we know, has not been suggested in the literature. It does not really involve a generalization of consumer's surplus but rather a notion of 'consumers' burden'.

Before we can go on to explain what is meant by consumers' burden, it is necessary to say something about the domain of the demand functions considered in the axiomatization. We look only at demand functions that are bounded. This implies the assumption that demand has a point of satiation in the sense that even at zero prices only finite quantities of all goods are consumed. We view this assumption as a reasonable idealization of real consumption behavior. The demand functions considered in this paper are not necessarily total demand functions. The axioms can be reasonably applied in a partial context too. This seems to be important since welfare judgements based on consumers' surplus usually are made in a partial context. Therefore we intentionally did not include income as a variable on which demand depends. In the case where income equals expenditure it is unnecessary anyhow to let income appear explicitly as an argument of the function.

Consumers' burden is the cost of being constrained by prices. This cost has two components. One is the expenditure for consumption, and the second is the 'loss' caused by not consuming as much as the satiation point. In Fig. 1 these two components are displayed for the single-commodity case.

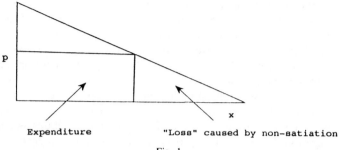

Fig. 1.

In the one-commodity case it makes no difference if one looks at the changes of consumers' surplus or changes of consumers' burden. The same is true for many commodities under the classical assumption that demand is derived from unconstrained maximization of a utility function which has the form of monetary consumption value minus expenditure. However, in the general multi-commodity case, consumers' surplus is difficult to define, whereas consumers' burden has a clear meaning under our assumption of satiation. Our axioms concern a concept of consumers' welfare gain. The notion of consumers' burden is not a point of departure but rather a result of the analysis.

2. Demand functions and consumption welfare

In this section we introduce a measure for consumers' welfare gain with its characterizing axioms. We will however start with some basic notations.

For a positive integer k, we denote by R_+^k (R_{++}^k) the set of all non-negative (positive) vectors in R^k. For vectors x, y in R^k, we denote by $x \cdot y$ the scalar $\sum_{i=1}^{k} x_i y_i$ and by xy the vector $(x_1 y_1, \ldots, x_k y_k)$. $x \geq y$ means $x_i \geq y_i$ for $1 \leq i \leq k$. Let M^k be the set of all continuous and bounded functions defined on the non-negative cone of all price vectors, i.e. $x: R_+^k \to R^k$. We will use the functions in M^k to represent the consumer behavior. $x_i(p_1, \ldots, p_k)$ is thus the demand[1] for commodity i under the price system $(p_1, \ldots, p_k)$. For every vector $\beta > 0$, let $\Delta_\beta^k = \{p \in R_+^k;\ p \leq \beta\}$. We will denote by M_β^k the set of all continuous and bounded functions x, defined on the box Δ_β^k. Finally, for $\beta \in R_{++}^k$ we will use $\|\cdot\|$ to denote the sup norm on M_β^k i.e. $\|x_i\| = \sup_{p \in \Delta_\beta^k} |x_i(p)|$.

A *consumers' welfare gain function* (W.G.F.) assigns to each integer k two price

---

[1] Note that we do not assume that the functions in M^k are non-negative. Negative demand can be interpreted as supply. Positive demand for leisure, for example, is a negative demand for labor, or a positive supply for labor. It should be remarked that the results of this paper are valid also when one restricts oneself to positive demand functions. Including demand functions which are not necessarily non-negative in our domain is merely for the purpose of using convergence theorems when the domain includes functions with zero values.

vectors, p^0 and p^1, in R_+^k, and a demand function x in M^k, a real number $c(p^0, p^1, x)$ which will be interpreted as the consumers' welfare gain induced by a price change between p^0 and p^1, under the consumption behavior x. The consumers' welfare gain function thus takes the demand functions as primitives and yields a value in money terms for every price change. We will focus our attention here on one such measure of welfare gain, which will then be justified through an axiomatic characterization. This W.G.F. in fact extends the Marshal–Dupuit measure for single-good consumption. To obtain our W.G.F. we first compute for each price vector p in R_+^k and a demand function x, the value (in money terms) of being able to get any amount from any good free of charge. This value, for which we will use the name consumers' burden, is obtained by integrating x_i (the ith coordinate of x), along the line connecting 0 and p, and then aggregating over all commodities. (Recall that the demand functions are bounded so that at price 0 the consumer still consumes finite quantities.) Denoting by $c(p, x)$ the consumers' burden for the price vector p and the demand function x, our W.G.F. is given by $c(p^0, p^1, x) = c(p^0, x) - c(p^1, x)$.

More precisely,

$$c(p^0, p^1, x) = \sum_{i=1}^{k} \left[p_i^0 \int_0^1 x_i(p^0 t)\, dt - p_i^1 \int_0^1 x_i(p^1 t)\, dt \right]. \tag{2.1}$$

Note that for the single good case, i.e. when $k = 1$, our W.G.F. coincides with the Marshal–Dupuit measure of welfare, which is the difference between the consumers' surplus at the two price vectors. A set of W.G.F. generalizing the Marshal–Dupuit measure, and based on path integration, was introduced by Silberberg (1972). We will now proceed with the axiomatic treatment for our W.G.F. The first five axioms will determine a W.G.F. up to the choice of the measure with respect to which the integration in (2.1) is taken. The last axiom will then determine the W.G.F. uniquely by fixing this measure to be the Lebesgue measure. Unless otherwise specified, p^0 and p^1 in the following axioms stand for arbitrary price vectors in R_+^k.

The first axiom asserts an additivity property of the W.G.F. with respect to the demand functions. It allows us to aggregate the demand of different consumers to obtain the total consumption welfare gain for a group of consumers. Alternatively, if the consumption behavior of a consumer changes over time, then his long-run welfare gain can be obtained by summing up his demand functions. Formally,

A_1. Additivity: for every two demand functions x_1 and x_2,
$c(p^0, p^1, x_1 + x_2) = c(p^0, p^1, x_1) + c(p^0, p^1, x_2)$.

It can be easily shown that additivity implies that we can obtain the W.G.F. by first computing the welfare gain for each commodity, independently of the demand for other goods, and then aggregating over all commodities, namely:

Lemma 1 *Under additivity we have,* $c(p^0, p^1, x) = \sum_{i=1}^{k} c_i(p^0, p^1, x)$, *where* $c_i(p^0, p^1, x)$ *is independent of* x_j *for all* $j \neq i$.

Proof. Let $x^i \in M^k$ be given by $x_i^j \equiv 0, \forall i \neq j$ and $x_i^i = x_i$, and write $c_i(p^0, p^1, x) = c(p^0, p^1, x^i) c_i$ is independent of x_j, for $j \neq i$, and $c(p^0, p^1, x) = c(p^0, p^1, \sum_{i=1}^{k} x^i) = \sum_{i=1}^{k} c_i(p^0, p^1, x)$. Q.E.D.

The second axiom requires independence with respect to intermediate price policies. Suppose that prices are given by the price vector $p^0 \in R_+^k$ at a certain day 0 which is followed by government tax policy at day 1 causing an increase in prices to p^1. If prices are reduced again to p^0 at day 2 due to subsidies, it is expected from a consistent measure to ignore these contradicting policies and to assign a zero value to the total welfare gain between days 0 and 2. This reads formally as follows:

A$_2$. Consistency: $c(p^0, p^1, x) + c(p^1, p^2, x) = c(p^0, p^2, x)$.

Note that consistency implies that

Lemma 2. *For every sequence of prices vectors* $p^0, \ldots, p^n$ *with* $p^0 = p^n$ *we have*

$$\sum_{i=0}^{n-1} c(p^i, p^{i+1}, x) = 0 . \tag{2.2}$$

Proof. First write $p^0 = p^1 = p^2 = p$, and observe $(c(p, p, x) = 0$. By induction on n we now obtain that for every price sequence $p^0, \ldots, p^n$, $\sum_{i=0}^{n-1} c(p^i, p^{i+1}, x) = c(p^0, p^n, x)$ and the result follows. Q.E.D.

The next axiom imposes the positivity of the welfare gain for non-negative demand functions and proportional price reduction (as in reducing the V.A.T. for example), i.e. if all prices decrease by the same proportion, then the consumers' welfare gain for such a decrease cannot be negative if his demand function is non-negative. Formally:

A$_3$. Positivity: for every $0 \leq \mu \leq 1$, and p in R_+^k, if $x(p') \geq 0$,[2] for all $p' \leq p$, then $c(p, \mu p, x) \geq 0$, and we have equality when $x(p') = 0$ for all $p' \leq p$.

We can now show that the axioms imposed up to this point imply the continuity of the W.G.F.

Lemma 3. *If c satisfies positivity, additivity and consistency, then for any* $\beta \in R_{++}^k$ *and any* $p^0, p^1 \leq \beta$, $c(p^0, p^1, \cdot)$ *is continuous in the sup norm on* M_β^k. *Namely, if*

---

[2] $x(p) \geq 0$ means $x^i(p) \geq 0$ for $1 \leq i \leq k$.

$\{x^n\}_{n=1}^{\infty}$ converges to x on Δ_β^k in the sup norm (i.e., $\forall 1 \leq i \leq k$ $\|x_i^n(p) - x_i(p)\| \to 0$ when $n \to \infty$), then, $c(p^0, p^1, x^n) \to c(p^0, p^1, x)$ as $n \to \infty$.

Proof. By additivity it is sufficient to show that if x^n converges to the zero function, then $c(p^0, p^1, x^n) \to 0$ as $n \to \infty$. Note first that for every rational number λ we have

$$\lambda c(p^0, p^1, x) = c(p^0, p^1, \lambda x). \tag{2.3}$$

For an integer λ, this is simply the additivity, and if $\lambda = r/s$ (r and s are integers), we have $c(p^0, p^1, (r/s)x) = rc(p^0, p^1, (1/s)x)$. Using now the additivity again for the demand functions $(1/s)x$, we get (2.3).

For each integer n let $\epsilon^n > 0$ be a rational number such that $\epsilon^n > \max_{i=\{1,\ldots,k\}} \sup_{p \in \Delta^k} x_i^n(p)$, and $\epsilon^n \to 0$ as $n \to \infty$. Let $\hat{\epsilon}^n$ be the demand function given by $\hat{\epsilon}_i^n(p) \equiv \epsilon^n$. Write $\hat{x}^n = \hat{\epsilon}^n - x^n$, and note that $\hat{x}^n \geq 0$. By the positivity axiom we have $c(p, 0, \hat{x}^n) \geq 0$, which by additivity means $c(p, 0, \hat{\epsilon}^n) \geq c(p, 0, x^n)$. By (2.3) we have $c(p, 0, \hat{\epsilon}^n) = \epsilon^n c(p, 0, 1) \geq c(p, 0, x^n)$, where 1 is the demand function which assigns 1 for each commodity at each price. If we now do the same for $-x^n$, we will get $\epsilon^n c(p, 0, 1) \geq -c(p, 0, x^n)$, which means that $c(p, 0, x^n) \to 0$ as $n \to \infty$. Now again by consistency we have $c(p^0, p^1, x^n) = c(p^1, 0, x^n) - c(p^0, 0, x^n)$, and thus $c(p^0, p^1, x^n) \to 0$ as $n \to \infty$. Q.E.D.

We proceed with the requirement under which the W.F.G. is independent of the specific unit in which quantities are measured. Suppose that x^1 is the demand function of a single commodity for which quantities are given in kilograms (i.e. price is issued per kilogram). If x^2 represents the same consumption behavior as x^1, but now with quantities measured in tons, then we must have $x^1(p) = 1000$ $x^2(1000p)$. According to the rescaling axiom it does not matter in which unit quantities are given as long as prices are adjusted accordingly. Namely, for price change (per kilogram) from p^0 to p^1 we have $c(p^0, p^1, x^1) = c(1000p^0, 1000p^1, x^2)$. Recall that c itself is always given in money terms. For the multi-commodity case this can be generalized as follows.

A$_4$. Rescaling: suppose that $x, \hat{x} \in M^k$ satisfy $\hat{x}(p) = \lambda x(\lambda p)$ for some $\lambda \in R_{++}^k$, then

$$c(p^0, p^1, \hat{x}) = c(\lambda p^0, \lambda p^1, x). \tag{2.4}$$

Note that λ is a vector, which means that we allow quantities of different commodities to be measured in different units. In particular, units of different commodities may not even refer to the same measure. One commodity may be measured by weight, for example, while another by time. For the single-commodity case the requirement of the rescaling axiom can also be interpreted as independence of the specific money unit in which prices are given. This interpretation, however, can hardly be extended to the multi-commodity case, since this would imply that prices of different commodities are given in different currencies. A different formulation for the rescaling axiom is the following.

Lemma 4. *Given* A_1 *and* A_3, *axiom* A_4 *is equivalent to*

$$c_i(p^0, p^1, \hat{x}) = \frac{1}{\lambda_i} c_i(\lambda p^0, \lambda p^1, x), \tag{2.5}$$

where $\hat{x}(p) = x(\lambda p)$.

Proof. First notice that (2.5) implies $c_i(p^0, p^1, \hat{x}) = c_i(\lambda p^0, \lambda p^1, (1/\lambda_i)x)$. This is shown for rational λ by additivity in the same manner as in Lemma 3, and for real λ it uses the continuity property. This shows that the requirement in this lemma implies A_4. For the converse use Lemma 1 with additivity. Q.E.D.

The last axiom in this section deals with the property of W.G.F. under which several commodities can be bundled to form a single commodity. This axiom is based on the notion of complimentary goods.

Consider a consumer who consumes two goods, shirts and ties. Suppose that any time he buys a tie, a suitable shirt will be immediately matched to it. Under this consumption behavior the demand for ties will be very much dependent on the prices of shirts, and we in fact can assume that this demand is a function of the total shirt–tie price. In other words, as far as the demand for ties is concerned, we can consider ties and shirts as a single commodity. The bundling axiom asserts that for such situations it should be possible to obtain at least the tie component of the welfare gain as a welfare gain of a single commodity, provided that all prices change in the same proportion. Formally,

A_5. Bundling: suppose that $x_i(p) = \hat{x}(\sum_{i=1}^{k} p_i)$, then for each $0 \neq p \in \Delta^k$ and $0 \leq \mu \leq 1$ we have

$$c_i(\mu p, p, x) = \frac{p_i}{\sum_{j=1}^{k} p_j} c\left(\sum_{j=1}^{k} \mu p_j, \sum_{j=1}^{k} p_j, \hat{x}\right).$$

Note that $x_i(p) = \hat{x}(\sum_{j=1}^{k} p_j)$ means that each commodity $j \neq i$ strictly complements i, that is, one unit of commodity i is consumed with one unit of each of the other commodities. This strict complimentary relation is however not symmetric. Namely, $x_i(p) = \hat{x}(\sum_{j=1}^{k} p_j)$ does not imply $x_i(p) = x_j(p)$ for all $i, j = (1, \ldots, k)$.

Obviously, in our foregoing example shirts strictly complement ties, but the converse is not true since shirts can be worn without ties. Finally, note that in A_5 the welfare gain of the bundled commodity is normalized by $p_i / \sum_{j=1}^{k} p_j$. This is due to the fact that this welfare gain is given in terms of the total price of the whole bundle while c_i is given in terms of the price of commodity i only.

In the next section we turn to the characterization of Consumers' Welfare Gain Functions.

3. Determining the W.G.F. by the axioms

According to Theorem 1 a W.G.F. is determined by A_1–A_5 up to the choice of the measure with respect to which integration is taken.

Theorem 1. *A W.G.F. c satisfies A_1–A_5 if and only if there exists a non-negative measure θ on $[0, 1]$ such that*

$$c(p^0, p^1, x) = \sum_{i=1}^{k} \left[p_i^0 \int_0^1 x_i(p^0 t) \, d\theta(t) - p_i^1 \int_0^1 x_i(p^1 t) \, d\theta(t) \right]. \quad (3.1)$$

We will start the proof by treating first the single-commodity case.

Lemma 5. *Let x be a demand function of a single commodity. If c satisfies A_1–A_5, then $c(0, p, x) = -p \int_0^1 x(pt) \, d\theta(t)$ for some non-negative measure θ which is uniquely determined.*

Proof. By the rescaling axiom it is sufficient to show the result for $p = 1$. By the positivity and additivity we first obtain that $c(0, 1, x)$ depends on x only through its behavior on prices smaller than 1, i.e. $c(0, 1, x_1) = c(0, 1, x_2)$ for x_1 and x_2 with $x_1(p) = x_2(p)$ for all $0 \leq p \leq 1$. This is shown by defining $x = x_2 - x_1$ and noting that we have zero welfare gain for x. We can therefore consider $-c(0, 1, \cdot)$ as a function on M_β^1 for $\beta = 1$. Owing to additivity and positivity (with Lemma 3), this operator is linear positive and continuous. Using the Riesz Representation Theorem, there exists a unique measure θ such that $-c(0, 1, x) = \int_0^1 x(t) \, d\theta(t)$. Q.E.D.

Proof of Theorem 1. Verifying the fact that c satisfies axioms A_1–A_5 is straightforward. We will show here therefore only the rescaling and the bundling axioms. Owing to the consistency, it is sufficient to show the rescaling only for $p^0 f = 0$. If $\hat{x}_i(p) = x_i(\lambda p)$ for some $\lambda \in R_{++}^k$, then

$$c_i(0, p, \hat{x}) = -p_i \int_0^1 \hat{x}_i(tp) \, d\theta(t) = -p_i \int_0^1 x_i(\lambda pt) \, d\theta(t)$$

$$= \frac{1}{\lambda_i} c_i(0, \lambda p, x).$$

For the bundling axiom, suppose that $x_i(p) = \hat{x}(\sum_{j=1}^k p_j)$, then

$$c_i(\mu p, p, x) = \mu p_i \int_0^1 \hat{x}\left(t \sum_j \mu p_j\right) d\theta(t) - p_i \int_0^1 \hat{x}\left(t \sum_j p_j\right) d\theta(t)$$

$$= \mu p_i \frac{1}{\sum_j \mu p_j} c\left(0, \sum_j \mu p_j, \hat{x}\right) - p_i \frac{1}{\sum_j p_j} c\left(0, \sum_j p_j, \hat{x}\right)$$

$$= \frac{p_i}{\sum_j p_j} c\left(\sum_j \mu p_j, \sum_j p_j, \hat{x}\right).$$

We will now show that c must have the property given in (3.1). By the consistency axiom (A_2), $c(p^0, p^1, x) = c(p, p^1, x) - c(p, p^0, x)$ for every price vector p, in particular for $p = 0$. It is therefore sufficient to show this part of the theorem for $p^0 = 0$. By Lemma 1 it is sufficient to show (3.1) only for demand functions with non-zero consumption only in one commodity (i.e. for some $1 \le i \le k$, $x_j \equiv 0$, $\forall j \ne i$). We will start with the case where the demand for this commodity is given by a polynomial, and then use Lemma 3 for general continuous demand functions. If $x_i(p)$ is a polynomial in $(p_1, \ldots, p_k)$, then $x_i(p)$ can be written as a linear combination of polynomials with the form

$$\left(\sum_{j=1}^k n_j p_j\right)^l, \qquad (3.2)$$

where n_j are non-negative. (For the proof of this claim see also Aumann and Shapley, 1974, p. 4.) Owing to additivity and continuity it is enough to prove (3.1) on polynomials of type (3.2) only. We will assume first that $n_j > 0$ for $1 \le j \le k$. Writing $\hat{x}(p) = p^l$, we have $x_i(p_1, \ldots, p_k) = \hat{x}(\sum_{j=1}^k n_j p_j)$. Using the rescaling axiom with the bundling for $\mu = 0$ we have

$$c_i(0, p, x) = \frac{1}{n_i} \frac{n_i p_i}{\sum_{j=1}^k n_j p_j} c(0, n \cdot p, \hat{x}),$$

where, $n = (n_1, \ldots, n_k)$. By Lemma 5,

$$c(0, n \cdot p, \hat{x}) = -n \cdot p \int_0^1 \hat{x}(n \cdot pt) d\theta(t),$$

for some non-negative measure θ, which yields

$$c_i(0, p, x) = -p_i \int_0^1 x(pt) d\theta(t)$$

as in (3.1). For the case where n_j are not all positive, write

$$x_i^\epsilon(p) = \left(\sum_{j=1}^k (n_j + \epsilon) p_j\right)^l, \quad \text{for } \epsilon > 0,$$

and obtain the result for $x_i^\epsilon(p)$, $\forall \epsilon > 0$. Since $\|x_i^\epsilon - x_i\| \to 0$ as $\epsilon \to 0$, we can use the continuity property of c_i to conclude the result also for polynomials with some zero coefficients. To consider the general case, take x in M^k and p^0 and p^1 in R_+^k. Let β be a vector in R_{++}^k such that $\beta_i > \max\{p_i^0, p_i^1\}$. Since the polynomials are dense in M_β^k with respect to the sup norm, there exists a sequence $\{x_n\}_{n=1}^\infty$ of polynomials such that $\|x_n - x\| \to 0$ as $n \to \infty$. Again, since c_i is continuous the result follows due to Lemma 5 for the general case. Q.E.D.

Theorem 1, although determining a specific form of consumers' welfare gain functions, allows a variety of such functions. Note in particular that if the measure θ is taken to have the total weight on 1, then the corresponding W.G.F. is just the difference between the expenditures at the two price vectors. A different special case is the one corresponding to total weight at 0. In this case the W.G.F. assigns 0 to every price change for every demand function. Note that this W.G.F. indeed satisfies all the axioms imposed so far. To obtain uniqueness one needs to determine the measure with respect to which integration is taken. We will do it by introducing the final axiom, which is based on one of the most basic ideas of the theory of consumers' welfare. According to this principle, which goes back to Dupuit's seminal work of 1844, the satisfaction obtained from a normal commodity (with a downward-sloping demand curve), is in excess of actual expenditure, and this excess vanishes only when the demand function is constant. Consequently, a consumer would be prepared to pay for a price reduction more than what he saves if his consumption level stays constant. Analogously, if prices increase, the consumer should typically be compensated by less than the increase in expenditure induced by keeping the consumption constant. This reads formally as follows:

A_6. Dupuit's principle: if x is a downward-sloping non-negative demand function of a single good, then

$$(p^0 - p^1)x(p^0) \leq c(p^0, p^1, x) \leq (p^0 - p^1)x(p^1).$$

We can now show that A_6 with the rest of the axioms determines the measure θ in (3.1) to be uniform (Lebesgue measure), and thus selects a unique W.G.F.

Theorem 2. *There exists a unique W.G.F. satisfying* A_1–A_6, *which is given by* (2.1).

Proof. As we have already argued, it is sufficient to show that A_6 determines the measure θ in (3.1) to be a Lebesgue measure. To show this it is sufficient to consider the single-good case for downward-sloping demand functions, since the measure is unique for all demand functions. First, note that by Lemma 5, for any two prices, p and $\bar{p}$, with $p \leq \bar{p}$, the welfare gain $c(\bar{p}, p, x)$ is just the area below the inverse demand curve between the two prices.

For $p^1, \ldots, p^m \in [p, \bar{p}]$, with $p = p^1 < p^2, \ldots, < p^m = \bar{p}$, we have by the consistency axiom, $c(\bar{p}, p, x) = \sum_{i=1}^m c(p^i, p^{i+1}, x)$. By Dupuit's principle (A_6) we have

$$\sum_{i=1}^{m-1} (p^i - p^{i+1})x(p^i) \le \sum_{i=1}^{m-1} c(p^i, p^{i+1}, x) \le \sum_{i=1}^{m-1} (p^i - p^{i+1})x(p^{i+1}).$$

Now sending m to infinity by taking a sequence of refining partitions of the interval $[p, \bar{p}]$, we obtain that $c(\bar{p}, p, x)$ coincides with the integral of the inverse demand function between the two prices. This means that the measure in (3.1) must be uniform, and that the W.G.F. must be the one given by (2.1). Q.E.D.

4. Utility maximization and the welfare gain

In our axiomatic treatment of the consumption measure we deliberately chose to take the demand function as a primitive on which the axioms are formulated. In classical consumption theory, however, the demand function is derived from the consumer's utility function on consumption bundles. In this short section we show that when the demand function is derived by utility maximization without budget constraints, i.e. when the consumer has linear (dis)utility for expenditure, then our measure for price change between p^0 and p^1 coincides with the difference between the indirect utility at these two prices.

Let U be a quasi-concave utility function on consumption bundles. $U(x_1, \ldots, x_k)$ is the utility from consuming x_i from the commodity i. Let W be the consumption utility function in which expenditure disutility is incorporated, i.e. $W(x_1, \ldots, x_k) = U(x_1, \ldots, x_k) - \sum_j p_j x_j$, where p_j is the price of commodity j. Let $x(p) = (x_1(p), \ldots, x_k(p))$ be the optimal consumption at price p, i.e. x maximizes $W(x)$ under the price vector p. Finally, write

$$W(p) = W(x(p)) = U(x_1(p), \ldots, x_k(p)) - \sum_i p_i x_i(p)$$

Our consumers' welfare gain can now be represented in terms of the indirect utility function W as follows:

Theorem 3. *For every two price vectors p^0 and p^1 we have*

$$c(p^0, p^1, x) = W(p^1) - W('p^0), \qquad (4.1)$$

provided that the left hand side of (4.1) is finite and well defined.

For the proof of Theorem 3 we make use of the following well-known mathematical Lemma. (For the proof see also Apostol, 1957, pp. 280–281 and p. 292.)

Lemma 6. *If $g(z)$ is a continuously differentiable function in an open convex subset B of R^n, then the line integral*

$$I = \int_{z^0}^{z^1} \sum_{i=1}^n g_i(z) \, dz_i, \qquad g_i \equiv \frac{\partial g}{\partial z_i}, \qquad (4.2)$$

connecting the points z^0 and z^1 in B by a piecewise smooth curve is independent of the path. Furthermore, $I = g(z^1) - g(z^0)$.

Proof of Theorem 2. By Lemma 6 we have

$$W(\hat{p}) - W(p) = \int_{T(p,\hat{p})} \sum_j \frac{\partial W}{\partial p_j} \, dp_j \,, \tag{4.3}$$

where $T(p, \hat{p})$ is any piecewise smooth path connecting p and $\hat{p}$. In particular we can take $p = 0$ and $\hat{p} = p^0$, and let $T(0, p^0)$ be the line connecting 0 and p^0, i.e. $p(t) = tp^0$, $0 \le t \le 1$, then

$$W(p^0) - W(0) = \int_0^1 \sum_j \frac{\partial W}{\partial p_j} (p^0 t) p_j^0 \, dt \,. \tag{4.4}$$

Consider now the derivatives of W with respect to the prices

$$\frac{\partial W}{\partial p_j} = \frac{\partial U}{\partial p_j} - \sum_i \frac{\partial x_i}{\partial p_j} p_i - x_j \,.$$

Furthermore,

$$\frac{\partial U}{\partial p_j} = \sum_i \frac{\partial U}{\partial x_i} \frac{\partial x_i}{\partial p_j} \,,$$

and so we obtain:

$$\frac{\partial W}{\partial p_j} = \sum_i \left(\frac{\partial U}{\partial x_i} - p_i \right) \frac{\partial x_i}{\partial p_j} - x_j \,,$$

which equals $-x_j$ by the first-order conditions of the optimization problem ($\partial U / \partial x_i = p_i$).

By (4.4) we therefore have

$$W(0) - W(p^0) = \int_0^1 \sum_j x_j(p^0 t) p_j^0 \, dt \,. \tag{4.5}$$

Obtaining an equation similar to (4.5) for the prices 0 and p^1 yields the result.

References

T.M. Apostol, Mathematical Analysis (Addison-Wesley, Reading, MA, 1957).
R.J. Aumann and L.S. Shapley, Values of Non Atomic Games (Princeton University Press, Princeton, NJ, 1974).
J. Dupuit, De la Mesure de l'utilite des travaux publics, Annales des Ponts et Chaussees, Memoires et documents relatifs a' l'art des constructions et au service de l'ingenieur (1884) 332–375. English translation: On the measurement of the utility of public works, in: K.J. Arrow and T. Scitovsky, eds., Readings and Welfare Economics (Irwin, Homewood, IL, 1969) 255–283.
J.R. Hicks, The rehabilitation of consumers' surplus, Rev. Econom. Studies 9 (1941) 108–116.
E. Silberberg, Duality and the many consumer's surpluses, Amer. Econom. Rev. 62 (1972) 942–952.

An Axiomatic Theory of a Risk Dominance Measure for Bipolar Games with Linear Incentives

REINHARD SELTEN

Rheinische Friedrich-Wilhelms-Universität Bonn, Wirtschaftstheoretische Abteilung I, Adenauerallee 24-42, D-53113 Bonn, Germany

Received August 27, 1993

Bipolar games are normal form games with two pure strategies for each player and with two strict equilibrium points without common equilibrium strategies. A normal form game has linear incentives, if for each player the difference between the payoffs for any two pure strategies depends linearly on the probabilities in the mixed strategies used by the other players. A measure of risk dominance between two strict equilibrium points of a bipolar game with linear incentives is characterized by 11 axioms. *Journal of Economic Literature* Classification Number: C72. © 1995 Academic Press, Inc.

1. INTRODUCTION AND FRAMEWORK

This paper has its roots in common work with John C. Harsanyi on equilibrium selection, which, however, never reached the stage of being written down. During the many years of our cooperation we developed three approaches to equilibrium selection. Only the third approach is described in our book (Harsanyi and Selten, 1988). The work presented here takes its point of departure from our first approach. An important part of this approach was a measure of risk dominance which expresses the strength of a risk dominance relationship by a real number. This measure, or more precisely a monotone transformation of it, restricted to a special class of games, will be axiomatized in this paper.

When we worked on our first approach to equilibrium selection Harsanyi and I made no attempt to develop an axiomatic theory of our risk dominance measure, not even for a limited class of games. One reason why we discarded our first approach was that the interpretation of the measure did not seem to be completely satisfactory. At least for the restricted class of games considered here the axiomatization leads to a deeper understanding and to a more convincing interpretation.

In our book (Harsanyi and Selten, 1988) we defined risk dominance as a binary relationship between equilibrium points. This binary relationship differs from the one induced by our first approach measure by the answer to the question in which direction the risk dominance between two equilibrium points is stronger.

The use of a risk dominance measure opens additional possibilities for selection criteria beyond the use of a binary relationship. Thus, in a case of circular risk dominance among three solution candidates, the one that is least dominated can be selected. Of course, this is just one example of the way in which a risk dominance measure may enter an equilibrium selection theory.

A *binary* game is a normal form game with two pure strategies for each player. A *bipolar* game is a binary game with two strict equilibrium points $\varphi = (\varphi_1, \ldots, \varphi_n)$ and $\psi = (\psi_1, \ldots, \psi_n)$ with $\varphi_i \neq \psi_i$ for $i = 1, \ldots, n$. For a selection of φ or ψ in such games a risk dominance measure does not really go beyond a risk dominance relationship. It is the purpose of a risk dominance measure to serve as a structural element of an equilibrium selection theory for a larger class of games. In the framework of such a theory risk dominance comparisons between strict equilibrium points must be reduced to comparisons in bipolar games. A very simple reduction method makes the comparison in a restricted bipolar game obtained by the elimination of all pure strategies other than those belonging to the two equilibrium points. However, this may not be the best method. Harsanyi and I proceeded differently in our first approach to equilibrium selection. The reduction method we used will not be explained here.

In the following the risk dominance measure of the first approach by Harsanyi and myself will be described, but only informally and not in full detail. I shall refer to this measure as the "first approach measure". Only later the connections to other measures proposed in the literature will be discussed.

1.1. *The First Approach Measure*

If one wants to examine the stability of one equilibrium point $\varphi = (\varphi_1, \ldots, \varphi_n)$ against deviations to another equilibrium point $\psi = (\psi_1, \ldots, \psi_n)$, it is natural to look at what happens, if every player i deviates to ψ_i with the same probability π and sticks to φ_i with the complementary probability $1 - \pi$. Such situations correspond to mixed strategy combinations $x^\pi = (x_1^\pi, \ldots, x_n^\pi)$. For each player i we may ask the question: What is the highest probability π_i such that φ_i is player i's best reply to x^π for all π with $0 \leq \pi \leq \pi_i$? This number $\pi_i(\varphi, \psi)$ is player i's *diagonal probability*.

(Harsanyi and I used this term since it was suggested by a stability diagram, in which the x^π form the diagonal.) The diagonal probability $\pi_i(\varphi, \psi)$ is a natural index of player i's individual stability at φ against deviations of other players to ψ.

The idea suggests itself that the $\pi_i(\varphi, \psi)$ should be combined to an overall stability index $s(\varphi, \psi)$ of φ against ψ. For this purpose we made use of *influence weights* $w_i(\varphi, \psi)$. The weight of a player reflects his importance for the stability of the other players. The computation of the weights is based on the influences that the players exert on each other by small deviations from strategy combinations x^π with π equal to one of the diagonal probabilities. The stronger the influences of a player and the higher the weights of the influenced players, the higher is his weight. This intuitive idea leads to a circular definition by a linear equation system. In Section 1.5 this definition will be explained in detail for the restricted class of games considered for the axiomatization presented here. The overall stability index $s(\varphi, \psi)$ is the product

$$s(\varphi, \psi) = \prod_{i=1}^{n} [\pi_i(\varphi, \psi)]^{w_i(\varphi, \psi)}. \tag{1}$$

The *first approach measure* of risk dominance is the quotient

$$R(\varphi, \psi) = \frac{s(\varphi, \psi)}{s(\psi, \varphi)}. \tag{2}$$

If player i has no other best replies to the strategy combinations x^π with $0 \le \pi \le 1$ than φ_i and ψ_i then we have $\pi_i(\varphi, \psi) + \pi_i(\psi, \varphi) = 1$. However, it may happen that on subintervals of positive length of $0 \le \pi \le 1$ neither φ_i nor ψ_i but other pure strategies are best replies to x^π. In the presence of such *foreign* best replies we have $\pi_i(\varphi, \psi) + \pi_i(\psi, \varphi) < 1$. In such cases also the weights $w_i(\varphi, \psi)$ are usually different from the weights $w_i(\psi, \varphi)$. However, in the restricted class of games considered for the axiomatization foreign best replies cannot occur, and we always have $\pi_i(\varphi, \psi) + \pi_i(\psi, \varphi) = 1$ and $w_i(\varphi, \psi) = w_i(\psi, \varphi)$ for $i = 1, \ldots, n$.

Actually the measure to be axiomatized will be $\ln R(\varphi, \psi)$, the natural logarithm of $R(\varphi, \psi)$, and not $R(\varphi, \psi)$ itself. The difference is not important, but also not completely insignificant, since the axioms will cardinally characterize the measure up to the unit of measurement.

1.2. *Measures Proposed in the Literature*

Güth and Kalkofen (1989) propose several measures which also take their point of departure from the mixed strategy combinations x^π explained at the beginning of 1.1. They use the term "resistance dominance" but in the following I shall also speak of risk dominance in connection with their measures. Their theory is based on *resistances* which are similar to diagonal probabilities. Player i's resistance $r_i(\varphi, \psi)$ is the highest probability π such that φ_i is player i's best reply to x^π. In the case that the set of all probabilities, such that φ_i is the best reply to x^π, is the union of several non-intersecting intervals, player i's resistance is different from his diagonal probability. However, in bipolar games we always have $r_i(\varphi, \psi) = 1 - \pi_i(\psi, \varphi)$.

The main difference between the first approach measure and the measures of Güth and Kalkofen lies in the way in which the individual stability indices are combined. Güth and Kalkofen, too, make use of player weights entering their formulas as exponents, but these weights are very different from ours.

In our first approach to equilibrium selection Harsanyi and I aimed at definitions which depend only on the best reply structure of the agent normal form. Güth and Kalkofen base their theory on the "complete agent normal form" which complements the usual agent normal form by a function which, depending on the strategy combination, specifies the probabilities with which the information sets of the agents are reached. These probabilities enter the construction of the weights. This results in a dependence on features of the extensive game not expressed by the agent normal form.

In a recent paper Güth (1992) considered three criteria of equilibrium selection by "unilateral deviation stability" which may be looked upon as risk dominance measures. These measures are based on stability indices for pairs of players and differ by the way in which the stability indices are aggregated. The measure favored by Güth in view of its properties is quite simple. It will be convenient to explain it in a way which is different from the exposition in Güth (1992). As before, the discussion is restricted to the comparison of two strict equilibrium points $\varphi = (\varphi_1, \ldots, \varphi_n)$ and $\psi = (\psi_1, \ldots, \psi_n)$ with $\varphi_i \neq \psi_i$ for $i = 1, \ldots, n$. For every player i let $L_i(\varphi)$ be player i's loss, if he deviates from φ_i to ψ_i whereas all other players stick to their strategies in φ. The *deviation losses* $L_i(\psi)$ are defined analogously. Let $\lambda_i(\varphi, \psi)$ be the quotient $L_i(\varphi)/L_i(\psi)$ and let $\lambda(\varphi, \psi)$ be the product of the $\lambda_i(\varphi, \psi)$ of all players. The measure is the $(n-1)$th power of $\lambda(\varphi, \psi)$. In this form the measure proposed by Güth (1992) has some relationship to the work presented here. In the restricted class of games considered in the axiomatization to be presented in this paper we always have $\lambda_i(\varphi, \psi) = \pi_i(\varphi, \psi)/\pi_i(\psi, \varphi)$ and $w_i(\varphi, \psi) = w_i(\psi, \varphi)$ for $i = 1, \ldots, n$. Therefore for these games the logarithmic version of the first approach measure can be written as follows:

$$R(\varphi, \psi) = \sum_{i=1}^{n} w_i(\varphi, \psi) \ln \lambda_i(\varphi, \psi). \tag{3}$$

The similarities between the measures proposed by Güth and Kalkofen (1989), Güth (1992) and the first approach measure are no coincidence. In all cases it was one of the aims of theory construction to obtain the "Nash property" which means that in "regular" unanimity games the equilibrium point with the highest "Nash product" is selected. The Nash product is the product $L(\varphi)$ of all deviation losses $L_i(\varphi)$ incurred by a single deviator i from φ, who deviates to a different pure strategy (in unanimity games it does not matter to which one). A unanimity game is regular, if the maximal Nash product is achieved by only one strict equilibrium. The Nash property puts some limits to the construction of risk dominance measures.

When Harsanyi and I worked on our first approach we also wanted to solve two-person unanimity games with incomplete information in a way which maximizes the generalized Nash product introduced in our paper on two-person bargaining under incomplete information (Harsanyi and Selten, 1972). Our first approach measure achieved this "generalized Nash property." (This will be shown in 3.2.) Güth and Kalkofen (1989) also aimed at the generalized Nash property and constructed measures which have it. Narrower limits than by the simple Nash property are imposed on measure construction by the generalized Nash property.

Later Harsanyi and I felt, that the generalized Nash property is not absolutely compelling. The theory of our book (Harsanyi and Selten, 1988) fails to have this property. A counterexample is given in 3.3. The measure proposed by Güth (1992) also does not have the generalized Nash property. If in an ordinary unanimity game one player is split into two identical types, both present with probability 1/2, this player's influence on the measure proposed by Güth (1992) is increased. It seems to be difficult to obtain the generalized Nash property without player weights. Therefore such weights enter the measures by Güth and Kalkofen (1989) as well as our first approach measure.

1.3. *Linear Incentives*

We say that a normal form game has *linear incentives*, if the difference between the payoffs for any two pure strategies of a player always depends linearly on the probabilities in the mixed strategies of the other players. The class of all games with linear incentives contains all two-person normal forms as a proper subset. It also contains all n-person normal forms arising from two-person normal form games with incomplete information in the sense of Harsanyi (1967/68) if the types are modelled as separate players.

The axiomatization of the first approach measure will be restricted to the class of bipolar games with linear incentives. Let $\varphi = (\varphi_1, \ldots, \varphi_n)$ and $\psi = (\psi_1, \ldots, \psi_n)$ be the two strict equilibrium points with $\varphi_i \neq \psi_i$ for

$i = 1, \ldots, n$ of a game of this kind. The measure will express the strength of risk dominance of φ over ψ. A generalization to more general games will not be discussed. It is the purpose of the measure to serve as a structural element in an equilibrium selection theory for games with linear incentives. Within a relatively short paper it is not possible to develop a full-fledged equilibrium selection theory of this kind. Therefore not much will be said about how the measure is meant to be used as a tool for equilibrium selection.

1.4. *Biforms*

The measure to be axiomatized depends only on the best reply structure of the game. However, this invariance with respect to transformations preserving the best reply structure will not be explicitly stated as an axiom. In order to simplify the formalism, classes of games with the same best reply structure will be represented by mathematical objects called "biforms" and the measure will be defined as a function defined on the set of all biforms.

The biform abstracts from the full information on payoffs and focuses on payoff differences. As before let $\varphi = (\varphi_1, \ldots, \varphi_n)$ and $\psi = (\psi_1, \ldots, \psi_n)$ be two strict equilibrium points with $\varphi_i \neq \psi_i$ for $i = 1, \ldots, n$ for a bipolar game with linear incentives. The difference of player i's payoffs for φ_i and ψ_i is a linear function of the probabilities p_j with which ψ_j is chosen by the other players. This function is called player i's *incentive function*. The best reply structure of the game is fully described by the n incentive functions. Moreover, the best reply structure is not changed if the incentive function is rescaled in such a way that for every player i the deviation losses $L_i(\varphi)$ and $L_i(\psi)$ add up to 1. The biform represents the n incentive functions normalized in this way. Formally a *biform* is a pair

$$B = (u, A) \qquad (4)$$

which consists of an n-dimensional column vector u and an $n \times n$-matrix $A = (a_{ij})$. For $i = 1, \ldots, n$ the components u_i of u satisfy $0 < u_i < 1$ and we have $a_{ii} = 0$. Moreover, the row sums of A are 1. These properties will become clear in the light of the interpretation of u and A.

The biform (u, A) represents the incentive function (after rescaling) as follows:

$$D_i = u_i - \sum_{j=1}^{n} a_{ij} p_j \qquad \text{for } i = 1, \ldots, n. \qquad (5)$$

Here, p_j is the probability with which ψ_j is used by player j. In matrix notation this can be expressed as follows:

$$D = u - Ap. \qquad (6)$$

Here D, u and p are the column vectors of the D_i, u_i and p_i, respectively.

The element a_{ij} of A is called player j's *influence* on player i and A is called the *influence matrix* of the biform. This terminology reflects the fact that the change of the payoff difference D_i caused by a change of p_j is proportional to a_{ij}. The diagonal elements a_{ii} vanish since the payoff difference D_i does not depend on p_i.

Obviously, u_i is nothing else than the deviation loss $L_i(\varphi)$ after the rescaling which leads to $L_i(\varphi) + L_i(\psi) = 1$. Therefore $1 - u_i$ is the deviation loss $L_i(\psi)$ after this rescaling. Since φ and ψ are strict equilibrium points we must have $0 < u_i < 1$ for $i = 1, \ldots, n$. It is clear that the *deviation loss ratio* $\lambda_i(\varphi, \psi) = L_i(\varphi)/L_i(\psi)$ before and after the rescaling has the value

$$\lambda_i(\varphi, \psi) = \frac{u_i}{1 - u_i} \qquad \text{for } i = 1, \ldots, n. \qquad (7)$$

For $p_j = u_i$ for $j = 1, \ldots, n$ the payoff difference D_i assumes the value zero. This shows that u_i coincides with the diagonal probability $\pi_i(\varphi, \psi)$ introduced in 1.1.

It will now be explained why the row sums of A must be 1. At ψ we have $p_i = 1$ for $i = 1, \ldots, n$. Therefore at ψ the payoff difference D_i is u_i diminished by the i-th row sum of A. The value of D_i at ψ is nothing else than $-L_i(\psi)$ after the rescaling. This shows that the i-th row sum of A must be 1.

It can be seen without difficulty that two games in the class considered here have the same best reply structure, if and only if they have the same biform. No formal proof of this fact will be given here.

The measure to be axiomatized will be defined as a function $R(u, A)$ which assigns real numbers to biforms. The determination of the measure of risk dominance of φ over ψ requires first a transition to the biform (u, A) and then a computation of $R(u, A)$. In this way invariance with respect to mappings which preserve the best reply structure is automatically guaranteed. The measure has the property that $R(u, A)$ is positive if φ risk dominates ψ and negative, if ψ risk dominates φ.

It is possible that a bipolar game with linear incentives has more than one pair of strict equilibrium points without common strategies. Therefore a biform is not associated to a bipolar game as such, but rather to a comparison between two strict equilibrium points $\varphi = (\varphi_1, \ldots, \varphi_n)$ and $\psi = (\psi_1, \ldots, \psi_n)$ with $\varphi_i \neq \psi_i$ for $i = 1, \ldots, n$ of a bipolar game.

The vectors $p = (p_1, \ldots, p_n)^T$ represent mixed strategy combinations in which player i uses ψ_i with probability p_i. (The symbol T indicates transposition; p is a column vector.) Obviously, φ and ψ are represented by $(0, \ldots, 0)^T$ and $(1, \ldots, 1)^T$, respectively. In view of this interpretation the symbols φ and ψ will be used for these vectors in the formalism connected to biforms.

1.5. Weights

It will now be explained how the weights $w_i(\varphi, \psi)$ introduced in 1.1 are defined for the restricted class of games to which the axiomatization of the measure is addressed. Since the measure $R(u, A)$ is defined for biforms and not directly for risk dominance comparisons in bipolar games and since the weights depend only on the influence matrix A, the notation $w_i(A)$ will mostly be used instead of $w_i(\varphi, \psi)$ for player i's influence weight.

The weight $w_i(A)$ expresses the extent to which player i influences other players j with their weights taken into account. Accordingly, $w_i(A)$ is the sum of all $a_{ij} w_j(A)$ with $i \neq j$. In view of $a_{ii} = 0$ this means that we have

$$w_i(A) = \sum_{j=1}^{n} a_{ji} w_j(A) \qquad \text{for } i = 1, \ldots, n \tag{8}$$

or in matrix notation

$$w(A) = A^T w(A), \tag{9}$$

where $w(A)$ is the column vector of the $w_i(A)$ and A^T is the transposed matrix of A. The sum of the $w_i(A)$ is normed to 1:

$$\sum_{i=1}^{n} w_i(A) = 1. \tag{10}$$

The problem arises, that the system formed by (9) and (10) does not always have a unique solution. In exceptional cases this system has either no solution or infinitely many solutions. In such cases we shall leave the weights undefined. The axiomatization will be restricted to biforms (u, A) with influence matrices for which the system formed by (9) and (10) yields a unique solution. For other biforms $R(u, A)$ the measure remains undefined.

Influence matrices A for which (9) and (10) together do not yield a unique solution for $w(A)$ are called *exceptional*. Biforms (u, A) with exceptional influence matrices are also called *exceptional*. Influence matrices and biforms which are not exceptional are called *regular*.

As we shall argue later it does not make sense to speak of risk dominance of one of the strict equilibrium points to be compared over the other if the associated biform is exceptional. In an equilibrium selection theory for games with linear incentives one may or may not wish to make a distinction between a lack of risk dominance for this reason or a lack of risk dominance in view of $R(u, A) = 0$ in the case of a regular biform (u, A). If one does not want to make the distinction (Harsanyi and I took this point of view in our first approach) one can define the weights for exceptional matrices A by $w_i(A) = 0$ for $i = 1, \ldots, n$. This yields $R(u, A) = 0$ for exceptional biforms.

In this paper the question will be left open whether one should distinguish between the two kinds of a lack of risk dominance. Accordingly, the measure $R(u, A)$ will not be extended to exceptional biforms.

The weight $w_i(A)$ of a player is not necessarily non-negative. This is due to the fact that some of the influences a_{ij} may be negative. If a_{ij} is negative, then player i's incentive to stick to φ is the higher the more player j deviates to ψ. Consider a player whose influences on other players are all negative. If the weights of all other players are positive, then this player's weight is negative.

1.7. The Measure

For every regular biform (u, A) the measure $R(u, A)$ is given by the following formula:

$$R(u, A) = \sum_{i=1}^{n} w_i(A) \ln \frac{u_i}{1 - u_i}, \tag{11}$$

where ln stands for the natural logarithm.

As Eq. (7) shows the quotient $u_i/(1 - u_i)$ is nothing else than the deviation loss ratio $\lambda_i(\varphi, \psi) = L_i(\varphi)/L_i(\psi)$. Moreover this ratio coincides with the diagonal probability ratio $\pi_i(\varphi, \psi)/\pi_i(\psi, \varphi)$.

In order to distinguish the measure (11) from other measures defined for regular biforms, the name *weighted average log measure* will be attached to it. The weighted average lot measure is defined for all regular n-player biforms with $n = 2, 3, \ldots$. As has been explained in 1.6 a biform (u, A) is regular, if (9) and (10) together uniquely determine the weights $w_i(A)$.

2. THE AXIOMS

The axioms characterize the weighted average log measure up to a constant positive factor. They refer to a risk dominance measure $R(u, A)$ which for every $n = 2, 3, \ldots$, is defined on a subset S_n of the set $\overline{S}_n$ of all n-player biforms (u, A). The regions S_n are not specified in advance, but are implicitly defined by the axioms together with R.

Some of the axioms do not refer to R itself but rather to the partial derivatives of $R(u, A)$ with respect to the components u_i of u. The symbol $R_i(u, A)$ will be used for $\partial R(u, A)/\partial u_i$. A special class of biforms has an important role in the axiomatization. A biform (u, A) is *equistable*, if all the components u_i of u have the same value. u_i is a natural index of player i's stability at φ. The idea suggests itself that therefore the common value of the u_i should be a natural collective stability index for φ in equistable games.

The weights $w_i(A)$ will also be implicitly defined by the axioms. They will be defined in terms of the partial derivatives $R_i(u, A)$ of the measure applied to equistable games.

Admittedly, the axioms differ with respect to their degree of plausibility. Some axioms seem to be unavoidable but others express little more than reasonable properties which hardly can be regarded as indispensable for a well-behaved measure of risk dominance.

In the following subsections the axioms will be introduced and discussed one after the other. As a first orientation a list of the names of the axioms follows:

1. Continuity, non-extendability, and the region where R is defined,
2. Differentiability with respect to u,
3. Symmetry,
4. Aggregability of identically motivated players,
5. Matrix independence for equistable biforms,
6. Incentive monotonicity for equistable biforms,
7. Vector independence of weights,
8. Weight equation for a single influence player,
9. Independence of R_i from u_j with $j \neq i$,
10. Inclusion of cyclical biforms in the region where R is defined,
11. Coordination game additivity.

The names of the axioms will also be used as titles of the subsections which deal with them. In most cases it is necessary to begin a subsection with some definitions and notations. Then the concerning axiom is stated and finally the motivation for this axiom is discussed under the heading "interpretation."

Each of the axioms is stated in a way which does not presuppose that other axioms are satisfied. This is necessary, if one does not want to lose the possibility to raise questions of independence. However, the interpretation of an axiom will sometimes assume the validity of other axioms discussed earlier.

2.1. Continuity, Non-extendability, and the Region Where R Is Defined

The set of all n-vectors u whose components satisfy $0 < u_i < 1$ for $i = 1, \ldots, n$ is denoted by U_n. The symbol $\overline{M}_n$ is used for the set of all $n \times n$-influence matrices or in other words, the set of all $n \times n$ matrices with vanishing diagonal elements and row sums 1. The set of all biforms (u, A) with $u \in U_n$ and $A \in \overline{M}_n$ is denoted by $\overline{S}_n$. The symbol $\overline{S}$ is used for the union of all $\overline{S}_n$ with $n = 2, 3, \ldots$.

In this paper the word "limit" will always be used in the sense of a finite limit, i.e., infinite limits are explicitly excluded from consideration. Consider a biform $(u', A') \in \overline{S}_n$. If we say, that the limit of $R(u, A)$ for $(u, A) \to (u', A')$ exists, we mean that for every sequence $(u^1, A^1), (u^2, A^2), \ldots$ of biforms in S_n with

$$\lim_{k \to \infty} (u^k, A^k) = (u', A') \tag{12}$$

the sequence $R(u^k, A^k)$ always converges to the same real number L. This number L is the limit of $R(u, A)$ for $(u, A) \to (u', A')$:

$$\lim_{(u, A) \to (u', A')} R(u, A) = L. \tag{13}$$

The values $+\infty$ and $-\infty$ are not considered to be permissible as limits.

AXIOM 1 (Continuity, Non-extendability, and the Region where R Is Defined). *For $n = 2, 3, \ldots$ let S_n be the subset of $\overline{S}_n$ for which R is defined. The following statements hold:*

(i) S_n *is open relative to* $\overline{S}_n$.

(ii) $\overline{S}_n$ *is the closure of* S_n.

(iii) *A subset M_n of $\overline{M}_n$ exists such that $(u, A) \in \overline{S}_n$ belongs to S_n, if and only if A belongs to M_n.*

(iv) R *is continuous on* S_n.

(v) If $(u', A') \in \bar{S}_n$ does not belong to S_n, then the limit of $R(u, A)$ for $(u, A) \to (u', A')$ does not exist.

Interpretation. (i) and (ii) make sure that almost all biforms in $\bar{S}_n$ belong to S_n. Condition (iii) requires that it depends only on A whether $R(u, A)$ is defined or not. This is a technical property without deeper significance. Continuity is a natural requirement. (iv) and (v) make sure that the measure is extended as far as this can be done continuously, but not to biforms at which continuity breaks down.

The lack of continuity at the excluded biforms is not a serious deficiency of the measure. Universal continuity cannot be expected in non-cooperative game theory. As payoffs vary continuously in normal form games, equilibrium points may suddenly appear or vanish. This does not diminish the central importance of the equilibrium point concept.

2.2. Differentiability with Respect to u

AXIOM 2 (Differentiability with Respect to u). *The measure $R(u, A)$ is differentiable with respect to u, wherever R is defined.*

Interpretation. This purely technical requirement hardly needs any comment.

2.3. Symmetry

An *n-player biform* is an element of $\bar{S}_n$. Let (u, A) with $u = (u_1, \ldots, u_n)^T$ and $A = (a_{ij})$ be an n-player biform and let τ be a permutation of $\{1, \ldots, n\}$. Let (u', A') with $u' = (u'_1, \ldots, u'_n)^T$ and $A' = (a'_{ij})$ be the biform which is related to (u, A) as follows:

$$u'_{\tau(i)} = u_i \quad \text{for } i = 1, \ldots, n \tag{14}$$

$$a'_{\tau(i)\tau(j)} = a_{ij} \quad \text{for } i, j = 1, \ldots, n. \tag{15}$$

We call A' the *τ-permuted* matrix of A and (u', A') the *τ-permuted* biform of (u, A). A *permuted* biform of (u, A) is a biform (u', A') such that for some permutation τ of $\{1, \ldots, n\}$ the biform (u', A') is the τ-permuted biform of (u, A).

AXIOM 3 (Symmetry). *If R is defined for (u, A) and (u', A') is a permuted biform of (u, A) then R is defined for (u', A') and*

RISK DOMINANCE MEASURE

$$R(u', A') = R(u, A) \tag{16}$$

holds.

Intepretation. The axiom postulates that the measure is invariant with respect to renamings of the players. This is an obvious and unavoidable requirement for any reasonable measure of risk dominance.

2.4. Aggregability of Identically Motivated Players

Let (u, A) be an n-player biform. The players k and m are *identically motivated* in (u, A) if the components $u_1, \ldots, u_n$ of u and the elements a_{ij} of A satisfy the following conditions:

$$u_k = u_m \tag{17}$$

$$a_{kj} = a_{mj} \quad \text{for } j = 1, \ldots, n. \tag{18}$$

Note that (18) implies $a_{km} = a_{mm}$ and $a_{kk} = a_{mk}$ and therefore

$$a_{mk} = a_{km} = 0. \tag{19}$$

In this sense two identically motivated players do not influence each other. Equations (17) and (18) also have the consequence that the incentive functions defined by (5) for players k and m coincide. For every strategy combination p the payoff differences D_k and D_m are equal.

Assume that in (u, A) the players $n - 1$ and n are identically motivated. An $(n - 1)$ player biform (u', A') will now be constructed in which the players $n - 1$ and n are "merged" to a new player $n - 1$. The components $u'_1, \ldots, u'_{n-1}$ of u' and the elements a'_{ij} of A' are as follows:

$$u'_i = u_i \quad \text{for } i = 1, \ldots, n - 1 \tag{20}$$

$$a'_{ij} = \begin{cases} a_{i,n-1} + a_{in} & \text{for } j = n - 1 \\ a_{ij} & \text{else.} \end{cases} \tag{21}$$

This means that u is shortened by the deletion of u_n, that the $(n - 1)$th column of A is replaced by the sum of the two last columns and that finally the last column and the last row is deleted. The biform (u', A') defined by (20) and (21) is called the biform, which *results by aggregation of players n and $n - 1$ from (u, A)*. In (u', A') player $n - 1$ has the same incentive function as players $n - 1$ and n in (u, A) and the combined influences of both players on other players.

AXIOM 4 (Aggregability of Identically Motivated Players). *Let (u, A) be an n-player biform with $n = 3, 4 \ldots$ in which players $n - 1$ and n are identically motivated and let (u', A') be the $(n - 1)$- player biform which results from (u, A) by the aggregation of players $n - 1$ and n. Then R is defined for (u', A') if and only if R is defined for (u, A) and*

$$R(u', A') = R(u, A) \tag{22}$$

holds, if R is defined for (u, A).

Interpretation. The axiom focuses on the special case that players $n - 1$ and n are identically motivated in (u, A) and not two arbitrary players k and m. This is only a technical convenience adjusted to the fact that according to our definitions players must be numbered from $1, \ldots, n - 1$ in an $(n - 1)$-player biform. One can aggregate two arbitrary identically motivated players in two steps. First a τ-permuted biform of (u, A) is formed with the help of a permutation τ which maps k and m to $n - 1$ and n. Then both players are aggregated. It follows by Axiom 3 (player symmetry) together with Axiom 4 that in this way the measure remains unchanged. Therefore the more general case need not be addressed directly.

In view of the common incentive function two identically motivated players $n - 1$ and n in the biform (u, A) described in the axiom always have the same pure best replies in the sense that φ_n is player n's best reply if and only if φ_{n-1} is the best reply of player $n - 1$. (The same holds for ψ_n and ψ_{n-1}). Therefore it is reasonable to suppose that the situation remains essentially the same, if player n delegates his strategic choice to player $n - 1$ in a way which restricts player $n - 1$ to the pure strategies $(\varphi_{n-1}, \varphi_n)$ and (ψ_{n-1}, ψ_n).

The requirement of aggregability of identically motivated players is a very reasonable one. It seems to be an almost unavoidable property of a well behaved measure of risk dominance defined for biforms.

2.5. *Matrix Independence for Equitable Biforms*

As has been explained before, a biform (u, A) is called *equitable*, if all components of u have the same value u_0.

AXIOM 5 (Matrix Independence for Equitable Biforms). *Let (u, A) and (u, A') be two equitable biforms, for which R is defined. Then*

$$R(u, A) = R(u, A') \tag{23}$$

holds.

Interpretation. In 1.4 it has been pointed out, that u_i coincides with the diagonal probability $\pi_i(\varphi, \psi)$ and $1 - u_i$ coincides with the diagonal probability $\pi_i(\psi, \varphi)$. This suggests that u_i (or a monotone transformation of u_i) is a natural index for player i's stability at φ. Since in an equistable biform all u_i are equal, their common value u_0 should be a natural collective index for the players' stability at φ. Therefore it is reasonable to require that for equistable biforms with a given number of players, the measure depends only on the common value u_0 of the u_i.

2.6. Incentive Monotonicity for Equistable Biforms

AXIOM 6 (Incentive Monotonicity for Equistable Biforms). *Let (u, A) and (u', A) be two equistable biforms for which R is defined and let u_0 and u_0' be the common values of the components of u and u', respectively. Then we have*

$$R(u', A) > R(u, A) \qquad \text{for } u_0' > u_0. \tag{24}$$

Interpretation. Assume that Axiom 1 holds. Then, according to (iii), it depends only on A whether $R(u, A)$ is defined or not. If A belongs to one of the sets M_n with $n = 2, 3, \ldots$ then $R(u, A)$ is defined for all $u \in U_n$. Axiom 6 requires that in this case the measure $R(u, A)$ of an equistable biform (u, A) is a monotonically increasing function of the common value u_0 of the u_i. In view of the interpretation of u_0 as a natural collective index of the players' stability at φ, this is a reasonable requirement.

It must be pointed out, however, that the argument in favor of Axiom 6 cannot be based on the idea, that generally $R(u, A)$ should be increased, if ceteris paribus one of the u_i is increased. This stronger form of incentive monotonicity does not hold for the weighted average log measure. The reason for this is, that a player may have a negative weight. If $w_i(A)$ is negative, then an increase of u_i results in a decrease of the measure. Consider the case that player i has negative influences on all other players and assume that their weights are positive. Then $w_i(A)$ is negative. An increase of u_i makes φ more attractive for player i, but the more attractive it is for player i, the less attractive is φ for the other players. Therefore it is not unreasonable that the measure is decreased by an increase of u_i.

The requirement of incentive monotonicity for equistable biforms means that the positive effects of an increase of the common value u_0 of the u_i on the overall attractivity of φ are stronger than the negative ones.

2.7. Vector Independence of the Weights

Assume that Axioms 1 and 2 hold and let (u, A) be an equistable biform with A in one of the sets M_n with $n = 2, 3, \ldots$. Moreover let u_0 be the common value of the u_i. The notation $R_0(u, A)$ will be used for the derivative

$$R_0(u, A) = \frac{\partial R(u, A)}{\partial u_0}. \tag{25}$$

Since $R(u, A)$ is differentiable with respect to u (this is more than differentiability with respect to each of the u_i separately), we have

$$R_0(u, A) = \sum_{i=1}^{n} R_i(u, A). \tag{26}$$

If Axiom 6 holds, then $R_0(u, A)$ is positive and we can form the quotient $R_i(u, A)/R_0(u, A)$. For a given player i, this quotient could depend on u_0 and A. The axiom will require that it depends only on A.

AXIOM 7 (Vector Independence of Weights). *Let $\hat{M}_n$ be the set of all $n \times n$-influence matrices A such that an equistable biform (u, A) exists, at which R is defined and differentiable with respect to u. For every given n with $n = 2, 3, \ldots$ and each $i = 1, \ldots, n$ a function w_i exists, which assigns a real number $w_i(A)$ to every $A \in \hat{M}_n$ such that*

$$R_i(u, A) = w_i(A) \sum_{j=1}^{n} R_j(u, A) \tag{27}$$

holds for every equistable (u, A) for which R is defined and differentiable with respect to u.

Interpretation. Assume that Axioms 1, 2, and 6 hold. Then $w_i(A)$ is the quotient $R_i(u, A)/R_0(u, A)$. Moreover $R_0(u, A)$ is positive and equal to the sum of the $R_i(u, A)$ as shown in (26). This means that $R_i(u, A)/R_0(u, A)$ can be looked upon as the "relative derivative" with respect to u_i. If u_i is interpreted as a stability index, then the relative importance of a change of player i's stability at φ for the overall stability of φ is expressed by $R_i(u, A)/R_0(u, A)$. The axiom requires that this relative importance does not depend on the vector u but only on A. This is the reason for the name "vector independence of weights". Of course, since (u, A) is equistable, vector independence really means independence of the common value u_0 of the components of u.

It is not unreasonable to assume that the relative importance expressed by $R_i(u, A)/R_0(u, A)$ depends only on A. Since (u, A) is equistable, differences among the players with respect to the relative importance of a change of their stability at φ cannot be based on differences among the u_i but only on differences with respect to influences received or exerted in A. There seems to be no reason why the way in which the relative importance $R_i(u, A)/R_0(u, A)$ are determined by A should depend on how high the common value u_0 of the u_i is.

Admittedly, the argument in favor of Axiom 7 is not very strong since it is based on the absence of a sufficient reason for a depencence on u_0, rather than a reason for independence.

2.8. Weight Equation for Single Influence Players

Let $A = (a_{ij})$ be an influence matrix. Player m is called a *single influence player* in A if player m has vanishing influences a_{ij} on all other players i with the possible exception of one specific player k; in other words $a_{im} = 0$ holds for all $i \neq k$ and a_{km} may or may not be different from zero. Assume that this is the case and that A belongs to one of the sets $\hat{M}_n$ defined in Axiom 7. Axioms 7 and 8 will have the consequence

$$w_m(A) = \rho a_{km} w_k(A), \tag{28}$$

where ρ is a positive constant, the same one for all cases which fit the description. Axiom 8 will postulate an analogous relationship between the derivatives $R_m(u, A)$ and $R_k(u, A)$ of the measure applied to equistable biforms (u, A). This analogous relationship together with Axiom 7 yields the "weight equation" (28).

AXIOM 8 (Weight Equation for Single Influence Players). *Let (u, A) be an equistable biform for which R is defined and differentiable with respect to u and assume that in $A = (a_{ij})$ player m is a single influence player with $a_{im} = 0$ for $i \neq k$, where k is a specific player with $m \neq k$. Then the following equation*

$$R_m(u, A) = \rho a_{km} R_k(A) \tag{29}$$

holds with the same positive constant ρ for all cases which fit the description.

Interpretation. Assume that Axioms 1, 2, 6, and 7 hold. In view of Axiom 6 the sum of all $R_j(u, A)$ is positive. Therefore (27) together with (29) yields (28). The interpretation will focus on this "weight equation" rather than the analogous relationship (29).

The weight $w_i(A)$ is the relative importance $R_i(u, A)/R_0(u, A)$ of player i's stability at φ for the overall stability in an equistable biform (u, A). These relative importances are thought of as forming an interdependent system in which $w_i(A)$ reflects the extent to which player i influences other players with their weights taken into account. In the special case considered in Axiom 9 player m influences nobody else than player k. It is therefore reasonable to assume that $w_m(A)$ should be increasing in a_{km} and in $w_k(A)$; moreover if $a_{km} = 0$ or $w_k(A) = 0$ holds, $w_m(A)$ should vanish. Equation (28) is the simplest possible specification of a relationship with these properties.

The general idea that $w_i(A)$ reflects the extent to which player i influences other players with their weights taken into account was also the basis of the intuition underlying the direct definition of $w(A)$ by (9) and (10) in 1.5. However, the direct definition makes much stronger assumptions on the form of interdependence among the weights than the weight Eq. (28). No more than proportionality of $w_m(A)$ with $a_{km}w_k(A)$ is assumed. The proportionality constant ρ is left unspecified.

If one accepts the general idea about the interdependence of the weights, it is natural to make the proportionality assumption (28). However, the argument in favor of axiom 8 rests on the plausibility of the general idea. Therefore it must be admitted that the requirement imposed by Axiom 8 is not absolutely compelling.

2.9. *Independence of R_i from u_j with $j \neq i$*

The requirements imposed by Axioms 5–8 concern the measure R and its derivatives R_i with respect to the u_i evaluated at equistable biforms. Axiom 9 imposes an independence requirement on the R_i evaluated at arbitrary biforms, at which R is defined and differentiable with respect to u.

AXIOM 9. *Let (u, A) and (u', A) be two biforms with the property, that one of the players, player i, has equal components $u_i = u_i'$ in u and u'. If R is defined and differentiable with respect to u, both at (u, A) and (u', A), then*

$$R_i(u, A) = R_i(u', A) \tag{30}$$

holds.

Interpretation. Axiom 9 requires that $R_i(u, A)$ does not depend on the components u_j of u with $j \neq i$. This is a convenient independence property. Independence assumptions tend to produce simpler structures and therefore recommend themselves wherever they are not in conflict with other considerations. Admittedly, this is a rather weak defense of Axiom 9. It cannot be argued that the required property is indispensable.

2.10. Inclusion of Cyclical Biforms in the Region Where R Is Defined

For every $n = 2, 3, \ldots$, a *cyclical* influence matrix $A = (a_{ij})$ is defined as follows:

$$a_{ij} = \begin{cases} 1 & \text{for } i = j + 1 \text{ and for } i = 1 \text{ and } j = n \\ 0 & \text{else.} \end{cases} \qquad (31)$$

One may think of the players $1, \ldots, n$ arranged on a circle like the numbers on the face of a clock. Influences are exerted only on the neighbor to the right.

AXIOM 10. *R is defined for every equistable biform (u, A) with a cyclical influence matrix A.*

Interpretation. The axiom imposes a technical requirement which hardly needs any comment.

2.11. Coordination Game Additivity

Axiom 12 will be concerned with equistable 2-player biforms. Influence matrices are defined by the properties that row sums are 1 and diagonal elements are zero. Therefore all 2-player biforms have the same influence matrix

$$A = \begin{pmatrix} 0 & 1 \\ 1 & 0 \end{pmatrix} \qquad (32)$$

Consequently two equistable 2-player biforms (u, A) and (u', A) differ only by the common values u_0 and u_0' of the components of u and u', respectively. The measure $R(u, A)$ of an equistable 2-player biform is a function of the common value u_0 of the components of u. Axiom 11 imposes a strong restriction on the structure of this function.

AXIOM 11 (Coordination Game Additivity). *Let (u, A), (u', A) and (u'', A) be three equistable 2-player biforms, for which R is defined. Let u_0, u_0' and u_0'' be the common values of the components of u, u', and u'', respectively, and assume that the condition*

$$\frac{u_0''}{1 - u_0''} = \frac{u_0}{1 - u_0} \cdot \frac{u_0'}{1 - u_0'} \qquad (33)$$

is satisfied. Then the equation

$$R(u'', A) = R(u, A) + R(u', A) \tag{34}$$

holds.

Interpretation. The explanation of the intuitive background of this axiom makes it necessary to go beyond the context of biforms and to look at *3 × 3-coordination games*. In such games both players have 3 pure strategies 1, 2, and 3. The payoff functions H_1 and H_2 of players 1 and 2 are as follows:

$$H_1(k, m) = H_2(k, m) = \begin{cases} x_k & \text{for } k = m \\ 0 & \text{for } k \neq m \end{cases} \tag{35}$$

with $x_k > 0$ for $k = 1, 2, 3$.

Consider the 2 × 2 games which result from a 3 × 3-coordination game G by taking away one of the three pure strategies, say strategy h, from the pure strategy sets of both players. G_{-h} denotes the 2 × 2 game which results from G in this way.

If in G one of the players does not use his pure strategy h with positive probability, then the other player's pure strategy h cannot be a best reply to what he does. In this sense the 2 × 2 games G_{-h} are substructures of G which are closed with respect to best replies. Such substructures are called *formations* in the book by Harsanyi and myself (Harsanyi and Selten, 1988, p. 198).

In this paper no detailed proposal is made with respect to the method by which risk dominance comparisons in more general games with linear incentives are reduced to risk dominance comparisons in bipolar games with linear incentives. However, the fact that the G_{-h} are closed with respect to best replies makes it unproblematic to perform the risk dominance comparison between two strict equilibrium points (k, k) and (m, m) of G in the 2 × 2 game G_{-h}, in which the two players are restricted to the pure strategies k and m. There seems to be no other reasonable possibility. In the following the biforms for the risk dominance comparisons between pairs of strict equilibrium points of G will be determined in this way.

For $k = 1, 2, 3$ the equilibrium point (k, k) of G will be denoted by φ^k. Let (u, A), (u', A), and (u'', A) be the 2-player biforms associated to risk dominance comparisons among the φ^k as follows:

(u, A) is associated to the comparison of φ^1 to φ^2

(u', A) is associated to the comparison of φ^2 to φ^3

(u'', A) is associated to the comparison of φ^1 to φ^3.

It can be seen immediately that the common values u_0, u_0' and u_0'' of u, u', and u'' are as follows:

$$u_0 = \frac{x_1}{x_1 + x_2} \tag{36}$$

$$u_0' = \frac{x_2}{x_2 + x_3} \tag{37}$$

$$u_0'' = \frac{x_1}{x_1 + x_3}. \tag{38}$$

This yields:

$$\frac{u_0}{1 - u_0} = \frac{x_1}{x_2} \tag{39}$$

$$\frac{u_0'}{1 - u_0'} = \frac{x_2}{x_3} \tag{40}$$

$$\frac{u_0''}{1 - u_0''} = \frac{x_1}{x_3}. \tag{41}$$

Condition (33) is satisfied for these values of u_0, u_0' and u_0''. Axiom 12 requires that (34) holds. In order to clarify the meaning of (34) we shall write $R(\varphi^1, \varphi^2)$, $R(\varphi^2, \varphi^3)$, and $R(\varphi^1, \varphi^3)$ instead of $R(u, A)$, $R(u', A)$, and $R(u'', A)$, respectively. With this notation (34) takes the following form:

$$R(\varphi^1, \varphi^3) = R(\varphi^1, \varphi^2) + R(\varphi^2, \varphi^3). \tag{42}$$

The name "coordination game additivity" attached to Axiom 12 now becomes clear. It is required that in 3×3-coordination games the risk dominance measure has the additivity property (42).

The interpretation of Axiom 11 in terms of risk dominance comparisons in 3×3-coordination games provides a basis for the discussion of the plausibility of the axiom. It is now necessary to ask the question, why the additivity property expressed by (42) should hold for 3×3-coordination games.

In view of the extreme simplicity of 3×3-coordination games one should expect that within such games risk dominance comparisons are especially easy. One condition for this being the case is the existence of a potential function $P(x_i)$ which permits us to write $R(\varphi^k, \varphi^m)$ as the difference between the potentials of x_k and x_m.

$$R(\varphi^k, \varphi^m) = P(x_k) - P(x_m). \tag{43}$$

The existence of a potential P greatly simplifies the task of risk dominance comparisons. The measure, which is a function of two arguments can be evaluated with the help of a function of only one argument. Obviously the existence of a potential P with (43) implies the additivity property (42). If (42) holds, then a potential can be constructed as follows:

$$P(x_3) = 0 \tag{44}$$

$$P(x_2) = R(\varphi^2, \varphi^3) \tag{45}$$

$$P(x_1) = R(\varphi^1, \varphi^3). \tag{46}$$

The additivity property (42) is equivalent to the existence of a potential P with (43).

It is a reasonable requirement that at least for very simple games with more than two strict equilibrium points, the risk dominance measure should be representable by some kind of potential difference. 3×3-coordination games are the simplest games of this type and the structure of these games suggests that the potential is a function of the common equilibrium payoff.

The interpretation in terms of the existence of a potential for the strict equilibrium points of a 3×3-coordination game shows that Axiom 11 imposes a reasonable requirement on a measure of risk dominance defined for biforms. However, it must be admitted, that the coordination game additivity property expressed by Axiom 11 is not compelling, even if it is intuitively appealing.

3. STATEMENT OF THE MAIN RESULT AND APPLICATION OF THE MEASURE TO TWO-PERSON UNANIMITY GAMES WITH INCOMPLETE INFORMATION

In this section the main result will be stated without proof. After the discussion of the main result an application of the weighted average measure to a special class of bipolar games will establish a connection to the theory of bargaining under incomplete information proposed by Harsanyi and myself (1972).

3.1. Statement of the Main Result

The weighted average log measure of risk dominance is defined by Eq. (11) for every regular biform, or in other words for every biform (u, A) with an influence matrix A for which the weight vector $w(A)$ is uniquely determined by (9) and (10). The axioms characterize the measure up to a positive factor and simultaneously determine the set of biforms for which it is defined.

THEOREM 1. *The weighted average log measure of risk dominance together with the set of biforms for which it is defined satisfies the Axioms 1–11. If a risk dominance measure R and the set S where it is defined satisfy Axioms 1–11, then R is the weighted average log measure multiplied by a positive constant and S is the set of all regular biforms.*

Discussion. The axioms determine the measure up to a positive factor. However, this result should not be overemphasized. It is probably possible to characterize some monotonous transformations of the weighted average log measure by very similar sets of axioms, e.g., the exponential function of this measure. This would lead to the first approach measure described by formula 1. In order to obtain the first approach measure one would have to change Eq. (11) in Axiom 3 as follows:

$$R(u, A)R(u', A) = 1. \tag{47}$$

Analogously Eq. (34) in Axiom 11 would have to be replaced by

$$R(u'', A) = R(u, A)R(u', A). \tag{48}$$

Of course, some additional adaptations at other places would also be necessary. No attempt will be made here to explore this possibility in more detail.

As has been said before, the axioms differ with respect to their plausibility. Maybe it would be worthwhile to try to find an intuitively more convincing axiom system which characterizes the weighted average log measure only up to monotonically increasing transformations.

The weighted average log measure is defined almost everywhere in a sense made precise by (i) and (ii) in Axiom 1. Moreover, in view of (v) in Axiom 1 this measure cannot be continuously extended to exceptional biforms. This is the reason why we take the point of view, that meaningful risk dominance comparisons are not possible in such cases. As has been said already in 1.5, in the framework of an equilibrium selection theory one may wish to treat this type of lack of risk dominance in the same way as the one connected to a risk dominance measure zero of a regular biform.

One of the properties of the measure, invariance with respect to mappings which preserve the best reply structure, has not been achieved by the axioms, but by the framework of the axiomatization which represents risk dominance comparisons by biforms. Probably it would not be difficult, but cumbersome, to axiomatize the measure directly for risk dominance comparisons in bipolar games with linear incentives by a system of axioms which explicitly requires invariance with respect to mappings which preserve the best reply structure.

This paper cannot claim to give a final answer to the problem of axiomatizing a measure of risk dominance. Theorem 1 goes further in this direction, than anything else which has been attempted before, but there is still ample room for improvement. It is necessary to generalize the theory beyond the context of linear incentives. One may want to change the axiom system in order to obtain a higher degree of plausibility. It may turn out, that another measure of risk dominance is intuitively more appealing than the weighted average log measure. Even if this should be the case, Theorem 1 may be useful as a result which helps to clarify the issue.

3.2. *Application of the Weighted Average Log Measure to Two-Person Unanimity Games with Incomplete Information*

In the following the weighted average log measure will be applied to risk dominance comparisons in games within a special class of games with linear incentives. A game in this class has n player $1, \ldots, n$ each of whom has K pure strategies $1, \ldots, K$, interpreted as proposed agreements. The extensive form is as follows:

1. An initial random move selects a pair $\{i, j\}$ of different players to be matched against each other. The probabilities of different pairs may be different.

2. The players i and j chosen by the initial random move simultaneously and independently both select one of the numbers $1, \ldots, K$. They do this without information about the identity of the opponent.

3. If i and j have selected the same number k then i receives x_i^k and j receives x_j^k. Otherwise both of them obtain zero payoffs. Players not chosen by the initial random move always receive zero payoffs. The numbers x_i^k are positive for all $i = 1, \ldots, n$ and $k = 1, \ldots, K$.

Such games are called *two-person unanimity games with incomplete information*. It can be seen without difficulty that these games have linear incentives.

The pair $\{i, j\}$ selected by the initial random move is unordered. This is important since otherwise player i could make his choice dependent on his position in the pair. The probability of $\{i, j\}$ is denoted by α_{ij} and, since the pair is unordered, also by α_{ji}. The probabilities α_{ij} form a symmetric probability matrix $\alpha = (\alpha_{ij})$ with $\alpha_{ii} = 0$ for $i = 1, \ldots, n$. The diagonal elements are zero, since only pairs of different players are chosen. For $i = 1, \ldots, n$ the i-th row sum is denoted by α_i. It will be assumed that we have

$$\alpha_i > 0 \quad \text{for } i = 1, \ldots, n. \tag{49}$$

Since each probability of a pair appears twice in α and since one pair is always chosen, the elements of α sum up to 2. The matrix α is called the *probability matrix* of the game.

The payoffs x_i^k, called *agreement payoffs* form an $n \times K$ matrix $X = (x_i^k)$ called the *agreement matrix*. The pair (α, X) is the *parameter constellation* of the game.

Consider the normal form G of a two-person unanimity game with incomplete information with the parameter constellation (α, X). It can be seen immediately that the pure strategy combinations $\varphi^k = (k, \ldots, k)$, in which all players choose the same number k, are strict equilibrium points of G.

For every pair of different numbers k and m among the numbers $1, \ldots, K$ let G_{km} be the game which results from G, if all pure strategies other than k and m are eliminated from the pure strategy sets of all players. The games G_{km} are formations of G (see the interpretation of axiom 12 in 2.12). In order to see this, consider a mixed strategy combination for G in which only k and m are used with positive probability by all players. It is clear that only k and m can be best replies to such combinations. As has been argued in 2.12, this makes it unproblematic to reduce the risk dominance comparison of φ^k with φ^m in G to the comparison of φ^k with φ^m in the bipolar game G_{km}. This method of reduction will be used in the following.

For fixed k and m we shall now determine the biform (u, A) which corresponds to the comparison of φ^k with φ^m in G_{km}. In this biform the strategy combination $\varphi = (0, \ldots, 0)$ corresponds to φ^k and $\psi = (1, \ldots, 1)$ corresponds to φ^m. The deviation losses of player i at φ^k and φ^m resp. are $\alpha_i x_i^k$ and $\alpha_i x_i^m$. A deviation of player j from φ^k to φ^m decreases player i's incentive to stick to φ^k by $\alpha_{ij}(x_i^k + x_i^m)$, since player i receives $\alpha_{ij} x_i^k$ less for choosing k and $\alpha_{ij} x_i^m$ more for choosing m. With the help of these simple facts it can be seen easily that the components u_i of u and the elements a_{ij} of A are as follows:

$$u_i = \frac{x_i^k}{x_i^k + x_i^m} \quad \text{for } i = 1, \ldots, n \tag{50}$$

$$a_{ij} = \frac{\alpha_{ij}}{\alpha_i} \qquad \text{for } i,j = 1,\ldots,n. \tag{51}$$

We now turn our attention to the determination of the weight vector $w(A)$. It is necessary to examine the question under which conditions A is regular. Since all elements of A are nonnegative, the system formed by (9) and (10) can be interpreted as connected to a stochastic chain, whose transition matrix for the states $1,\ldots,n$ is A^T. Looked upon in this way $w(A)$ has the character of a stationary distribution. It follows by the well known theorem by Perron–Frobenius (see, e.g. Gantmacher, 1986, p. 398) that $w(A)$ is uniquely determined by (9) and (10) if and only if A is indecomposable. In the following it will be assumed that this is the case. A necessary, but by no means sufficient condition for indecomposibility is satisfied if all probabilities α_{ij} are positive.

In view of (51) the Eq. (9) assume the following form

$$w_i(A) = \sum_{j=1}^{n} \frac{\alpha_{ji}}{\alpha_j} w_j(A) \qquad \text{for } i = 1,\ldots,n. \tag{52}$$

This system is solved by weights proportional to the row sums α_i, which in view of the symmetry of α are also the column sums of α:

$$w_i(A) = \mu \alpha_i \qquad \text{for } i = 1,\ldots,n. \tag{53}$$

Since the sum of the α_i is 2 it follows by (10) that μ is equal to 1/2:

$$w_i(A) = \frac{\alpha_i}{2} \qquad \text{for } i = 1,\ldots,n. \tag{54}$$

In view of (50) we have:

$$\frac{u_i}{1 - u_i} = \frac{x_i^k}{x_i^m} \qquad \text{for } i = 1,\ldots,n. \tag{55}$$

Equations (54) and (55) yield the following value of the weighted average log measure:

$$R(\varphi^k, \varphi^m) = R(u, A) = \frac{1}{2} \sum_{i=1}^{n} \alpha_i \ln \frac{x_i^k}{x_i^m}. \tag{56}$$

In the same way as in the case of the 3 × 3-coordination game (see 2.12) a potential $P(\varphi^k)$ can be associated to every strict equilibrium point of G with the property that

$$R(\varphi^k, \varphi^m) = P(\varphi^m) \tag{57}$$

always holds. This potential is as follows:

$$P(\varphi^k) = \frac{1}{2} \sum_{i=1}^{n} \alpha_i \ln x_i^k. \tag{58}$$

The exponential function of $2P(\varphi^k)$

$$e^{2P(\varphi^k)} = \prod_{i=1}^{n} [x_i^k]^{\alpha_i} \tag{59}$$

has the form of a *generalized Nash product* as introduced in the paper by Harsanyi and myself (1972) on bargaining under incomplete information. In the special case considered there the players $1, \ldots, n$ were partitioned into two subsets $1, \ldots, n_1$, and $n_1 + 1, \ldots, n$, interpreted as types of two players I and II. In this case the right hand side of (59) is the generalized Nash product associated to the equilibrium payoff vector $x^k = (x_1^k, \ldots, x_n^k)$ in the sense of the paper of 1972, with $(0, \ldots, 0)$ as the conflict point.

In the paper of 1972 the generalized Nash product was maximized over the convex hull of the x^k and $(0, \ldots, 0)$. However, if the φ^k are considered to be the only candidates for equilibrium selection, and one of the strict equilibrium points has a higher potential $P(\varphi^k)$ than all the others, then the maximization of the generalized Nash product (59) is equivalent to the selection of that strict equilibrium point which risk dominates all others in the sense of the weighted average log measure.

A final remark in this section concerns the case of a decomposable matrix A. In this case the set of players can be partitioned into several subsets with the property that no player in one of the subsets influences any player in another subset. This means that the game can be decomposed into separate games for the subsets. In such cases a reasonable equilibrium selection theory should reduce the task of equilibrium selection in the whole game to equilibrium selection in the smaller games for the subsets. If the game is decomposed as much as possible, these smaller games will give rise to indecomposable influence matrices.

3.3. *A Numerical Counterexample*

In the following a numerical example of a two-person unanimity game with incomplete information will show, that the weighted average log measure can yield a risk dominance relationship which is opposite to the one obtained on the basis of the definition in the book by Harsanyi and myself (1988). For $n = 2$ something similar cannot happen, since in bipolar 2×2 games one strict equilibrium point risk dominates the other, if it has the higher deviation loss product. The same is true for the definition based on the weighted average log measure. Therefore one needs an example with three players. We shall look at the two-person unanimity game with incomplete information described by the following parameters:

$$n = 3, \quad K = 2 \tag{60}$$

$$\alpha_{12} = \alpha_{13} = \frac{1}{2} \tag{61}$$

$$\alpha_{23} = 0 \tag{62}$$

$$x_1^1 = 12, \ x_2^1 = 15, \quad x_3^1 = 2 \tag{63}$$

$$x_1^2 = 8, \ x_2^2 = 5, \quad x_3^2 = 18. \tag{64}$$

It can be seen immediately that the biform associated to the comparison of φ^1 with φ^2 is as follows:

$$u_1 = .6 \tag{65}$$

$$u_2 = .75 \tag{66}$$

$$u_3 = .1 \tag{67}$$

$$A = \begin{bmatrix} 0 & .5 & .5 \\ .5 & 0 & 0 \\ .5 & 0 & 0 \end{bmatrix}. \tag{68}$$

The generalized Nash products Π_1 and Π_2 of φ^1 and φ^2 are as follows:

$$\Pi_1 = 12 \cdot \sqrt{15} \cdot \sqrt{2} = 65.7 \tag{69}$$

$$\Pi_2 = 8 \cdot \sqrt{5} \cdot \sqrt{18} = 75.9. \tag{70}$$

This shows, that according to the definition based on the weighted average log measure φ^2 risk dominates φ^1. The measure $R(\varphi^1, \varphi^2)$ has the value $-.098$.

Without going into detail, it will now be indicated why, according to the definition in the book by Harsanyi and myself (1988), φ^1 risk dominates φ^2. This definition will not be repeated here. It requires, that the tracing procedure is applied to the bicentric prior. It will be assumed that the reader is familiar with these concepts.

In our case the bicentric prior specifies probability u_1 for player i's strategy φ_i^1 in φ^1 and $1 - u_i$ for his strategy φ_i^2 in φ^2. It can be seen easily that the combination of best replies to the bicentric prior is $(\varphi_1^1, \varphi_2^1, \varphi_3^2)$. The best reply to this combination is φ^1. This has the consequence that in the trace a constant segment at the best reply to the bicentric prior is followed by a jump segment at which player 3 shifts from φ_3^2 to φ_3^1. Finally a constant segment at φ^1 leads to φ^1 as the result of the tracing procedure. Therefore φ^1 risk dominates φ^2.

As we have seen in 3.2 the weighted average log measure has the property that in bipolar two-person unanimity games risk dominance (where it is defined) is in the direction of the higher generalized Nash product. The counterexample shows that the risk dominance concept of the book by Harsanyi and myself (1988) does not have this "generalized Nash property."

4. Properties of the Weights

The notion of an influence matrix A has been introduced in 1.5. An influence matrix can be described as an $n \times n$ matrix whose diagonal elements are zero and whose row sums are 1. As has been said in 1.7 an influence matrix A is regular if the system formed by (9) and (10) permits a unique solution for $w(A)$. Otherwise A is exceptional. The *weight function* w assigns the weight vector $w(A)$ determined by (9) and (10) to every regular $n \times n$ matrix with $n = 2, 3, \ldots$. As before M_n stands for the set of all regular $n \times n$-influence matrices. Later it will be shown that this is the set M_n characterized by the axioms (see (iii) in Axiom 1).

4.1. A Regularity Criterion

In the following it will be shown that an influence matrix A is regular if and only if a certain determinant connected to it is different from zero.

LEMMA 1. *Let A be an $n \times n$-influence matrix with $n = 2, 3, \ldots$, let I be the $n \times n$-unity matrix whose elements are 1 on the diagonal and 0 everywhere else, and let $\psi = (1, \ldots, 1)^T$ be the n-dimensional column vector, whose elements are all equal to 1. Then A is regular if and only if we have:*

$$|A^T - I + \psi\psi^T| \neq 0. \tag{71}$$

Proof. We first show that $w(A)$ satisfies (9) and (10) if and only if it is the solution of the following system:

$$(A^T - I + \psi\psi^T)w(A) = \psi. \tag{72}$$

This system is obtained from (9) by adding (10) to each of the n equations of (9). Therefore $w(A)$ satisfies (72) if it satisfies (9) and (10). We now show that (72) implies (9) and (10). One obtains (10) from (72) by adding all equations and dividing by $n - 1$. This can be seen without difficulty if one remembers that the column sums of A^T are all equal to 1 and that $\psi\psi^T$ is the $n \times n$ matrix whose elements are all equal to 1. Since (9) can be obtained by subtracting (10) from each of the equations of (72), it follows that (9) is implied by (72), too.

We can conclude that A is regular, if and only if (72) has a uniquely determined solution $w(A)$. Obviously this is the case if and only if (71) holds.

Remark. The proof of Lemma 1 has shown that $w(A)$ is the solution of the system formed by (9) and (10) if and only if it is the solution of (72) regardless whether A is regular or not. If A is exceptional, then (72) may have no solution or infinitely many solutions.

4.2. The Cases of Two and Three Players

As has been shown in 2.11 there is only one 2×2-influence matrix namely the matrix shown by (32). It is an immediate consequence of Lemma 1, that this matrix is regular. (72) takes the form

$$\begin{bmatrix} 0 & 2 \\ 2 & 0 \end{bmatrix} \begin{bmatrix} w_1(A) \\ w_2(A) \end{bmatrix} = \begin{bmatrix} 1 \\ 1 \end{bmatrix}. \tag{73}$$

This yields

$$w_1(A) = w_2(A) = \tfrac{1}{2}. \tag{74}$$

In the following we shall look at the case $n = 3$. Since the row sums are 1, a 3×3-influence matrix is fully determined by the three parameter a_{13}, a_{21}, and a_{32}. If $A = (a_{ij})$ is a 3×3-influence matrix we have:

$$A = \begin{bmatrix} 0 & 1 - a_{13} & a_{13} \\ a_{21} & 0 & 1 - a_{21} \\ 1 - a_{32} & a_{32} & 0 \end{bmatrix}. \tag{75}$$

This yields:

$$A^T - I + \psi\psi^T = \begin{bmatrix} 0 & 1 + a_{21} & 2 - a_{32} \\ 2 - a_{13} & 0 & 1 + a_{32} \\ 1 + a_{13} & 2 - a_{21} & 0 \end{bmatrix}. \tag{76}$$

By Lemma 1 the influence matrix A is regular if and only if the determinant of the matrix (76) does not vanish. An easy computation yields:

$$|A^T - I + \psi\psi^T| = 3(3 - a_{13} - a_{21} - a_{32} + a_{13}a_{21} + a_{13}a_{32} + a_{21}a_{32}). \tag{77}$$

In view of

$$a_{13}a_{31} = a_{13}(1 - a_{32}) = a_{13} - a_{13}a_{32} \tag{78}$$
$$a_{21}a_{12} = a_{21}(1 - a_{13}) = a_{21} - a_{13}a_{21} \tag{79}$$
$$a_{32}a_{23} = a_{32}(1 - a_{21}) = a_{32} - a_{21}a_{32} \tag{80}$$

we have:

$$|A^T - I + \psi\psi^T| = 3(3 - a_{13}a_{31} - a_{21}a_{12} - a_{32}a_{23}). \tag{81}$$

If this determinant does not vanish we can solve the system (73) by Cramer's rule. After some computations omitted here one obtains the following result:

LEMMA 2. *Let A be a regular 3×3-influence matrix. Then we have*

$$w_i(A) = \frac{1 - a_{jk}a_{kj}}{3 - a_{13}a_{31} - a_{21}a_{12} - a_{23}a_{32}} \tag{82}$$

for every permutation (i, j, k) of $(1, 2, 3)$.

EXAMPLE OF AN EXCEPTIONAL INFLUENCE MATRIX. Consider the influence matrix

$$A = \begin{bmatrix} 0 & 2 & -1 \\ 2 & 0 & -1 \\ -1 & 2 & 0 \end{bmatrix}. \tag{83}$$

This influence matrix is exceptional, since the right hand side of (81) takes the value zero. Moreover it can also be seen without difficulty that for this matrix A the system formed by (9) and (10) admits no solution.

4.3. Aggregation of Identically Influenced Players

Let $A = (a_{ij})$ be an $n \times n$-influence matrix with the property that for some integer m with $1 < m < n$ we have:

$$a_{ij} = a_{mj} \quad \text{for } i = m+1, \ldots, n \text{ and } j = 1, \ldots, n. \tag{84}$$

If this is the case, we say that the players $m, \ldots, n$ are *identically influenced* in A. Note that (84) implies

$$a_{ij} = 0 \quad \text{for } i, j \in \{m, \ldots, n\}. \tag{85}$$

In this sense two identically influenced players do not influence each other. We now construct the matrix $\overline{A}$ which results from A by the *aggregation* of the identically influenced players $m, \ldots, n$. This matrix $\overline{A}$ is an $m \times m$-influence matrix whose elements $\overline{a}_{ij}$ are as follows:

$$\overline{a}_{ij} = \begin{cases} \sum_{k=m}^{n} = a_{ik} & \text{for } j = m \\ a_{ij} & \text{else.} \end{cases} \tag{86}$$

The definition is essentially the same as in the case of identically motivated players (see (21) in 2.4) with the only difference that now more than two players may be aggregated. The distinction between identically motivated and identically influenced players is made, since in the latter case we deal with influence matrices only and not with biforms.

LEMMA 3. *Let A be an $n \times n$ matrix in which the players $m, \ldots, n$ are identically influenced and let $\overline{A}$ be the matrix which results by the aggregation of players $m, \ldots, n$ from A. Then $\overline{A}$ is regular if and only if A is regular. In the case that A and $\overline{A}$ are regular we have*

$$w_i(\overline{A}) = w_i(A) \quad \text{for } i = 1, \ldots, m-1 \tag{87}$$

and

$$w_m(\overline{A}) = \sum_{i=m}^{n} w_i(A). \tag{88}$$

Proof. Consider the system formed by (9) and (10) and assume that $w(A)$ is a solution of this system. It can be seen without difficulty that the system formed by (9) and (10) applied to $\overline{A}$ instead of A and with n replaced by m is solved by $w(\overline{A})$ as defined by (87) and (88). In order to see this, it is sufficient to add the $n - m + 1$ last equations in the system (9) applied to A and to make the substitution indicated by (88). It follows that $w(\overline{A})$ is uniquely determined if A is regular. Consequently $\overline{A}$ is regular if A is regular.

Now assume that $\overline{A}$ is regular. Then $w(\overline{A})$ is uniquely determined. In view of (85) the $w_m(A), \ldots, w_n(A)$ are uniquely determined by the $w_1(A), \ldots, w_{n-1}(A)$ and the $w_1(A), \ldots, w_{n-1}(A)$ depend only on each other and the sum on the right side of (88). Therefore the system formed by (9) and (10) applied to A has one and only one solution $w(A)$ if $\overline{A}$ is regular. Consequently A is regular if $\overline{A}$ is regular.

4.4. *Feasible Weight Vectors*

In this section we shall investigate the question which vectors can be obtained as solutions of the system formed by (9) and (10). We first ask this question for arbitrary influence matrices, regardless of whether they are regular or exceptional.

LEMMA 4. *Let A be an $n \times n$-influence matrix (not necessarily a regular one) with $n = 2, 3, \ldots,$ and let $w = (w_1, \ldots, w_n)$ be a solution for $w(A)$ in the system formed by (9) and (10). Then w has at least two non-vanishing components.*

Proof. In view of (10) at least one component of w must be unequal to zero. The only vectors excluded by the assertion are those with one component equal to 1 and all others equal to zero. Let w be a vector of this kind and assume $w_k = 1$. If w is a solution of the system formed by (9) and (10) we must have

$$w_i = a_{ki} w_k = a_{ki} = 0 \qquad \text{for } i \neq k. \tag{89}$$

This follows by $w_j = 0$ for $j \neq k$. In view of $a_{kk} = 0$, Eq. (89) has the consequence that the k-th row sum A is zero, contrary to the definition of an influence matrix, which requires that all row sums are 1.

LEMMA 5. *Let $w = (w_1, w_2, w_3)$ be a vector with at least two non-vanishing components and with*

$$w_1 + w_2 + w_3 = 1. \tag{90}$$

Then a regular 3×3-influence matrix $A = (a_{ij})$ exists, such that $w = w(A)$ holds.

Proof. It will be assumed that w_2 and w_3 are unequal to zero. This entails no loss of generality, since the validity of the assumption can be achieved by a renumbering of the players. Obviously such a renumbering does not change the regularity status of an influence matrix. It will be shown that the following matrix A has the required properties:

$$A = \begin{bmatrix} 0 & 1 & 0 \\ 1 - \dfrac{w_3}{w_2} & 0 & \dfrac{w_3}{w_2} \\ 1 - \dfrac{w_2 - w_1}{w_3} & \dfrac{w_2 - w_1}{w_3} & 0 \end{bmatrix}. \tag{91}$$

Easy computations show that (9) is satisfied with w inserted for $w(A)$. It remains to show that A is a regular influence matrix. Obviously A is an influence matrix. In view of (81) the influence matrix A is regular if the denominator on the right hand side of (82) is unequal to zero. An easy computation yields

$$3 - a_{12}a_{21} - a_{13}a_{31} - a_{23}a_{32} = 1 + \frac{w_1 + w_3}{w_2}. \tag{92}$$

In view of (90) we can replace $w_1 + w_3$ by $1 - w_2$. We obtain

$$3 - a_{12}a_{21} - a_{13}a_{31} - a_{23}a_{32} = \frac{1}{w_2}. \tag{93}$$

This shows that A in (91) is a regular influence matrix.

LEMMA 6. *Let $w = (w_1, \ldots, w_n)$ with $n = 3, 4, \ldots$ be a vector with at least two non-vanishing components and with*

$$w_1 + \cdots + w_n = 1. \tag{94}$$

Then a regular $n \times n$ matrix exists, such that $w = w(A)$ holds.

Proof. Define $\bar{w} = (\bar{w}_1, \bar{w}_2, \bar{w}_3)$ by

$$\bar{w}_i = w_i \quad \text{for } i = 1, 2 \tag{95}$$

and

$$\overline{w}_3 = \sum_{k=3}^{n} w_k. \tag{96}$$

As we shall see later, it can be assumed without loss of generality that $\overline{w}_2$ and $\overline{w}_3$ are unequal to zero. In view of Lemma 5 a regular 3×3-influence matrix $\overline{A}$ with $\overline{w} = w(\overline{A})$ can be found. It is the plan of the proof to construct a regular $n \times n$-influence matrix A with $w = w(A)$ by a disaggregation of player 3 in $\overline{A}$ into identically influenced players $3, \ldots, n$ with the right weights. In this way Lemma 3 can be used to reduce the assertion to the case $n = 3$, already covered by Lemma 5.

We now show that by a suitable renumbering of the players we can always obtain a vector w with w_2 and $\overline{w}_3$ unequal to zero. As we shall see it is sufficient to renumber the players in such a way that w_2 becomes one of the smallest non-vanishing components and w_1 becomes one of the smallest among the other components. Obviously we have

$$w_2 < 1, \tag{97}$$

since at least two components are non-vanishing. This yields

$$1 - w_2 > 0 \tag{98}$$

$1 - w_2$ is the sum of all w_i with $i \neq 2$ and w_1 is one of the smallest among them. Hence

$$w_1 \leq \frac{1 - w_2}{n - 1}. \tag{99}$$

This permits the conclusion

$$\overline{w}_3 = 1 - w_2 - w_1 \geq \frac{n - 2}{n - 1}(1 - w_2) > 0. \tag{100}$$

In the following we shall assume that w_2 and $\overline{w}_3$ are unequal to zero. As we have seen this entails no loss of generality.

Let $\overline{A}$ be a regular 3×3-influence matrix with $\overline{w} = w(\overline{A})$. In view of Lemma 5 such a matrix $\overline{A}$ exists. Define

$$\alpha_i = \frac{w_i}{\overline{w}_3} \quad \text{for } i = 3, \ldots, n. \tag{101}$$

We now construct an $n \times n$-influence matrix $A = (a_{ij})$ whose elements are as follows:

$$a_{ij} = \begin{cases} \bar{a}_{ij} & \text{for } i = 1, 2 \text{ and } j = 1, 2 \\ \alpha_i \bar{a}_{i3} & \text{for } i = 1, 2 \text{ and } j = 3, \ldots, n \\ \bar{a}_{3j} & \text{for } i = 3, \ldots, n \text{ and } j = 1, 2 \\ 0 & \text{for } i = 3, \ldots, n \text{ and } j = 3, \ldots, n. \end{cases} \quad (102)$$

It can be seen without difficulty that A is an influence matrix and that players $3, \ldots, n$ are identically influenced in A. Moreover it is clear that $\bar{A}$ results by aggregation of players $3, \ldots, n$ from A. It follows by Lemma 3 that A is regular and that (87) and (88) hold. The second and fourth lines of (102) together with (101) permit the conclusion $w_i(A) = w_i$ for $i = 3, \ldots, n$. Consequently $w(A) = w$ holds.

Feasibility. For $n = 2, 3, \ldots$ a vector $w = (w_1, \ldots, w_n)$ is called a *feasible* weight vector, if a regular $n \times n$-influence matrix A with $w(A) = w$ exists.

LEMMA 7. *For $n = 2$ the vector $w = (.5, .5)$ is the only feasible weight vector. For $n = 3, 4 \ldots$ a weight vector is feasible if and only if it has at least two non-vanishing components and the sum of all its components is 1.*

Proof. For $n = 2$ only one influence matrix exists, namely the one shown by (32). The weight vector of this influence matrix is $(.5, .5)$. For $n = 3, 4 \ldots$ the assertion follows by Lemma 6.

Remark. Lemma 4 shows that only feasible weight vectors can be solutions of the system formed by (9) and (10) regardless of whether A is regular or exceptional.

4.5. The Weight Function in the Vicinity of an Exceptional Influence Matrix

In this section it will be shown that the weight function does not approach a limit, if the influence matrix comes nearer and nearer to an exceptional one. Therefore the weight function cannot be continuously extended to exceptional influence matrix. This does not yet imply that the weighted average log measure cannot be continuously extended to at least some exceptional biforms. Therefore some additional results about the behavior of the weight function near exceptional influence matrices need to be derived in this section.

For any two $n \times n$-influence matrices $A = (a_{ij})$ and $A' = (a'_{ij})$ we define a *distance* $|A - A'|$:

$$|A - A'| = \sum_{i=1}^{n} \sum_{j=1}^{n} |a_{ij} - a'_{ij}|. \tag{103}$$

The *norm* $|w|$ of a vector $w = (w_1, \ldots, w_n)$ is defined by

$$|w| = \sum_{i=1}^{n} |w_i|. \tag{104}$$

It will be shown that close to an exceptional influence matrix regular influence matrices can be found whose weight vectors have arbitrarily high norms. For this purpose, it will be necessary to make a distinction between two types of exceptional matrices. An influence matrix A is a *no-weight matrix*, if the system formed by (9) and (10) has no solutions for $w(A)$ and A is a *multiweight matrix*, if this system has more than one solution.

LEMMA 8. *Let A be a multiweight influence matrix and let w be a solution of the system formed by (9) and (10). Then for every $\varepsilon > 0$ a regular influence matrix A' with $w(A') = w$ and $|A' - A| < \varepsilon$ exists.*

Proof. It follows by the remark after Lemma 7 that w is a feasible weight vector, which means that a regular influence matrix $\overline{A}$ exists, such that $w = w(\overline{A})$ holds. Let $\overline{A}$ be a matrix of this kind. For every real number α let A_α be the matrix

$$A_\alpha = (1 - \alpha)\overline{A} + \alpha A. \tag{105}$$

It can be seen immediately that for every α the matrix A_α is an influence matrix with the property that w solves the system formed by (9) and (10) with A replaced by A_α. We have $w = w(A_\alpha)$ if A_α is regular. We shall now show that A_α can be exceptional at finitely many values of α only. In view of Lemma 1 the influence matrix A_α is exceptional if and only if we have:

$$|A_\alpha^T - I + \psi\psi^T| = 0. \tag{106}$$

This is an algebraic equation in α, which has only finitely many solutions unless it holds for every α. The latter case cannot arise, since $\overline{A}$ is regular and therefore (106) does not hold for $\alpha = 0$. Therefore A_α can be exceptional for finitely many values of α only. Obviously a regular matrix A_α arbitrarily near to A can be found.

LEMMA 9. *Let A be a multiweight influence matrix. Then for every $K > 0$ the system formed by (9) and (10) has a solution w with $|w| > K$.*

Proof. Let w^1 and w^2 be two different solutions of the system formed by (9) and (10). Let $v = (v_1, \ldots, v_n)$ be the vector $w^2 - w^1$. Obviously we have

$$Av = 0 \qquad (107)$$

and

$$\sum_{i=1}^{n} v_i = 0. \qquad (108)$$

Therefore every w of the form

$$w = w^1 + \lambda v \qquad (109)$$

is a solution of the system formed by (9) and (10). The norm $|w|$ of such a solution satisfies the following inequality:

$$|w| \geq \lambda |v| - |w^1|. \qquad (110)$$

Since w^2 and w^1 are different, the norm $|v|$ is positive. Obviously the right hand side of (110) can be made as great as desired by an appropriate choice of λ.

LEMMA 10. *Let A be an exceptional influence matrix. Then for every $\varepsilon > 0$ and every $K > 0$ a regular influence matrix A' with $|A' - A| < \varepsilon$ and $|w(A')| > K$ can be found.*

Proof. For the case of a multiweight matrix the assertion is an immediate consequence of the two preceding lemmata. In the following we shall assume that A is a no-weight matrix. Let $A^1, A^2, \ldots$ be a sequence of regular influence matrices converging to A, such that $|A^m - A| < \varepsilon$ holds for every $m = 1, 2, \ldots$. Suppose that for some $K > 0$ the sequence contains no member A^m with $|w(A^m)| > K$. Then the weight vectors $w(A^m)$ remain in a compact set bounded by $|w(A^m)| \leq K$ and it follows that the sequence $w(A^1), w(A^2), \ldots$ has a converging subsequence. Let w be the limit of such a subsequence. It is clear that w must be a solution of the system formed by (9) and (10), since the subsequence of the matrices involved converges to A. However, this is impossible, since A is a no-weight matrix. Therefore, for every K the sequence $A^1, A^2, \ldots$ contains a member with the required properties.

Remark. It is a consequence of Lemma 10 that the limit of $w(A')$ for $A' \to A$ does not exist, if A is exceptional.

5. Properties of the Weighted Average Log Measure

It is the goal of this section to prove that the weighted average log measure satisfies Axioms 1–11. As before M_n will denote the set of all regular $n \times n$-influence matrices and S_n stands for the set of all regular biforms. The symbol R will be used for the weighted average log measure.

5.1. Continuity Properties

LEMMA 11. *The weighted average log measure satisfies Axiom 1.*

Proof. In view of the remark after the proof of Lemma 1 the weight vector $w(A)$ of a regular influence matrix A is the unique solution of the system (72). It follows that for $n = 2, 3, \ldots$ the weight function $w(A)$ is continuous in M_n. This has the consequence that (iv) in Axiom 1 is satisfied for the weighted average log measure. It can also be seen without difficulty that (i), (ii), and (iii), in Axiom 1 holds for the weighted average log measure.

Let (u', A') be an exceptional n-player biform. In view of Lemma 10 the limit of $w(A)$ for $A \to A'$ does not exist. Nevertheless, the limit of $R(u', A)$ for $A \to A'$ may exist. It can be seen easily that this limit exists if (u', A') is equitable. Nothing remains to prove for the case that the limit of $R(u', A')$ for $A \to A'$ does not exist. Therefore it will be assumed in the following that this limit does exist. Define

$$\lim_{A \to A'} R(u', A) = L. \tag{111}$$

In view of Lemma 10 we can construct a sequence of regular $n \times n$-influence matrices $A^1, A^2, \ldots$ converging to A' such that $|w(A^k)| > k$ holds for $k = 1, 2, \ldots$. Let $A^1, A^2, \ldots$ be a sequence of this kind. In view of (111) we have

$$\lim_{k \to \infty} R(u', A^k) = L. \tag{112}$$

In order to show that the limit of (u, A) for $(u, A) \to (u', A')$ does not exist we shall construct a sequence $u^1, u^2, \ldots$ of vectors such that $(u^1, A^1), (u^2, A^2), \ldots$ forms a sequence of regular biforms converging to (u', A') with the property that the limit of $R(u^k, A^k)$ for $k \to \infty$ is different from L. Define

$$y_k = \frac{y}{|w(A^k)|} \quad \text{for } k = 1, 2, \ldots, \tag{113}$$

where y is an arbitrary fixed real number different from zero. In view of $|w(A^k)| > k$ the sequence $y_1, y_2, \ldots$ converges to zero. The components u_i^k of the members of the sequence $u^1, u^2, \ldots$ are defined as follows

$$u_i^k = \begin{cases} \dfrac{u_i' e^{y_n}}{1 - u_i' + u_i' e^{y_n}} & \text{for } w_i(A^k) \geq 0 \\[2mm] \dfrac{u_i' e^{-y_n}}{1 - u_i' + u_i' e^{-y_n}} & \text{for } w_i(A^k) < 0 \end{cases} \tag{114}$$

for $i = 1, \ldots, n$ and $k = 1, 2, \ldots$. An easy computation shows that this definition has the following consequence

$$\frac{u_i^k}{1 - u_i^k} = \begin{cases} \dfrac{u_i'}{1 - u_i'} e^{y_n} & \text{for } w_i(A^k) \geq 0 \\[2mm] \dfrac{u_i'}{1 - u_i'} e^{-y_n} & \text{for } w_i(A^k) < 0 \end{cases} \tag{115}$$

for $i = 1, \ldots, n$ and $k = 1, 2, \ldots$. Equation (115) yields a relationship between the weighted average log measures $R(u^k, A^k)$ and $R(u', A')$:

$$R(u^k, A^k) = R(u', A^k) + \sum_{i=1}^{n} |w_i(A)| y_k. \tag{116}$$

In view of (113) we obtain

$$R(u^k, A^k) = R(u', A^k) + y \tag{117}$$

for $k = 1, 2, \ldots$ and therefore

$$\lim_{k\to\infty} R(u^k, A^k) = L + y. \tag{118}$$

Consequently the limit of $R(u, A)$ for $(u, A) \to (u', A')$ does not exist.

Remark: The proof has shown that any real number can be reached as the limit of the measures of a sequence $(u^1, A^1), (u^2, A^2), \ldots$ converging to (u', A') if A' is exceptional and the limit of $R(u', A)$ for $A \to A'$ exists.

Comment. Consider an equistable biform (u', A') and let u'_0 be the common value of the components of u'. It may seem to be natural to define the weighted average log measure of (u', A') as the logarithm of $u'_0/(1 - u'_0)$ regardless of whether A' is exceptional or regular. The remark shows that this extension of the weighted average log measure to exceptional equistable biforms is not reasonable in view of the discontinuity with respect to u'. Since positive as well as negative numbers can be reached as the limit of the measures associated to the members of a sequence of regular biforms converging to (u', A'), it makes no sense to speak of risk dominance in either direction at (u', A').

5.2. The Validity of the Axioms

LEMMA 12. *The weighted average log measure together with the set of biforms for which it is defined satisfies Axioms 1–11.*

Proof. In view of Lemma 11 it remains to show that Axioms 2–11 are satisfied. It can be seen immediately that Axiom 2 is satisfied. The derivatives with respect to u_i are as follows:

$$R_i(u, A) = w_i(A) \frac{1}{u_i(1 - u_i)} \quad \text{for } i = 1, \ldots, n. \tag{119}$$

It is clear that the weighted average log measure is invariant with respect to a renumbering of the players, Axiom 3 is satisfied. The fact that the weighted average log measure satisfies Axiom 4 is an easy consequence of Lemma 3. Axiom 5 is satisfied, since the weighted average log measure of a regular equistable biform is the logarithm of $u_0/(1 - u_0)$, where u_0 is the common value of all components of u; the weights do not matter. Axiom 6 holds since this logarithm is monotonically increasing in u_0.

With the help of (119) it can be seen immediately that (27) holds for the weighted average log measure in the case of a regular equistable biform (u, A). Consequently the components w_i of the weight function have the properties required by Axiom 7. This axiom is satisfied. In the case of a single influence player m with a non-zero influence on a player k equation

(29) holds with $\rho = 1$. This is an immediate consequence of (9). Therefore Axiom 8 is satisfied. It follows by (119) that $R_i(u, A)$ depends only on u_i and A. Axiom 9 is satisfied.

For a cyclical influence matrix A, as defined in 2.11, the system (9) takes the following form:

$$w_k(A) = w_{k+1}(A) \quad \text{for } k = 1, \ldots, n - 1 \tag{120}$$

$$w_n(A) = w_1(A). \tag{121}$$

Together with (10) this has the consequence that we have

$$w_i = \frac{1}{n} \quad \text{for } i = 1, \ldots, n. \tag{122}$$

The weights are uniquely determined by (9) and (10). Therefore cyclical influence matrices are regular. Axiom 10 is satisfied.

Axiom 11 is an immediate consequence of the fact mentioned above, that the weighted average log measure of a regular equistable biform (u, A) is the logarithm of $u_0/(1 - u_0)$ where u_0 is the common value of all components of u.

6. Proof of the Main Result

Lemma 12 has established the first part of Theorem 1. In this section it will be shown that the axioms determine the measure up to a positive constant.

Proof of Theorem 1. Let R be a risk dominance measure which, together with the set S of biforms for which it is defined, satisfies Axioms 1–11. It remains to show that R must be the weighted average log measure multiplied by a positive constant and defined for all regular biforms. The proof will be subdivided into the proof of a sequence of nine assertions.

ASSERTION 1. *The measure R is defined for all two-player biforms. For equistable 2-player biforms (u, A) with $u = (u_0, u_0)$ we have*

$$R(u, A) = c \ln \frac{u_0}{1 - u_0}, \tag{123}$$

where c is a positive constant, the same one for every equistable 2-player biform.

RISK DOMINANCE MEASURE

Proof. Only one 2×2-influence matrix exists, namely the matrix A shown in (32). This matrix is cyclical. It follows by Axiom 10 that the measure is defined for all two player biforms. In view of Axiom 6 the measure $R(u, A)$ of an equistable 2-player biform (u, A) with $u = (u_0, u_0)$ is a monotonically increasing function $f(u_0)$ of u_0. Let g be the function defined by

$$g\left(\ln \frac{u_0}{1 - u_0}\right) = f(u_0) \tag{124}$$

Since f is defined for $0 < u_0 < 1$, the function g is defined on the set of all real numbers. In view of Axiom 2 the function f is differentiable. Therefore g is differentiable, too. Axiom 11 has the consequence that g satisfies the following functional equation

$$g(x + y) = g(x) + g(y) \tag{125}$$

for any two real members x and y. It can be seen immediately that (125) yields the conclusion that the derivative g' of g is constant. Therefore g must be of the form:

$$g(x) = b + cx. \tag{126}$$

Obviously (125) does not hold unless at least one of the constants b and c in (126) is equal to zero. It follows by Axiom 6 that g is monotonically increasing. Therefore we must have $b = 0$ and $c > 0$. This shows that the assertion of the lemma holds.

ASSERTION 2. *Let A be an $n \times n$-influence matrix with $n = 2, 3, \ldots$ which belongs to the set M_n (see Axiom 1). Then we have:*

$$\sum_{i=1}^{n} w_i(A) = 1. \tag{127}$$

Proof. Let (u, A) be an equistable biform. It follows by (iii) in Axiom 1 that $R(u, A)$ is defined. In view of Axiom 6 the derivative $R(u, A)$ with respect to the common value u_0 of the components of u is positive. This derivative is nothing else than the sum of all $R_i(u, A)$. Therefore we can derive (127) by first adding the n Eq. (27) from Axiom 7 and dividing by the sum of all $R_i(u, A)$.

ASSERTION 3. *Let (u, A) be an equistable biform, for which R is defined and let u_0 be the common value of the components of u. Then we have*

$$R(u, A) = c \ln \frac{u_0}{1 - u_0}, \tag{128}$$

where c is the same positive constant as in Assertion 1.

Proof. In view of Assertion 1 Eq. (128) holds for the case that (u, A) is an equistable 2-player biform. We shall show by induction on the number of players n that the assertion is generally true. Suppose that it holds up to n players. In view of Axiom 10 and of (iii) in Axiom 1 an $n \times n$-influence matrix A exists, such that R is defined for every equistable biform (u, A) with this influence matrix. Let A be a matrix of this kind. For every equistable biform (u, A) one can easily construct an equistable $(n + 1)$-player biform (u', A') with the property that in (u', A') the players n and $n + 1$ are identically motivated and that (u, A) results from (u', A') by the aggregation of players n and $n + 1$. It follows by Axiom 4 that we have:

$$R(u', A') = R(u, A) = c \ln \frac{u_0}{1 - u_0}. \tag{129}$$

In view of Axiom 5 the measure of an equistable biform depends only on the common value u_0 of the components of the incentive vector. This together with (129) permits the conclusion that the assertion holds for $n + 1$.

ASSERTION 4. *Let A be an $n \times n$-influence matrix with $n = 2, 3, \ldots$ which belongs to the set M_n. Then $R(u, A)$ is defined for every $u = (u_1, \ldots, u_n)$ in U_n and we have*

$$R(u, A) = c \sum_{i=1}^{n} w_i(A) \ln \frac{u_i}{1 - u_i}, \tag{130}$$

where c is the same positive constant as in Assertion 1.

Proof. It follows by (iii) in Axiom 1 that $R(u, A)$ is defined for every $u \in U_n$. For every $u = (u_1, \ldots, u_n)$ let $\mu(u)$ be the greatest number m such that we have

$$u_k = u_1 \quad \text{for } k = 1, \ldots, m. \tag{131}$$

If u_2 is different from u_1 we have $\mu(u) = 1$ and in the case of an equistable biform (u, A) the number $\mu(u)$ takes the value n. We shall prove (128) by induction on $n-\mu(u)$.

It is an immediate consequence of Assertion 3 that (130) holds in the equistable case $\mu(u) = n$. Assume that (130) holds for $\mu(u) = m + 1$. It has to be shown that (130) holds for $\mu(u) = m$. Let $u = (u_1, \ldots, u_n)$ be a vector with $\mu(u) = m$. Let $z^m = (z_1^m, \ldots, z_n^m)$ be the vector with

$$z_i^m = \begin{cases} 1 & \text{for } i = m \\ 0 & \text{else.} \end{cases} \tag{132}$$

As before ψ denotes the n-dimensional vector whose components are all equal to 1. Let $u' = (u_1', \ldots, u_n')$ with

$$u_i' = \begin{cases} 0 & \text{for } i = m \\ u_i & \text{else.} \end{cases} \tag{133}$$

Assume $u_m > u_{m-1}$. In view of Axiom 2 the measure $R(u, A)$ can be determined as follows:

$$R(u, A) = R(u' + u_{m+1}z^m, A) + \int_{u_{m+1}}^{u_m} R_m(u' + xz^m, A) \, dx. \tag{134}$$

Since we have

$$\mu(u' + u_{m+1}z^m) = m + 1, \tag{135}$$

we can evaluate $R(u' + u_{m+1}z^m, A)$ by (130). This yields

$$R(u' + u_{m+1}z^m, A) = c \sum_{i=1, i \neq m}^{n} w_i(A) \ln \frac{u_i}{1 - u_i} \\ + c w_m(A) \ln \frac{u_{m+1}}{1 - u_{m+1}}. \tag{136}$$

In order to evaluate the second term on the right hand side of (134) we make use of the fact that by Axiom 9 we have:

$$R_m(u' + xz^m, A) = R_m(x\psi, A). \tag{137}$$

The right hand side of (137) can be evaluated with the help of Assertion 1:

$$R_m(x\psi, A) = w_m(A)\frac{1}{x(1-x)}. \tag{138}$$

We can replace the integrand in (134) by the right hand side of (138). It can be seen without difficulty that this together with (136) yields (130).

Now assume $u_m < u_{m+1}$. In this case we have

$$R(u, A) = R(u' + u_{m+1}z^m, A) - \int_{u_m}^{u_{m+1}} R_m(u' + xz^m, A)\, dx. \tag{139}$$

In the same way as in the case $u_m > u_{m+1}$ it follows that (130) holds.

ASSERTION 5. *The constant ρ in Eq. (29) of Axiom 8 has the value*

$$\rho = 1 \tag{140}$$

Proof. Let A be the cyclical $n \times n$-influence matrix ($n = 2, 3, \ldots$) defined in 2.11 and for $m = 1, \ldots, n$ let (u^m, A) be the biform with

$$u^m = \psi/2 + xz^m. \tag{141}$$

Here x is a constant with $0 < x < \frac{1}{2}$. As before ψ is the n-dimensional vector whose components are all equal to 1. As in the proof of Assertion 3 the vector z^m is defined by (132). In view of Axiom 10 and (iii) in Axiom 1 the measure R is defined at each (u^m, A). It follows by Assertion 4 that we have:

$$R(u^m, A) = cw_m(A) \ln\frac{x}{1-x}. \tag{142}$$

It can be seen without difficulty that each of the (u^m, A) with $m = 2, 3, \ldots$ is a permuted biform of (u^1, A). Therefore Axiom 3 yields

$$R(u^m, A) = R(u^1, A) \quad \text{for } m = 2, \ldots, n. \tag{143}$$

This together with (142) permits the conclusion that all weights $w_i(A)$ are equal. We have

$$w_i(A) = \frac{1}{n}. \tag{144}$$

Axiom 8 implies

RISK DOMINANCE MEASURE

$$w_i(A) = \rho w_2(A). \tag{145}$$

Therefore (140) holds.

ASSERTION 6. *Let A be an $n \times n$-influence matrix with $n = 2, 3, \ldots$ which belongs to the set M_n. Let τ be a permutation of $\{1, \ldots, n\}$ and let A' be the τ-permuted matrix of A. Then we have:*

$$w_{\tau(i)}(A') = w_i(A') \quad \text{for } i = 1, \ldots, n. \tag{146}$$

Proof. Let u^m be defined as in (141). As in the proof of Assertion 5, Assertion 4 yields (142). The τ-permuted biform of (u^m, A) is $(u^{\tau(m)}, A')$. By Axiom 3 the measure R is defined for this biform and we have:

$$R(u^{\tau(m)}, A') = R(u^m, A). \tag{147}$$

Assertion 3 yields

$$R(u^{\tau(m)}, A') = c w_{\tau(m)}(A') \ln \frac{x}{1-x}. \tag{148}$$

This together with (140) and (145) permits the conclusion that (144) holds.

ASSERTION 7. *Let A be an $n \times n$-influence matrix with $n = 3, 4, \ldots$ with the property that for some m with $1 < m < n$ the players $m, \ldots, n$ are identically influenced (see 4.3). Let $\overline{A}$ be the $m \times m$-influence matrix which results from A by the aggregation of players $m, \ldots n$. Then A belongs to M_n if and only if $\overline{A}$ belongs to M_m. If A belongs to M_n we have:*

$$w_m(\overline{A}) = \sum_{i=1}^{n} w_i(A). \tag{149}$$

Proof. For $k = m, \ldots, n$ let A^k be defined as follows: A^n is the matrix A and for $k = m, \ldots, n - 1$ the matrix A^k results from A^{k+1} by the aggregation of the identically influenced players k and $k + 1$. It is clear that this is a valid recursive definition, since after the aggregation the players $m, \ldots, k$ remain identically influenced. Obviously A^m is the matrix $\overline{A}$.

For $k = m, \ldots, n$ let $u^k = (u_1^k, \ldots, u_k^k)$ be the k-dimensional vector with the following components

$$u_i^k = \begin{cases} \dfrac{1}{2} & \text{for } i = 1, \ldots, m-1 \\ x & \text{for } i = m, \ldots, k, \end{cases} \quad (150)$$

where x is a number with $1/2 < x < 1$. The biforms (u^k, A^k) have the property that for $k = m+1, \ldots, n$ the players $m, \ldots, k$ are identically motivated. For $k = m+1, \ldots, n-1$ the biform (u^k, A^k) results from (u^{k+1}, A^{k+1}) by the aggregation of the identically motivated players k and $k+1$. Repeated application of Axiom 4 together with (iii) in Axiom 1 permits the conclusion that $\overline{A}$ belongs to M_m if and only if A belongs to M_n. Moreover it follows that we have

$$R(u^m, \overline{A}) = R(u^n, A) \quad (151)$$

if A belongs to $\hat{M}_n$. Assertion 4 yields

$$R(u^n, A) = c \sum_{i=m}^{n} w_i(A) \ln \frac{x}{1-x} \quad (152)$$

and

$$R(u^m, \overline{A}) = c\, w_m(\overline{A}) \ln \frac{x}{1-x}. \quad (153)$$

In view of (151), (152), and (153), it is clear that (149) holds.

ASSERTION 8. *Let A be an $n \times n$-influence matrix with $n = 2, 3, \ldots$ which belongs to M_n. Then $w(A) = (w_1(A), \ldots, w_n(A))$ is a solution of the system formed by (9) and (10).*

Proof. In view of Assertion 2 Eq. (10) holds. It remains to show that (9) holds. As we shall see it is sufficient to prove the last equation of (9):

$$w_n(A) = \sum_{i=1}^{n} a_{in} w_i(A) \quad (154)$$

The analogous equation for $w_j(A)$ with $j \neq n$ can be reduced to (154) with the help of assertion 6. For this purpose one can look at a τ-permuted matrix of A with $\tau(j) = n$.

We show now that (154) holds. Consider first the case $n = 2$. In this case (154) is a direct consequence of Axiom 8 together with Assertion 5. In the following we assume $n = 3, 4, \ldots$. We shall construct a $(2n-2) \times (2n-2)$-influence matrix $A' = (a'_{ij})$ with the property that in

A' the $n-1$ players $n, \ldots, 2n-2$ are identically influenced. The elements of A' are as follows:

$$a'_{ij} = \begin{cases} a_{ij} & \text{for } i = 1, \ldots, n-1 \text{ and } j = 1, \ldots, n-1 \\ a_{in} & \text{for } i = 1, \ldots, n-1 \text{ and } j = n-1+i \\ a_{nj} & \text{for } i = n, \ldots, 2n-2 \text{ and } j = 1, \ldots, n-1 \\ 0 & \text{else.} \end{cases} \quad (155)$$

The third and the fourth lines show that the players $n, \ldots, 2n-2$ are identically influenced in A'. The second and the fourth lines show that in A' each of the players $j = n, \ldots, 2n-2$ influences only a single other player, namely the player $i = j - n + 1$. It is clear, that A results from A' by the aggregation of the identically influenced players $n, \ldots, 2n-2$. Therefore it follows by Assertion 7 that A' is in M_{2n-2} and that

$$w_n(A) = \sum_{i=1}^{n-1} w_{n-1+i}(A') \quad (156)$$

holds. Since for $i = 1, \ldots, n-1$ the player $j = n - 1 + i$ influences only player i, it follows by Axiom 8 together with Assertion 5 that we have

$$w_{n-1+i}(A') = a_{in} w_i(A) \quad \text{for } i = 1, \ldots, n-1. \quad (157)$$

This together with (156) yields (154).

ASSERTION 9. *For $n = 2, 3, \ldots$ the set M_n contains no exceptional influence matrices.*

Proof. It follows by Axiom 8 together with (iii) in Axiom 1 that a weight vector $w(A)$ is defined for every $A \in M_n$. In view of Assertion 8 this weight vector satisfies (9) and (10). Therefore an exceptional influence matrix $A \in M_n$ cannot be a no-weight matrix. It remains to show that M_n does not contain any multiweight influence matrix.

Assume that $A \in M_n$ is a multiweight influence matrix. It follows by (i) and (iii) in Axiom 1 that M_n is open relative to $\overline{M}_n$. Therefore a whole ε-neighborhood of A belongs to M_n. It follows by Lemma 8 that for every solution w of the system formed by (9) and (10) we can construct a sequence $A^1, A^2, \ldots$ of regular influence matrices in this neighborhood such that for $k \to \infty$ the sequence $A^1, A^2, \ldots$ converges to A and the sequence $w(A), w(A), \ldots$ of the weight vectors uniquely determined by (9) and (10) converges to w. In view of Assertion 8 these are the weight vectors determined by the weight functions whose existence is required by Axiom 8.

Let $w = (w_1, \ldots, w_n)$ and $\overline{w} = (\overline{w}_1, \ldots, \overline{w}_n)$ be two different solutions of the system formed by (9) and (10). Let j be a player with $w_j \neq \overline{w}_j$. Let $u = (u_1, \ldots, u_n)$ be the n-dimensional vector with the following components

$$u_i = \begin{cases} \dfrac{1}{2} & \text{for } i \neq j \\ x & \text{for } i = j, \end{cases} \tag{158}$$

where x is a real number with $1/2 < x < 1$. Let $A^1, A^2, \ldots$ and $\overline{A}^1, \overline{A}^2, \ldots$ be sequences of regular influence matrices in M_n converging to A such that $w(A^1), w(A^2), \ldots$ converges to w and $w(\overline{A}_1), w(\overline{A}_2), \ldots$ converges to $\overline{w}$. In view of Axiom 1 and Assertion 4 the continuity of R requires

$$R(u, A) = cw_j \ln \frac{x}{1-x} \tag{159}$$

and

$$R(u, A) = c\overline{w}_j \ln \frac{x}{1-x} \tag{160}$$

contrary to the assumption $w_j \neq \overline{w}_j$.

ASSERTION 10. *For $n = 2, 3, \ldots$ the set M_n is the set of all regular $n \times n$-influence matrices.*

Proof. In view of Assertion 9 it is sufficient to show that M_n contains every regular $n \times n$-influence matrix. Assume that this is not the case. Let A be a regular $n \times n$-influence matrix which does not belong to M_n. In view of (ii) and (iii) in Axiom 1 the closure of M_n is $\overline{M}_n$. Therefore A is at the border of M_n. Every sequence $A^1, A^2, \ldots$ of matrices in M_n converging to A has the property that the weight vectors $w(A^1), w(A^2), \ldots$ converge to the unique solution of the system formed by (9) and (10). This has the consequence that for every $u \in U_n$ the limit of $R(u', A')$ for $(u', A') \to (u, A)$ exists, contrary to (v) in Axiom 1.

Completion of the Proof of Theorem 1. Assertion 10 has the consequence that R is defined wherever the weighted average log measure is defined and vice versa. Assertions 4 and 8 permit the conclusion that R is the weighted average log measure multiplied by a positive constant.

References

GANTMACHER, F. R. (1986). *Matrizentheorie*. Berlin/Heidelberg/New York/Tokyo: Springer.

GÜTH, W. (1992). "Equilibrium Selection by Unilateral Deviation Stability," in *Rational Interaction, Essays in Honor of John C. Harsanyi* (R. Selten, Ed.), pp. 161–189. Berlin/Heidelberg/New York/London/Tokyo/Hong Kong/Barcelona/Budapest: Springer.

GÜTH, W., AND KALKOFEN, B. (1989). "Unique Solutions for Strategic Games." *Lecture Notes in Economics and Mathematical Systems*. Berlin/Heidelberg/New York/London/Paris/Tokyo: Springer.

HARSANYI, J. C. (1967/1968). "Games with Incomplete Information Played by "Bayesian Players," *Manage. Sci.* **14,** Part I, 159–182, Part II, 320–334, Part III, 486–502.

HARSANYI, J. C., AND SELTEN, R. (1972). "A Generalized Nash Solution for Two-Person Bargaining Games with Incomplete Information," *Manage. Sci.* **18,** No. 5, Part II, 80–106.

HARSANYI, J. C., AND SELTEN, R. (1988). *A General Theory of Equilibrium Selection in Games*, MA/London: The MIT Press. Cambridge.

PART III

LEARNING

Evolution, Learning, and Economic Behavior*

REINHARD SELTEN

Department of Economics, University of Bonn, Adenauerallee 24-42, D-5300 Bonn 1, Germany

Received August 7, 1990

This is the entire text of the 1989 Nancy L. Schwartz Lecture delivered by the author at the J. L. Kellogg Graduate School of Management, Northwestern University, Evanston, Illinois. *Journal of Economic Literature* Classification Numbers: 011, 036, 215. © 1991 Academic Press, Inc.

It is doubtful whether I shall be able to meet the high standards set by previous speakers. I shall not prove deep theorems. I shall not present astonishing new results. Instead of this I shall try to catch your attention by a fictitious dialogue. I shall employ the help of imaginary discussants like the "Bayesian" or the "experimentalist." A "chairman" will determine who speaks next, but he shall also make his own remarks.

Chairman: I open the discussion with a question: What do we know about the structure of human economic behavior?

One of our participants, the Bayesian, has signalled his willingness to answer this question. I give the floor to the Bayesian.

Bayesian: As far as economic activities are concerned, it is justified to assume that man is a rational being. Since Savage (1954) simultaneously axiomatized utility and subjective probability we know what rational economic behavior is. Rational economic behavior is the maximization of subjectively expected utility.

Chairman: Among us is an economist. I would like to ask him whether this is the agreed upon opinion in economic theory.

Economist: Yes, to a large extent this is the agreed upon opinion. Most

---

* We thank Dean Donald P. Jacobs of the J. L. Kellogg Graduate School of Management for permission to publish this lecture.

of microeconomics takes Bayesianism for granted. However, there are exceptions. Some theorists have different views.

As an example let me mention. Allais. Since 1953 he insists that in the evaluation of risky decisions not only the expectation, but also the variance of utility must be taken into account. He has shown that his theory is in better agreement with observed behavior than Bayesianism is.

Chairman: One of our discussants is an experimentalist. His background is in experimental economics. I would like to ask the experimentalist for his opinion on Allais and Bayesian decision theory and their agreement with observed behavior.

Experimentalist: If I understand the opening question correctly, we are here to discuss *human* economic behavior, not the behavior of a mythical hero called "rational man," a mythical hero whose powers of computation and cogitation are unlimited. For this mythical hero it is easy to form consistent probability and preference judgments, but not for ordinary people like you and me.

People are not consistent. I would like to mention just one of many experimental results which show this. In a paper published by Tyszka (1983) he describes an experiment in which the subjects had to make choices from triples, say $\{A, B, C\}$ or $\{A, B, D\}$. Tyszka succeeded in constructing triples such that

95% of the subjects chose A from $\{A, B, C\}$ and

95% of the subjects chose B from $\{A, B, D\}$.

The choice between A and B is influenced by the irrelevant alternatives C and D. This should not be the case, if consistent preference judgments are formed.

The theory of Allais (1953) is in better agreement with behavior than Bayesianism is, but the agreement is only slightly better. The violations of ideal rationality observed in experiments are much more basic than the theory of Allais suggests. Rejecting Bayesianism in favor of Allais's theory is like going up to the top of a skyscraper in order to be nearer to the moon!

Chairman: I agree with the experimentalist. One must make a distinction between normative and descriptive theory. My opening question was meant descriptively. Decision theory and game theory have made tremendous progress on the clarification of the concept of ideal rationality. In this discussion we are not concerned with ideal rationality, but with actual human decision behavior.

However, we should not be too quick in the rejection of the optimization approach as nondescriptive. Evolutionary game theory—started by Maynard Smith and Price (1973)—is successful in biology and biology is thoroughly descriptive. The book by Maynard Smith (1982), *Evolution and the Theory of Games,* provides many examples. Game theory has been created as a theory of rational behavior, but it is now applied to animals and plants.

Among us is a biologist whom we call the "adaptationist" since he strongly beliefs in adaptation in the biological sense. The adaptationist thinks that the principle of fitness maximization is applicable to human behavior. I would like to ask him to explain his views.

Adaptationist: Let me explain what adaptation means in biology. Adaptation means fitness maximization. Fitness is reproductive success—roughly speaking the expected number of offspring in the next generation. Natural selection drives organisms toward fitness maximization. Fitness maximization is a powerful explanatory principle also for human behavior.

However, as far as human decisions are concerned—the same holds for animal decisions by the way—we must be aware of the fact that near to the optimum, selective pressures are weak. Therefore we often observe nearly optimal rules of thumb instead of truly optimal behavior.

The experimentalist has told us about deviations from rationality in experiments. I think that these deviations cannot be of great practical importance. Otherwise natural selection would have eliminated them long ago.

Chairman: Now several discussants want to say something. The next speaker is the Bayesian.

Bayesian: I find the remarks of the adaptationist very interesting. The principle of fitness maximization permits us to construct new kinds of theories in economics. Preferences can now be explained as a result of evolution. People like what is good for their fitness.

Let me make an additional remark on rules of thumb. What seems to be only nearly optimal may be truly optimal, if decision costs are taken into account. I think that with decision costs taken into account, many rules of thumb may turn out to be truly optimal upon closer inspection.

Chairman: I would like to ask the experimentalist for his opinion on rules of thumb and decision costs.

Experimentalist: As many people have pointed out, decision problems tend to become more difficult with decision costs taken into account. If one tries to save decision costs by taking them into account, one may easily end up with higher decision costs.

Adaptationist (*interrupting*): No! I must interrupt here. Excuse me for doing this. Remember that I think of rules of thumb as inherited. They are not made up on the spot. Evolution has already solved the optimization problem. The decision maker does not have to solve it any more.

Experimentalist: I concede this point. However, I wonder whether all rules of thumb which people use in everyday life are inherited. Sometimes rules of thumb have to be made up on the spot.

I would like to point out that nearly optimal rules of thumb can be far away from truly optimal policies. This may not matter for the decision maker very much, but it may be important for other people. I would like to tell you about an example in which the structure of experimentally observed behavior is dramatically different from that of the optimal policy.

Claus Berg (1973, 1974) published the results of an investment experiment. The subjects could invest in cash or a risky asset. In every period they had to divide total asset into cash and the risky asset. The end was decided by a stopping probability of 1%.

The risky asset yielded a positive interest, say 25%, or a negative interest, say -25%. The percentage was the same in both cases. It was fixed and known to the subject. There was a fixed probability p for the positive interest. This probability was not known to the subject. Profits and losses changed total assets from period to period.

Bayesian decision theory yields the following prediction about changes of the proportion of the risky asset in total assets:

After a positive interest the risky asset proportion is increased.

After a negative interest the risky asset proportion is decreased.

This is due to an increase of the posterior probability for a positive interest after a positive interest and to a decrease of this posterior probability after a negative interest. However, in 75% of all cases the subjects changed the risky asset proportion in the direction opposite to the Bayesian prediction!

A closer look at the data revealed the reason for this phenomenon. The subjects tend to form an aspiration level for the total assets they want to obtain at the end of the period. Often this aspiration level is total assets at the beginning of the last period. This has the following consequences: After a loss as much is risked as necessary in order to recuperate the loss; this requires a raise of the risky asset proportion. After a gain not more is risked than has been won; this requires a decrease of the risky asset proportion.

Berg ran computer simulations with an idealized description of the aspiration guided behavior observed in the experiments. He compared the results with those for Bayesian optimal policies starting from beta-distributed priors with expected values near to the true probability. Strangely enough the results for the behavioral theory were often better than those for the optimal policies, particularly in the parameter range of the experiments. Maybe the priors were not appropriate. The success of a Bayesian policy may crucially depend on the prior. Only asymptotically the prior does not matter. But how should we choose the prior?

In any case Berg's theory performs very well. However, its structure of aspiration guided behavior is very different from that of the optimal Bayesian policy. The Bayesian prediction of the change of the risky asset proportion goes in the wrong direction!

Chairman: I permit the Bayesian to make a short comment.

Bayesian: I only want to say that aspiration guided behavior may be truly optimal, if decision costs are taken into account.

Chairman: Well, this is a possibility, but I would say a remote one. I would like to come back to the statement of the adaptationist: *Natural selection drives organisms toward fitness maximization.* Among us is a population geneticist. I would like to ask him, what can be said about this statement from the point of view of population genetics?

Population geneticist: Fitness maximization is thought of as the result of a dynamic process of natural selection. We need a justification of this idea within explicit dynamic models. Such models are the subject matter of population genetics.

Ronald Fisher, the great population geneticist and econometrician proved a "fundamental theorem," which under certain conditions shows that natural selection increases fitness until a maximum is reached. Unfortunately Fisher's conditions are rarely satisfied in genetic systems.

In the following I shall rely heavily on a very illuminating paper by Ilan Eshel (1988) which has been made available as a preprint. Maynard Smith and Price (1973) have introduced the concept of an evolutionarily stable strategy as an attempt to give a static description of a dynamically stable result of natural selection in game situations. We may say that evolutionary stability is the generalization of fitness maximization to game situations. Eshel's paper is concerned with the dynamic foundations of evolutionary stability.

We must distinguish two mechanisms of natural selection:

1. adaptation of genotype frequencies without mutation,
2. gene substitution by mutation.

Mendelian inheritance and selective pressures combine to change the frequencies with which genotypes are represented in the population. This is the process called "adaptation of genotype frequencies without mutation."

Moran (1964) has shown that under realistic conditions adaptation of genotype frequencies without mutation does not necessarily maximize fitness, it may even decrease fitness, until a local minimum is reached. This result holds even without any game interaction. The explanation lies in the combined effects of Mendelian inheritance and linkage. Linkage is the phenomenon that genes near to each other on the same chromosome are likely to be inherited together.

After Moran's result the idea of fitness maximization fell into disrepute among population geneticists. Only recently a new picture emerged, first in a paper by Eshel and Feldman (1984). They showed for two-locus systems that gene substitution by mutation works in the direction of evolutionary stability. This result has been generalized to an arbitrary number of gene loci by Liberman (1988).

This is very good news for all those who work on evolutionary game theory. Evolutionary game theory now has a more solid dynamic foundation.

However, successful mutants are very rare. Gene substitution by mutation is very slow. Therefore fitness maximization or evolutionary stability can only be expected as a long run equilibrium phenomenon. It is dubious whether any mutations have changed human economic behavior in the relatively short time since the beginning of the dispersion of agriculture about 10,000 years ago. This means that biologically man may still be a hunter and gatherer not very well adapted to the necessity of long run planning. This may be the reason why some Ph.D. dissertations take much longer than planned. In any case it would be silly to expect that man is genetically adapted to modern industrial society.

Chairman: The remarks of the population geneticist throw doubt on the near optimality of human economic behavior. At least we can say that natural selection did not necessarily produce this result. Among us is a naturalist, a man who knows animals and plants. He wants to make a comment.

Naturalist: We rely heavily on the principle of fitness maximization in the explanation of field phenomena. This principle has tremendous explanatory power. However, it must be used with care. Fitness maximization does not work absolutely but only under structural constraints. Let me give you an example in order to make it clear what I mean by structural constraints.

The example is the giraffe. As you all know the giraffe has a very long neck, but only 7 neck bones like every other mammal. This is very inconvenient for the giraffe. It has difficulties lying down to sleep and standing up quickly in danger. It actually sleeps very little. Why did evolution fail to increase the number of neck bones? The answer is simple: *Evolution cannot change many things at once*. A change of the number of neck bones alone would be disastrous. Muscles, nerves, and other things would have to be adjusted. This would require many simultaneous mutations. Therefore the number of neck bones acts as a *structural constraint* on the evolution of the giraffe.

A related problem is that of the correct strategy space in biological game models. Hammerstein and Riechert (1988) have modelled the fighting behavior of *agelonopsis aperta* (a spider). In this model the spiders ignore some useful information like the number of days passed in the season. From the point of view of fitness maximization the spider's strategy should depend on this information. Hammerstein and Riechert do not permit this in their model. I have no objection! The data justify the restriction of the strategy space. Structural constraints on the spiders' behavior must be working here.

I am convinced of the principle of fitness maximization, but under structural constraints. Fitness maximization alone is not sufficient to explain natural phenomena. A thorough knowledge of nature cannot be replaced by abstract principles.

Chairman: We have heard three reasons against the biological deduction of human economic behavior:

1. the slowness of gene substitution by mutation,

2. the fact that genotype frequency adaptation without mutation does not necessarily optimize fitness,

3. structural constraints.

We have to gain empirical knowledge. We cannot derive human economic behavior from biological principles. I would like to ask the Bayesian what he thinks about this.

Bayesian: Well, maybe the discussion has overemphasized natural selection. Rationality needs training. Small children are not yet rational. Untrained grown-ups still make many mistakes. Maybe we all are not yet sufficiently trained. We have to change this. Bayesian methods should be taught to future executives much more than this is done now. In this way Bayesian methods will become more and more widespread in business and government. Haphazard natural decision behavior will be replaced by superior Bayesian methods. This process of cultural evolution will establish descriptive relevance of Bayesian decision theory, at least where it matters, in business and government.

Chairman: Thank you for mentioning cultural evolution. Up to now this topic has been neglected in our discussion. Two population geneticists, Cavalli-Sforza and Feldman, have created a fascinating mathematical theory of cultural evolution. In 1981 they published a book on the subject. Another useful systematic exposition can be found in the book of Boyd and Richerson (1985). The population geneticist is familiar with this literature. I would like to ask him to describe the basic ideas underlying the mathematical theory of cultural evolution.

Population geneticist: This is a difficult task. I shall give a highly simplified picture. The theory of cultural evolution focuses on cultural traits. A cultural trait is something like the use of a dialect or the adherence to a religion. We think of cultural traits as acquired in the formative years of childhood and adolescence and not changed later in life. Of course this is a simplification.

One can model a cultural trait as absent or present, but it is often more adequate to think of a cultural trait as a quantifiable variable measured on a scale. Somebody may more or less adhere to a religion. He may go to church every Sunday or only occasionally. In the following I shall restrict my attention to the case of a quantifiable cultural trait.

Models of cultural evolution with quantifiable cultural traits are similar to models of quantitative inheritance. The theory of quantitative inheritance was initiated by Galton (1889) long before the beginnings of population genetics. Quantitative inheritance theory does not make use of Mendelian genetics. Nevertheless models of quantitative inheritance continue to be useful in animal and plant breeding.

Consider a trait like "height." Quantitative inheritance of height is as follows. Three components determine the height of an individual:

1. the parents' average height,
2. the population mean,
3. a random component.

First a weighted average with fixed coefficients is formed of the first two components and then the random component is added. In this way the height of an individual is determined.

Models of cultural evolution are similar. However, there may be many cultural parents. The biological parents may or may not be among them. The cultural parents exert their influence by teaching or setting an example. They transmit an average in which different cultural parents may have different weights reflecting differences of importance.

I shall now give a sketch of a possible model which is meant to be an illustration and should not be taken too seriously. We shall look at the cultural trait "conformance to work ethics." We have to make an assumption on who becomes a cultural parent of whom. We make the simplest possible assumption. The cultural parents of an individual are a random sample of fixed size taken from the previous generation. In the transmitted average the cultural parents have weights which increase with prestige. Economic success has a positive influence on prestige and conformance to work ethics has a positive influence on economic success. It can be seen how in a model along these lines a high level of conformance to work ethics can evolve in the population, even if in the beginning this level is very low.

I am now at the end of my short exposition. Of course much more could be said about the theory of cultural evolution.

Chairman: I can see that models of this type may be useful in the explanation of economic development. Cultural traits like values, ambitions, and left styles influence economic behavior and thereby economic conditions. Economic conditions exert selective pressure on the cultural traits. In this way we obtain a feedback loop. Obviously the application of cultural evolution theory to economics offers some interesting possibilities.

However, we should not forget evolutionary theories which have their origin within economies. Already Schumpeter (1934) created an evolutionary theory of innovation and imitation. Nelson and Winter (1982) present formal models in their book on "an evolutionary theory of economic change." The economist is familiar with this work and I would like to ask him to explain the approach of Nelson and Winter.

Economist: Nelson and Winter focus on firms rather than individuals. In their models firms do not maximize profits. Instead of this they adapt to success and failure in a trial and error fashion. I think that the best way to explain the approach of Nelson and Winter is with a short sketch of a particular model described by Sidney Winter (1971).

I shall sketch a slightly simplified version of the model. A finite number of goods is produced by many firms. Production methods connect inputs and outputs by fixed proportions. Inputs and outputs differ from method to method. An output of one firm may be an input of another.

The market as a whole ends up with a surplus for some commodities and with a deficit for others. We may think of surpluses as sold to consumers and of deficits as imported. The surpluses and deficits determine the prices by a relationship technically named "inverse demand function."

I now come to the behavioral assumptions. After a profit, a firm expands by one unit. After a loss a firm contracts by one unit and in addition to this starts a search for a new production method which is found and adopted with a positive probability. A firm whose production method yields zero profits does not change anything. Essentially the same rules apply to potential firms which do not produce anything. These firms also have a production method. They enter in case of profitability. They search for a new production method if the present one is not profitable at current prices.

Sidney Winter has shown that under appropriate assumptions on the inverse demand function the stochastic process defined by the model converges to an absorbing state with the properties of a competitive equilibrium. In this way he provided a new foundation to the theory of competitive equilibrium without profit maximization assumptions.

I admire the work of Nelson and Winter and in particular the model which I just described. However, I cannot see a close analogy to biological and cultural evolution. Firms do not reproduce and do not die. There is no cultural transmission from firm to firm—at least not in this model. I have no objection! Nelson and Winter have created an evolutionary theory in its own right, much better adapted to economics than any literal translation from biology could be.

Chairman: Maybe there is a closer analogy than one thinks at first glance. It is necessary to change the perspective. Not the firms but the production methods are the animals under selective pressure. The behavior of the firms is the environment of the production methods. Production methods are born and may die.

In equilibrium all active production methods have zero profitability. This is analogous to biological models with asexual reproduction where in equilibrium all surviving genotypes have fitness 1. Obviously profitability in one case has the same role as fitness in the other case. I would like to know whether a similar analogy can be established between cultural and biological evolution. Is it possible to define a cultural fitness? Maybe the population geneticist can answer this question.

Population geneticist: The idea of cultural fitness sometimes appears in the literature, for example, in the book of Cavalli-Sforza and Feldman (1981), but I cannot remember having seen a precise definition. Nevertheless it seems to be easy to give a meaning to the term. Cultural fitness could be defined as a measure of the expected influence exerted as a cultural parent on the next generation. However, it is unclear whether a cultural fitness concept could be useful.

In the models I described, cultural inheritance is similar to quantitative inheritance in biology and it is unclear whether the biological fitness concept is useful in models of quantitative inheritance, unless very special assumptions are made. Two parents of optimal height usually have children of nonoptimal height—due to the random component and maybe the influence of the population mean. In equilibrium not everybody will be of optimal height. The equilibrium height distribution has a positive variance. Generally there will not even be a monotonic relationship between fitness and representation in the equilibrium population. Only under very special assumptions on the way in which fitness depends on height do we obtain such a monotonic relationship. This indicates that a cultural fitness concept is not very useful in models of cultural evolution like those I have described. This may be different for other types of models.

Chairman: The adaptationist wants to make a remark.

Adaptationist: I want to say that cultural evolution tends to the maximization of biological fitness. We do not need a concept of cultural fitness. Mechanisms of cultural evolution are shaped by biological evolution. There must be fitness advantages for transmitters and receivers. Consider the following example:

Parents teach their children which fruits are edible and which are poisonous.

It is advantageous for the children to accept the transmission and it is advantageous for the parents to transmit—after all the children are their fitness.

The shakers provide another example. The shakers have the cultural trait of not having children. Therefore they are dying out. In the long run a cultural trait which reduces biological fitness cannot persist.

Chairman: Is this really true? I would like to hear the opinion of the population geneticist.

Population Geneticist: My exposition of cultural evolution was highly simplified. I did not talk about the interaction of cultural and biological evolution. Actually this interaction is emphasized in the literature. Cavalli-Sforza and Feldman (1981) present a model in which family size is the cultural trait under consideration. In spite of the interaction with biological evolution, in this model cultural evolution stabilizes a small family size.

Adaptationist (interrupting): No! No! I must interrupt here. Excuse me. The assumptions of the model must be wrong. In the long run natural selection would favor a tendency not to accept the cultural transmission of a small family size. If a bigger family size offers a fitness advantage, a small family size cannot persist.

Population geneticist: Maybe in the very long run this is true, but not on the time scale of cultural evolution processes in recorded history. In the explanation of such processes we can safely ignore the interaction with biological evolution, at least if one avoids extreme examples like the shakers.

In my simplified exposition of cultural evolution I also omitted another kind of interaction—the interaction of cultural evolution with individual learning. It seems to me that the interaction with individual learning is much more important than the interaction with natural selection. Actually the interaction with individual learning is also emphasized in the literature. Boyd and Richerson (1985) describe the psychological learning model by Bush and Mosteller (1955) and they also explain Bayesian updating. However, they do not make any explicit use of these modelling possibilities. They only model the effect of learning. The precise mechanism is left open.

Boyd and Richerson assume that learning is guided by a criterion of success, for example, money income. Moreover they assume that at any point of time there is exactly one value of the trait under consideration, which is optimal with respect to the criterion of success. The value of an individual's trait is influenced by three components:

1. a culturally inherited value,
2. the optimal value,
3. a random error added to the optimal value.

The idea is that the optimal value is not correctly perceived. This is expressed by the random error. Learning shifts the culturally inherited value in the direction of the misperceived optimal value. The value of the trait is a weighted average with fixed coefficients for the culturally inherited value and the optimal value modified by the random error. The optimal value is endogenous rather than exogenous; it may depend on the distribution of the trait.

Chairman: This reminds me of Arrow's (1962) theory of learning by doing. In this theory, too, only the effect of learning is modelled. It is assumed that experience results in a downward shift of the cost curve, but how the firms learn to save costs is left open.

If we want to describe economic behavior, it is not enough to model the effects of learning. We must ask the question: What is the structure of learning? Maybe the experimentalist can help us to answer this question.

Experimentalist: It is necessary to distinguish at least three kinds of learning:

1. rote learning,
2. imitation,
3. belief learning.

In rote learning success and failure directly influence the choice probabilities. Rote learning does not require any insight into the situation. It is based on a general trust in the stability of the environment: What was good yesterday will be good today.

Imitation of successful others is similar to rote learning with the difference being that it is the success of others which directly influences choice probabilities.

Belief learning is very different. Here experiences strengthen or weaken beliefs. Belief learning has only an indirect influence on behavior.

We do not know very much about the structure of learning, but we know more about rote learning than belief learning. Up to now we do not yet have a sufficient understanding of belief learning. Therefore I shall restrict my comments to rote learning.

Psychological learning models like the one by Bush and Mosteller (1955) describe rote learning. These models have only two possible reward levels, success and failure. This is reasonable for some animal experiments: The rat either finds food in the maze or it does not find food. For the description of learning by an economic agent motivated by profits one needs a continuum of possible reward levels. In his book from 1983 John Cross has generalized the model of Bush and Mosteller to a continuum of reward levels. As far as I know the generalized model has not been confronted with data.

I would like to tell you about an experimental investigation by Malawski described in his unpublished Ph.D. dissertation (1989). He looked at game situations with minimal information. The subjects did not even know that they played games. They only knew that they had to choose one of several alternatives, say A, B, C. After each choice they were informed about their own payoff (sometimes they obtained additional information which will not be described here). They experienced the same decision situation about 70 times.

Malawski developed a theory of "learning through aspiration," which is in good agreement with his data.

His theory assumes that aspiration levels on payoffs are formed in the beginning of the session (the first 12 decisions). After this initial phase only two responses to the experience made last time are possible:

(a) the same choice as last time or
(b) a randomly picked different alternative.

Depending on the situation one of these responses is "normal" and the other one is "exceptional." The normal response is taken more often.

Under the condition that the aspiration level has been reached or surpassed last time, the normal response is the same choice as last time. Under the condition that the aspiration level has not been reached, the normal response is a randomly picked different alternative.

The normal response is taken with a probability $p > \frac{1}{2}$ and the exceptional response is taken with probability $1 - p$. The probability p is a parameter which varies from subject to subject.

Once more the good fit of this theory shows the importance of aspiration levels in economic behavior. Of course Malawski's theory is not yet firmly established. More experiments have to be made, but his results look very promising.

Chairman: I would like to hear the opinion of the economist on the structure of learning.

Economist: I am surprised about the remark of the experimentalist, that we cannot say very much about belief learning. After all Bayesian updating is a plausible model of belief learning. Bayesian updating can be descriptively right, even if subjectively expected utility maximization is descriptively wrong. Somebody who does not maximize utility can still adjust his subjective probabilities by Bayes' rule.

We can replace subjectively expected utility maximization by alternative theories in the literature, for example, prospect theory by Kahneman and Tversky (1984) or regret theory by Loomes and Sugden (1982), Bell (1982), and Fishburn (1982). However, I cannot see any alternative to Bayesian updating. The experimentalist did not even sketch an alternative model of belief learning.

Chairman: The experimentalist should answer this remark. Is it really true that we have no alternative to Bayesian updating as a model of belief learning?

Experimentalist: The trouble with Bayes' rule is that people do not obey it. It is, for example, well known that people overvalue the information content of small samples. In this connection Tversky and Kahneman speak of a "law of small numbers" (1982a).

I have been accused of not having sketched an alternative model of belief learning. Therefore I shall do this now. What I shall tell you is highly speculative, even if there is some experimental support. I refer to an oligopoly experiment described by Selten (1967).

If one wants to model belief learning, one first has to model a belief system. Bayes' rule operates on a belief system which is a probability distribution over all possible states of the world. However, belief systems of human decision makers may have a completely different structure. In my sketch of an alternative model of belief learning, beliefs are formed on causal links like

Advertising increases sales.

The belief structure has the form of a causal diagram composed of such causal links. The term causal diagram has been used by Selten (1967). Later very similar structures were called "cognitive maps" by Axelrod (1976). Axelrod has done interesting empirical research on cognitive maps of politicians expressed in speeches and writings.

A causal diagram shows chains which connect a decision variable like advertising with a goal variable like profits. Suppose that there are two causal chains in the diagram:

1. *Advertising increases sales and sales increase profits.*
2. *Advertising increases cost and cost decreases profit.*

Subjects reason qualitatively on the basis of such chains. The first chain is an argument for more advertising and the second chain is an argument for less advertising. Such conflicts are resolved by judgments on the relative importance of causal links. Suppose that the causal link from advertising to cost is judged to be the least important one. On the basis of this judgment the second chain is neglected and a decision to increase advertising is based on the first chain. Of course, this determines only the direction of change. The amount of change has to be determined in some other way.

Belief learning exerts its influence on importance judgments. Experience may show that the influence of advertising on cost is more important than the influence of advertising on sales. The mechanism of belief learning can be modelled in a fashion similar to that of the mechanism of rote learning.

Chairman: These remarks show that belief learning is closely connected to boundedly rational reasoning. Not only belief systems must be modelled but also their use in reasoning. We need to know more about bounded rationality. In my view the development of a theory of bounded rationality is one of the most important tasks of economics in our time. Bayesian updating has been mentioned favorably and unfavorably. I think that the Bayesian should comment on Bayes' rule.

Bayesian: I would rather like to comment on bounded rationality. I may have no opportunity to do this later. Herbert Simon (1957) introduced the idea of satisficing. His view of bounded rationality did not exert a strong influence on economic theory. The work of Herbert Simon inspired the book by Cyert and March (1963) on the "behavioral theory of the firm." Many people were very impressed by this book, in particular by the surprising empirical success of the model which describes the behavior of a department store manager with high predictive accuracy. However, the book did not start a revolution of economic theory. What is the reason for this? Let me give my answer to this question.

A typical piece of research in the behavioral theory of the firm is a simulation study, based on a complex model, sometimes with hundreds of parameters and with many behavioral ad hoc assumptions. Generally no clear conclusion can be drawn from such simulation studies. What we see here is a *theory without theorems*. A theory without theorems cannot succeed.

In the behavioral theory of the firm and the evolutionary approach by Nelson and Winter, behavior is described by ad hoc assumptions, which vary from model to model. This is very unsatisfactory. Economic theory needs a description of economic behavior based on a few general principles which can be applied to every conceivable decision situation. Bayesian decision theory meets this requirement. Bayesian decision theory should not be thrown away in favor of ad hoc explanations of experimental phenomena.

Experiments are often too quickly interpreted as evidence against Bayesian decision theory. I would like to mention the example of the finitely repeated prisoner's dilemma. This game has only one equilibrium outcome, namely noncooperation in every period. Nevertheless one observes cooperation until shortly before the end. This seems to refute the rationality assumptions of game theory. However, a rational explanation has been given by Kreps *et al.* (1982). They introduced a small amount of incomplete information on payoffs of the other players. In the slightly modified game the usual pattern of behavior is an equilibrium outcome.

Even the limits of computational capabilities permit a Bayesian treatment. Repeated games with limited memory or limited complexity have been analyzed by Neyman (1985), Aumann and Sorin (1989), Kalai and

Stanford (1987), and others. Another approach to problems connected to bounded rationality is the relaxation of common knowledge assumptions explored by Neyman. We see here the beginnings of a Bayesian theory of bounded rationality. It is not necessary to construct a theory of bounded rationality outside the Bayesian framework. It is much more fruitful to do this within the Bayesian framework.

Chairman: I would like to ask the experimentalist what he thinks about Bayesian bounded rationality.

Experimentalist: Let me first make a comment on the paper by Kreps *et al.* (1982). Game theoretically their work is very interesting, but behaviorally they miss the mark! Selten and Stoecker (1986) describe an experiment where each subject plays 25 supergames of 10 periods each against anonymous opponents changing from supergame to supergame. In these experiments the typical pattern of cooperation until shortly before the end did not emerge in the first supergame, but only after a considerable amount of learning. In the beginning behavior is chaotic. Only slowly cooperation is learned and after cooperation the end effect. According to Kreps *et al.*, the typical pattern is due to thinking rather than learning and therefore should emerge immediately. In their theory there is no room for chaotic behavior in the beginning and for slow learning afterward.

I now want to comment on infinite supergames with restricted memory. In such games only the operating memory is restricted, or in other words, the storage space available for the execution of a strategy. The computational capabilities for the analysis of the game remain unrestricted. In fact, the analysis of the game tends to become more difficult by constraints on the operating memory. I cannot see any contribution to a theory of bounded rationality in this kind of work.

Let me now say something about common knowledge or the lack of it. Consider a chain of the following kind:

I know, that he knows, that I know, that he knows, ...

Roughly speaking, common knowledge means, that such chains can be continued indefinitely. Does it really matter in practical decision situations whether I have common knowledge or whether I have to break off such chains after stage 4? I do not think so. As far as human decision behavior is concerned I dare say: A lack of common knowledge is not important; what often is important is a very common lack of knowledge.

I highly appreciate the behavioral theory of the firm and the evolutionary approach by Nelson and Winter. I do not accept the criticism against the use of ad hoc assumptions. Look at human anatomy and physiology: bones, muscles, nerves, and so on. Human anatomy and physiology cannot be derived from a few general principles.

Let me also say something else in defense of ad hoc assumptions.

Experiments show that human behavior is ad hoc. Different principles are applied to different decision tasks. Case distinctions determine which principles are used where. Successful explanations of experimental phenomena have been built up along these lines, for example, the theory of equal division payoff bounds for three-person games in characteristic function (Selten, 1987). Let me conclude my comments with a final remark: It is better to make many empirically supported ad hoc assumptions, than to rely on a few unrealistic principles of great generality and elegance.

Chairman: I would like to ask the economist whether he thinks that economic theory should abolish the optimization approach in favor of a more realistic description of economic behavior.

Economist: Many economic theorists are uneasy with the usual exaggerated rationality assumptions—but they continue to use them. They do not see a clear alternative. In recent years the interest in experimental economics has increased tremendously. This offers a hope for a new foundation of microeconomics. However, as long as this new foundation has not yet been established, we have to go on relying on exaggerated rationality assumptions. I do not think that present-day microeconomics will become completely obsolete. Market experiments by Smith, Plott, and others reviewed in the literature (Smith, 1980; Plott, 1982) confirm competitive equilibrium theory. What is now derived as a result of optimization may later be explained as a result of learning.

Bayesianism may be wrong descriptively, but this does not touch its great normative significance. Moreover as has been pointed out earlier in the discussion, teaching of Bayesian methods in universities will increase their use in business and government and thereby establish descriptive relevance for Bayesian decision theory.

Chairman: Yes, we did not yet sufficiently discuss the idea that in the future economic behavior will become more Bayesian than it is now due to the influence of teaching. I would like to ask the experimentalist what he thinks of the prospects of a cultural evolution toward a widespread use of Bayesian methods in business and government.

Experimentalist: The application of Bayesian methods makes sense in special contexts. For example, a life insurance company may adopt a utility function for its total assets; subjective probabilities may be based on actuarial tables. However, a general use of Bayesian methods meets serious difficulties. Subjective probabilities and utilities are needed as inputs. Usually these inputs are not readily available.

There is no probability and utility book in the brain which can be looked up like a telephone directory. Probability and preference judgments require information processing in the brain. They are outputs rather than inputs. There is no reason to suppose that information processing in the brain yields consistent probability and preference judgments. There is much experimental evidence to the contrary. Let me introduce an example concerning probability judgments, the conjunction effect, described by Tversky and Kahneman (1982b). They asked their subjects to rank a number of statements on some future events with respect to likelihood. One of the events was a tennis match involving Borg. Among the statements were the following two:

Statement A: Borg will lose the first set.

Statement B: Borg will lose the first set, but he will win the match.

Subjects tend to judge statement B as more likely than statement A, in spite of the fact that statement B describes a subcase of statement A. This shows that probability judgments do not even have one of the most basic consistency properties, namely monotonicity of the probability measure with respect to set inclusion.

The phenomenon is a "representativeness effect." The statement "Borg will loose the first set" is not representative of the image of Borg in the mind of the subjects. Borg was a winner, not a loser. The additional detail "but he will win the match" is representative and therefore improves the impression of credibility.

Preference judgments are as inconsistent and unreliable as probability judgments. Martin Weber *et al.* (1988) have published an experimental investigation of "dimension splitting in multiattribute utility measurements." They show that the weight of an attribute is increased if it is split into several subattributes. This is important, since multiattribute utility measurement is recommended as an instrument of decision aid by Bayesian decision theorists interested in practical applications. We see here that the result of the method heavily depends on the way in which the problem is presented to those who have to make the preference judgments.

We can conclude: Normative Bayesianism is dubious in view of the unreliability of its inputs. If you ask people to be consistent, you ask for too much. Imagine a normative theory of sports which commands: Athlete, jump 100 m high!—It cannot be done.

I do not see an unavoidable cultural evolution toward a more and more widespread use of Bayesian methods in business and government.

Chairman: It seems to be necessary to make a distinction between a *practical normative theory* and an *ideal normative theory*. A practical normative theory can be used in order to help people to improve their decisions. Ideal normative theory has the purpose of clarifying the concept of rationality independent of the limitations of real persons. Even if Bayesianism may fail as a practical normative theory it still remains an ideal normative theory of great philosophical importance.

We now must come to the end of the discussion. I shall try to summarize the results as I see them. I opened the discussion with a question: What do we know about the structure of human economic behavior? I must admit that the answer is disappointing. We know very little.

We know that Bayesian decision theory is not a realistic description of human economic behavior. There is ample evidence for this, but we cannot be satisfied with negative knowledge—knowledge about what human behavior fails to be. We need more positive knowledge on the structure of human behavior. We need quantitative theories of bounded rationality, supported by experimental evidence, which can be used in economic modelling as an alternative to exaggerated rationality assumptions.

We have identified a hierarchy of dynamic processes which shape economic behavior. I name these processes in the order of increasing speed:

1. (the slowest process) gene substitution by mutation,
2. adaptation of genotype frequencies without mutation,
3. cultural transmission from generation to generation,
4. learning (including imitation).

The speed differences are so great that for many purposes an adiabatic approximation seems to be justified. Adiabatic approximation means that if we look at one of the four processes, results of slower processes can be taken as fixed and quicker processes can be assumed to reach equilibrium instantly.

Learning, the quickest process, is the most important one for economics. Day to day price movements on the stock exchange and other competitive processes involve learning and imitation. For slower dynamic phenomena like economic development, cultural transmission from generation to generation is also important.

The two processes of biological evolution have shaped the inherited components of economic behavior. Gene substitution by mutation maximizes fitness, but slowly and under structural constraints. Adaptation of genotype frequencies without mutation is nonoptimizing.

It is interesting to speculate on the evolution of behavioral tendencies. One may, for example, construct theories on the influence of prehistoric or even prehuman environmental factors on mechanisms of cultural evolution. However, such speculations, as interesting as they may be, are no substitute for empirical research. It makes no sense to speculate on the evolution of unicorns unless unicorns have been found in nature. Biological theory cannot be used as an instrument to discover facts by armchair reasoning.

We have to do empirical research if we want to gain knowledge on the structure of human economic behavior. In order to replace unrealistic rationality assumptions, we need theories of bounded rationality. As I have already said, we need quantitative theories which can replace the usual rationality assumptions in economic models.

In the near future theories of limited range which apply to restricted areas of experimental research have to be expanded. A number of such theories already can be found in the literature. Some of them have been mentioned in the discussion. It is hoped that eventually many theories of limited range will grow together and evolve into a comprehensive picture of the structure of human economic behavior. Only painstaking experimental research can bring us nearer to this goal.

I close the discussion now. Whoever wants to add something must do this in private conversations.

References

ALLAIS, M. (1953). "Le Comportement de l'Homme Rationnel Devant le Risque Critique des Postulates et Axiomes de l'Ecole Americaine," *Econometrica* **21**, 503–546.

ARROW. K. J. (1962). "The Economic Implications of Learning by Doing," *Rev. Econ. Stud.* **29**, 155–173.

AUMANN, R. I., AND SORIN, S. (1989). "Cooperation and Bounded Recall," *Games Econ. Behav.* **1**, 5–39.

AXELROD, R. (1976). *Structure of Decision: The Cognitive Maps of Political Elites*. Princeton, NJ: Princeton Univ. Press.

BELL, D. E. (1982). "Regret in Decision Making under Uncertainty," *Oper. Res.* **30**, 961–981.

BERG, C. C. (1973). "Individuelle Entscheidungsprozesse: Laborexperimente und Computersimulation," Wiesbaden.

BERG, C. C. (1974). "Individual Decision Concerning the Allocation of Resources for Projects with Uncertain Consequences," *Manag. Sci.* **21**, 98–105.

BOYD, R., AND RICHERSON, P. I. (1985). *Culture and the Evolutionary Process*. Chicago/London: Univ. of Chicago Press.

BUSH, R. R., AND MOSTELLER, F. (1955). *Stochastic Models of Learning*. New York: Wiley.

CAVALLI-SFORZA, L. L., AND FELDMAN, M. W. (1981). *Cultural Transmission and Evolution: A Quantitative Approach*. Princeton, NJ: Princeton Univ. Press.

CROSS, J. G. (1983). *A Theory of Adaptive Economic Behavior*. Cambridge/London/New York: Cambridge Univ. Press.

CYERT, R. M., AND MARCH, J. G. (1963). *A Behavioral Theory of the Firm*. Englewood Cliffs, NJ: Prentice–Hall.

ESHEL, I. (1988). "Game Theory and Population Dynamics in Complex Genetical Systems: The Role of Sex in Short Term and in Long Term Evolution," Working Paper No. 24, Game Theory in the Behavioral Sciences, ZiF, Universität Bielefeld.

ESHEL, I., AND FELDMAN, W. M. (1984). "Initial Increase of New Mutants and Some Continuity Properties of ESS in Two Locus Systems," *Amer. Naturalist* **124**(5), 631–640.

FISHBURN, P. C. (1982). "Nontransitive Measurable Utility," *J. Math. Psychol.* **26**, 31–67.

GALTON, F. (1989). *Natural Inheritance.* New York: Macmillan Co.

HAMMERSTEIN, P., AND RIECHERT, S. E. (1988). "Payoffs and Strategies in Territorial Contests of Two Ecotypes of Spider, *Agelonopsis Aperta*," *Evol. Ecol.* **2**, 115–138.

KAHNEMAN, D., AND TVERSKY, A. (1984). "Prospect Theory: An Analysis of Decision under Risk," *Econometrica* **47**, 263–291.

KALAI, E., AND STANFORD, W. (1987). "Finite Rationality and Interpersonal Complexity in Finitely Repeated Games," *Econometrica* **56**, 397–410.

KREPS, D., MILGROM, P., ROBERTS, J., AND WILSON, R. (1982). "Rational Cooperation in the Finitely Repeated Prisoner's Dilemma," *J. Econ. Theory* **27**, 245–252.

LIVERMAN, V. (1988). "External Stability and ESS: Criteria for Initial Increase of New Mutant Allele," *Math. Biol.* **26**, 477–485.

LOOMES, G., AND SUGDEN, R. (1982). "Regret Theory: An Alternative Theory of Rational Choice under Uncertainty," *Econ. J.* **92**, 805–824.

MALAWSKI, M. (1989). "Some Learning Processes in Population Games," Unpublished dissertation, University of Bonn.

MAYNARD SMITH, J. (1982). *Evolution and the Theory of Games.* Cambridge: Cambridge Univ. Press.

MAYNARD SMITH, J., AND PRICE, G. R. (1973). "The Logic of Animal Conflict," *Nature* **246**, 15–18.

MORAN, P. A. P. (1964). "On the Nonexistence of Adaptive Topographgies," *Amer. Human Genet.* **27**, 343–383.

NELSON, R., AND WINTER, S. G. (1982). *An Evolutionary Theory of Economic Change.* Cambridge, MA/London: Belknap Press of Harvard Univ. Press.

NEYMAN, A. (1985). "Bounded Complexity Justifies Cooperation in the Finitely Repeated Prisoner's Dilemma," Manuscript.

PLOTT, C. R. (1982). "Industrial Organization Theory and Experimental Economics," *J. Econ. Lit.* **20**, 1485–1527.

SAVAGE, L. J. (1954). *The Foundation of Statistics.* New York: Wiley.

SCHUMPETER, J. A. (1934). *The Theory of Economic Development.* Cambridge, MA: Harvard Univ. Press.

SELTEN, R. (1967). "Invetitionsverhalten im Oligopolexperiment," in *Beiträge zur experimentellen Wirtschaftsforschung* (H. Sauermann, Ed.), Vol. 1, pp. 60–102. Tübingen: Mohr (Paul Siebeck).

SELTEN, R. (1987). "Equity and Coalition Bargaining," in *Experimentation in Economics* (A. Roth, Ed.). Cambridge/New york: Cambridge Univ. Press.

SELTEN, R., AND STOECKER, R. (1986). "End Behavior in Sequences of Finite Prisoner's Dilemma Supergames," *J. Econ. Behav. Organ.* **7**, 47–70.

SIMON, H. A. (1957). *Models of Man.* New York: Wiley.

SMITH, V. L. (1980). "Microeconomic Systems as a Science," *Amer. Econ. Rev.* **5**, 923–955.

TVERSKY, A., AND KAHNEMAN, D. (1982a). "Belief in the Law of Small Numbers," in *Judgment and Uncertainty, Heuristics and Biases* (D. Kahneman, P. Slovic, and A. Tversky, Eds.), pp. 23–31. Cambridge/London/New York: Cambridge Univ. Press.

TVERSKY, A., AND KAHNEMAN, D. (1982b). "Judgements of and by Representativeness," in *Judgment and Uncertainty, Heuristics and Biases* (D. Kahneman, P. Slovic, and A. Tversky, Eds.), pp. 84–98. Cambridge/London/New York: Cambridge Univ. Press.

TYSZKA, T. (1983). "Contextual Multiattribute Decision Rules," in *Human Decision Making* (L. Sjöberg, T. Tyszka, and J. Wise, Eds.). Bodafors: Doxa.

WEBER, M., EISENFÜHR, F., AND VON WINTERFELDT, D. (1988). "The Effects of Splitting Attributes on Weights in Multiattribute Utility Measurement," *Manag. Sci.* **34,** 432–445.

WINTER, S. G. (1971). "Satisficing, Selection and the Innovating Remnant," *Quart. J. Econ.* **85,** 237–261.

[7]

ANTICIPATORY LEARNING IN TWO-PERSON GAMES

by

Reinhard Selten

Summary

A learning process for 2-person games in normal form is introduced. The game is assumed to be played repeatedly by two large populations, one for player 1 and one for player 2. Every individual plays against changing opponents in the other population. Mixed strategies are adapted to experience. The process evolves in discrete time.

All individuals in the same population play the same mixed strategy. The mixed strategies played in one period are publicly known in the next period. The payoff matrices of both players are publicly known.

In a preliminary version of the model, the individuals increase and decrease probabilities of pure strategies directly in response to payoffs against last period's observed opponent strategy. In this model, the stationary points are the equilibrium points, but genuinely mixed equilibrium points fail to be locally stable.

On the basis of the preliminary model an anticipatory learning process is defined, where the individuals first anticipate the opponent strategies according to the preliminary model and then react to these anticipated strategies in the same way as to the observed strategies in the preliminary model. This means that primary learning effects on the other side are anticipated, but not the secondary effects due to anticipations in the opponent population.

Local stability of the anticipatory learning process is investigated for regular games, i.e., for games where all equilibrium points are regular. A stability criterion is derived which is necessary and sufficient for sufficiently small adjustment speeds. This criterion requires that the eigenvalues of a matrix derived from both payoff matrices are negative.

It is shown that the stability criterion is satisfied for 2×2-games without pure strategy equilibrium points, for zero-sum games and for games where one player's payoff matrix is the unit matrix and the other player's payoff matrix is negative definite. Moreover, the addition of constants to rows or columns of payoff matrices does not change stability.

The stability criterion is related to an additive decomposition of payoffs reminiscent of a two way analysis of variance. Payoffs are decomposed into row effects, column effects and interaction effects. Intuitively, the stability criterion requires a preponderance of negative covariance between the interaction effects in both players' payoffs.

The anticipatory learning process assumes that the effects of anticipations on the other side remain unanticipated. At least for completely mixed equilibrium points the stability criterion remains unchanged, if anticipations of anticipation effects are introduced.

1. Introduction

A loose description of the anticipatory learning process investigated here has already been given in the summary. The aim of the paper is a speculative attempt to develop an idealized descriptive theory. The effort remains speculative since the task of experimental validation is left to the future. The theory is idealized since in real laboratory situations populations cannot be very large and individual differences must be expected to matter. Nevertheless, our assumption on the structure of behavior may be basically correct and the stability criterion derived here may have some predictive relevance. At least it may serve to guide the planning of experiments.

1.1 Decision Emergence

The modelling efforts of this paper are based on a picture of limited rationality which involves "decision emergence" rather than "decision making". Mixed strategies are thought of as behavioral dispositions which do not arise from rational calculations. It is assumed that the individual has no direct understanding of the way in which decisions come about. A person without special training does not know how the body performs physiological functions like digestion. Similarly, the decision process has important parts which are *inaccessible* in the sense that introspection is unable to reveal the mechanism. *Decisions are not made, decisions emerge*. The conscious mind supplies inputs to the black box containing the inaccessible parts of the decision mechanism and observes the emergence of decisions as an output.

Anticipation also fits into this picture. An individual can imagine to be in the role of the other player. This amounts to the use of one's own decision mechanism and its inaccessible parts for the prediction of the opponent's behavior. Prediction does not involve conscious calculations, *anticipations emerge* in the same way as decisions do.

A lack of understanding of the internal and the external environment prevents optimization. This explains the possibility of non-optimal behavior. Computational complexity is not the immediate reason for the failure to optimize, even if it may explain the evolutionary adaptiveness of limited rationality.

Of course, an individual who is in the possession of a numerically specified correct model of the learning process should optimize against it. Therefore, the anticipations in our theory should not be misunderstood as representing approximatively correct conscious calculations. In order to emphasize this point, we use the word "anticipation" rather than the customary term "expectation".

1.2 Remarks on the Literature

The notion of an equilibrium point has been introduced as a normative concept (Nash, 1951). Nevertheless, predictions based on equilibrium points (or, more precisely, saddlepoints) are surprisingly successful in some experiments (O'Neill, 1987). Encouraged by the predictive success of a learning theory approach in a different experimental context (Selten and Stoecker, 1986), the author thinks that learning processes are a promising modelling tool for the description of game playing behavior.

The process of fictitious play (Brown, 1951; Robinson, 1951) has not been proposed as a serious model of game learning but rather as an algorithm for the computation of saddlepoints. The usual interpretation as a process of repeated play between the same two individuals does not look plausible. However, a reinterpretation in the population framework employed here is less objectionable.

Shapley has shown instability of fictitious play for a class of 3×3-games; Miyasawa has proved convergence for all 2×2-games (Miyasawa, 1961). Rosenmüller has established conditions for limit cycles (Rosenmüller, 1967). Due to a lack of continuity, it seems to be difficult to obtain general results on stability conditions for fictitious play.

Since Bush and Mosteller published their path-breaking book, many learning models have been presented in the psychological literature (Bush and Mosteller, 1955). Cross considers generalizations of such models which could be applied to game learning (Cross, 1983).

A dynamic process considered in the biological literature as a model of natural selection in symmetric two-person games (Taylor and Jonker, 1978) leads to evolutionary stability (Maynard Smith and Price, 1973; Maynard Smith, 1982) as a sufficient condition for dynamic stability. A reinterpretation as a learning process is possible. This interpretation involves only one population for both players. The symmetry restriction is important. Asymmetric games can be symmetrized by a random assignment of player roles, but evolutionarily stable strategies of the symmetrized game must be pure (Selten, 1980). Further references to the literature can be found in a recent survey article (Bomze, 1986).

Experiments performed by Milinski show that fishes learn to distribute themselves on two food sources in proportion to the amount of nutrition offered per time unit (Milinski, 1979). Game learning processes are of interest for the description of animal behavior. Harley has developed an interesting biological learning model (Harley, 1981). A more careful examination of the convergence properties of this model would be desirable (Selten and Hammerstein, 1984). Crawford has examined a class of reasonable learning models for which he has shown that genuinely mixed equilibrium points are almost always locally unstable (Crawford, 1985).

None of the theories mentioned above involves an element of anticipation. It seems to be possible to construct anticipatory modifications of some of the models in the literature, but no attept in this direction will be made here.

1.3 Linearity Properties

The modelling approach taken in this paper is based on the idea that as much linearity as possible should be imposed on the functional form of dynamic relationships. Non-negativity constraints on probabilities prevent full linearity in the space of mixed strategy combinations. However, it is possible to achieve linearity in every region of the space where a given set of constraints is binding.

Unfortunately, it is not possible to concentratge on the interior of the space of mixed strategies and to ignore border problems unless one is satisfied with the investigation of local stability properties of completely mixed equilibrium points. Border problems matter for other mixed equilibrium points.

As long as we do not have any empirical reason to prefer one functional form over another, it seems to be reasonable to impose linearity properties for the sake of mathematical convenience. Alternatively, one might want to avoid to specify functional forms. It seems to be difficult to obtain explicit stability criteria if one takes this approach. Therefore, a specific functional form is used for the anticipatory learning process proposed here.

2. Regular Equilibrium Points

In the following definitions and notations concerning two person games in normal form will be introduced. The concept of a regular equilibrium point and its properties will be discussed.

2.1 Basic Definitions and Notations

A *two-person game* in normal form can be described as a pair (A,B) of two n×m-matrices. Player 1 has the pure strategies 1,...,n and player 2 has the pure strategies 1,...,m. The matrices

$$A = (a_{ij})_{n \times m} \qquad (1)$$

and

$$B = (b_{ij})_{n \times m} \qquad (2)$$

are the *payoff matrices* of players 1 and 2, respectively. a_{ij} and b_{ij} are the *payoffs* for players 1 and 2, respectively, if player 1 plays i and player 2 plays j. The set $\{1,...,n\}$ of player 1's pure strategies is denoted by N and the set $\{1,...,m\}$ of player 2's pure strategies is denoted by M.

In this paper, all definitions and all statements in lemmata and theorems will be relative to an arbitrary but fixed two-person game (A,B) unless something else is said explicitly. Whenever the word "game" is used without further specification, we mean a two-person game in normal form.

Mixed strategies are described by column vectors with non-negative components summing up to 1. The number of components is n for player 1 and m for player 2. The k-th component is the probability of choosing the pure strategy k. If p denotes a mixed strategy, then p_k denotes the k-th component of p. The same convention is also applied if other letters q, r, or s denote mixed strategies of player 1 or 2. The set of all mixed strategies of player 1 is denoted by P and the set of all mixed strategies of player 2 is denoted by Q. A *mixed strategy pair* (p,q) is defined by $p \in P$ and $q \in Q$.

The *payoffs* H(p,q) and K(p,q) of players 1 and 2 for a mixed strategy pair (p,q) are as follows:

$$H(p,q) = p^T A q \qquad (3)$$

$$K(p,q) = p^T B q. \qquad (4)$$

Here and in the remainder of the paper, the upper index T indicates transposition.

Pure strategies are looked upon as special mixed strategies of the same player. The pure strategy k is identified with that mixed strategy, whose k-th component is 1. Accordingly, we use the notation H(i,q) for player 1's payoff obtained, if player 1 plays i and player 2 plays q.

The *support* of a mixed strategy $p \in P$ is the set I of all $i \in N$ with $p_i > 0$. Analogously, the *support* of $q \in Q$ is the set J of all $j \in M$ with

$q_j > 0$. If I is the support of p and J is the support of q, then (I,J) is called the *support* of the mixed strategy pair (p,q).

A mixed strategy is called *genuinely mixed* if its support has at least two elements. A mixed strategy pair (p,q) is called *genuinely mixed* if p and q are genuinely mixed. A mixed strategy is called *completely mixed* if its support is the set of all pure strategies of the player concerned. A mixed strategy pair (p,q) is called *completely mixed* if p and q are completely mixed.

2.2 Best Replies

A mixed strategy $r \in P$ of player 1 is a *best reply* of player 1 to $q \in Q$ if we have:

$$H(r,q) = \max_{p \in P} H(p,q). \tag{5}$$

Analogously, a mixed strategy $s \in Q$ is a *best reply* of player 2 to $p \in P$ if we have

$$K(p,s) = \max_{p \in P} K(p,q). \tag{6}$$

It is well known that $r \in P$ is a best reply to $q \in Q$ if and only if every pure strategy i with $r_i > 0$ is a best reply to q. An analogous assertion holds for best replies of player 2.

2.3 Equilibrium Point

An equilibrium point is a mixed strategy pair (r,s) with the property that r is a best reply to s and s is a best reply to r.

Let (r,s) be an equilibrium point and let $E = H(r,s)$ and $F = K(r,s)$ be the corresponding equilibrium payoffs of both players. Let (I,J) be the support of (r,s). With the help of the property of best replies mentioned above, it can be seen that the following equations must be satisfied:

$$\sum_{j \in J} a_{ij} s_j = E \quad \text{for } i \in I \tag{7}$$

$$\sum_{i \in I} b_{ij} r_i = F \quad \text{for } j \in J \tag{8}$$

$$\sum_{i \in I} r_i = 1 \tag{9}$$

$$\sum_{j \in J} s_j = 1. \tag{10}$$

We shall use the notation $|S|$ for the number of elements of a finite set S. Eqs. (7), (8), (9) and (10) can be looked upon as a system of $|I|+|J|+2$ variables r_i with $i \in I$ and s_j with $j \in J$ and E and F. By assumption, the system has at least one solution, but it may happen that there are infinitely many solutions which fill a linear subspace of the $(|I|+|J|+2)$-dimensional vector space for these variables.

An equilibrium point (r,s) is calles *isolated* if an open neighborhood V of (r,s) can be found such that V contains no other equilibrium points. It is clear that the system (7), (8), (9) and (10) has exactly one solution if (r,s) is isolated.

An equilibrium point (r,s) is called *quasistrong* (Harsanyi, 1973) if in addition to (7), (8), (9) and (10) the following inequalities are satisfied:

$$\sum_{j \in J} a_{ij} s_j < E \qquad \text{for } i \in N \setminus I \qquad (11)$$

$$\sum_{i \in I} b_{ij} r_i < F \qquad \text{for } j \in M \setminus J. \qquad (12)$$

These inequalities mean that at the equilibrium point pure strategies with zero probabilities fail to be best replies.

2.4 Regularity

A game (A,B) is called *regular* if it is isolated and quasistrong. A game (A,B) is called *regular* if all equilibrium points of (A,B) are regular. It is clear that in the space of all two-person games the irregular games form a set of lower dimension. In this sense, one can say that only degenerate cases are excluded by the definition of regularity.

It is an interesting property of equilibrium points that the system (7), (8), (9) and (10) is composed of two independent subsystems, namely (7) and (10) on the one hand and (8) and (9) on the other hand. The payoffs of player 1 determine the strategy of player 2 and vice versa. We refer to this property as the *heterodependence* of equilibrium strategies. Not a player's own payoffs determine his equilibrium probabilities of pure strategies, but those of the other player.

If (r,s) is regular, then we must have $|I| = |J|$; otherwise one of both subsystems would have more unknowns than equations and therefore could not have a unique solution. It also follows by the definition of regularity that two regular equilibrium points cannot have the same support. Consequently, a regular game (A,B) has only a finite number of equilibrium points.

2.5 The Restricted Game

Let I be a non-empty subset of N and let J be a non-empty subset of J. Let $\underline{A}$ and $\underline{B}$ be the submatrices of A and B respectively, resulting from A and B by the removal of all rows with indices not in I and all columns with indices not in J. We call the game $(\underline{A},\underline{B})$ the *restriction* of (A,B) to (I,J). In $\underline{A}$ and $\underline{B}$ rows have new numbers $1,\ldots,|J|$ and columns have new numbers $1,\ldots,|J|$. In the transition from (A,B) to $(\underline{A},\underline{B})$ pure strategies are renumbered accordingly. the elements of $\underline{A}$ are denoted by $\underline{a}_{ij}$ and those of $\underline{B}$ by $\underline{b}_{ij}$.

Let (I,J) be the support of a regular equilibrium point (r,s). In this case, the restriction $(\underline{A},\underline{B})$ to (I,J) is also called the *restricted game* of (r,s). If in (r,s) the positive components of r and s are renumbered in the same way as the pure strategies in I and J respectively, and the other components are left out, one obtains a pair of mixed strategies (r,s) for the restricted game. Obviously, $(\underline{r},\underline{s})$ is a regular equilibrium point of the restricted game $(\underline{A},\underline{B})$.

A game in which both players have the same number of pure strategies is called *quadratic*. The restricted game of a regular equilibrium point is quadratic.

3. The Preliminary Model

As a preparation for the anticipatory learning process to be defined later, a preliminary model is introduced in this section. The interpretation is based on a population framework common to both models.

3.1 Population Framework

As has been explained in the summary, learning is supposed to take place in two large populations, one for player 1 and the other for player 2. The same individual always plays in the role of the same player. Time is a succession of discrete periods. In every period, the individuals of both populations are randomly matched into pairs playing the game. The same game (A,B) is played in every period.

The exposition will be based on the assumption that all individuals in the same role behave in exactly the same way. In every period all individuals of the same population play the same mixed strategy, also referred to as the mixed strategy of the concerned player. Of course, this is a very restrictive assumption. In view of the linearity properties of our models, a more generous interpretation seems to be possible which permits some strategic diversity within a population. However, this idea will not be made precise.

The populations are thought of as infinite. This means that no distinction is made between the relative frequency of a pure strategy and its probability of being played. We assume that these frequencies are observed, or, in other words, mixed strategies are observed before the beginning of the next period.

It is assumed that all individuals know both payoff matrices. Knowledge of own payoffs would be sufficient for the preliminary model, but anticipation requires knowledge of the opponent's payoffs.

The learning process is described by a piecewise linear system of difference equations. The interpretation is based on the decision emergence view explained in the introduction.

3.2 Notation

Time is modelled as a succession of periods $t = 0,1,2,\ldots$. The common mixed strategy in population 1, or, shortly, player 1's mixed strategy at time t is denoted by $p(t)$. Similarly, player 2's mixed strategy at time t is denoted by $q(t)$. The notation $p_i(t)$ is used for the i'th component of $p(t)$. Similarly, $q_j(t)$ is the j-th component of $q(t)$. Often we shall suppress the dependence on t and simply write p and q instead of $p(t)$ and $q(t)$. In such cases, p_+ and q_+ will be used instead of $p(t+1)$ and $q(t+1)$, respectively, while the notation p_{i+} and q_{j+} is used for components. The same conventions will be applied to other time dependent variables.

We shall also use the notation Δp for the first difference vector $p_+ - p$ and Δq for $q_+ - q$. The difference operator Δ is also applied to components; thus Δp_i stands for $p_{i+} - p_i$. The same notation will also be used for other time dependent variables.

3.3 A Constrained Proportionality Principle

In the following, we shall present intuitive arguments for the specific way in which the preliminary model deals with non-negativity constraints. The formulas of the model are hard to interpret directly and may seem to be arbitrary without an intuitive justification. Therefore, we first introduce an intuitively appealing *constrained proportionality principle* which then completely determines the structure of the model.

The game being played is (A,B). We focus on player 1. It will be convenient to use a shorter notation for player 1's payoffs obtained for his pure strategies i against player 2's current strategy q. Define:

$$E_i = H(i,q) \qquad \text{for } i = 1,\ldots,n. \tag{13}$$

The quantities E_i measure the current success of player 1's pure strategies. We want to construct a model where the increase Δp_i of player 1's probability of choosing i is determined by these success measures $E_1,\ldots,E_n$. It is natural to require that ceteris paribus Δp_i should be the greater, the greater E_i is, at least as long as non-negativity constraints do not prevent this. Accordingly, for $p_{i+} > 0$ and $p_{k+} > 0$ we should have:

$$\Delta p_i > \Delta p_k \quad \text{for} \quad E_i > E_k. \tag{14}$$

Since we want to achieve as much linearity as possible, (14) suggests the following *proportionality principle*:

$$\Delta p_i - \Delta p_k = \alpha(E_i - E_k) \quad \text{for} \quad p_{i+} > 0 \quad \text{and} \quad p_{k+} > 0 \tag{15}$$

where α is a positive parameter. (15) can be rewritten as follows:

$$\Delta p_i = \Delta p_k + \alpha(E_i - E_k) \quad \text{for} \quad p_{i+} > 0 \quad \text{and} \quad p_{k+} > 0. \tag{16}$$

Obviously, we must have

$$\Delta p_i \geq - p_i. \tag{17}$$

Otherwise, p_{i+} would be negative. If (16) yields a smaller vlaue than $-p_i$ for Δp_i, then (16) cannot apply. In this case, it is natural to require that Δp_i should assume its minimum value $-p_i$. If one wants to achieve as much linearity as possible, deviations from proportionality should not be permitted for any other reason. This leads to the following *constrained proportionality principle*:

$$\Delta p_i = \max[-p_i, \Delta p_k + \alpha(E_i - E_k)] \quad \text{for} \quad p_{k+} > 0 \quad \text{and} \quad i = 1,\ldots,n. \tag{18}$$

Assume that the constrained proportionality principle holds. Consider the case $p_{i+} > 0$ and $p_{k+} > 0$ where (18) assumes the form (16). Since the same value of Δp_i is obtained for every k with $p_{k+} > 0$, the expression $\alpha E_k - \Delta p_k$ must assume the same value u for all these k: A number u exists such that the following is true:

$$u = \alpha E_k - \Delta p_k \quad \text{for all} \quad k \quad \text{with} \quad p_{k+} > 0. \tag{19}$$

With the help of (19), we obtain an equivalent reformulation of the constrained proportionality principle: A number u exists for which the following is true:

$$\Delta p_i = \max[-p_i, \alpha E_i - u] \quad \text{for} \quad i = 1,\ldots,n. \tag{20}$$

In the transition from (19) to (20), we have made use of the obvious fact that we must have $p_{k+} > 0$ for at least one $k \in N$. The p_{i+} must sum up to 1. In view of the definition of Δp_i, Eq. (20) can be rewritten as follows:

$$p_{i+} = \max[0, p_i + \alpha E_i - u] \quad \text{for} \quad i = 1,\ldots,n. \tag{21}$$

This is the reformulation of the constrained proportionality principle to be used in the model. Of course, the same principle is also applied to the Δq_j.

The number u can be thought of as an *auxiliary variable*. It depends on $p_1,\ldots,p_n$ abd $E_1,\ldots,E_n$. As we shall see, u is uniquely determined by the condition that the p_{i+} sum up to 1.

3.4 Definition of the Preliminary Model

In the preliminary model, the mixed strategies p_+ and q_+ for period $t+1$ depend as follows on the mixed strategies p and q for period t:

$$p_{i+} = \max[0, p_i + \alpha H(i,q) - u] \quad \text{for } i = 1,\ldots,n \tag{22}$$

$$q_{j+} = \max[0, q_j + \beta K(p,j) - v] \quad \text{for } j = 1,\ldots,m \tag{23}$$

with

$$\sum_{i=1}^{n} \max[0, p_i + \alpha H(i,q) - u] = 1 \tag{24}$$

$$\sum_{j=1}^{m} \max[0, q_j + \beta K(p,j) - v] = 1 \tag{25}$$

where α and β are positive parameters. The condition that the p_{i+} must sum up to 1 is expressed by (24). The interpretation of (25) is analogous.

Usually, a system of difference equations explicitly describes the dynamic relationships. Eqs. (22) - (25) provide only an implicit description. Nevertheless, the description is complete. This is shown by the following lemma.

<u>Lemma 1</u>: Let α and β be positive constants. Then, for given p and q next period's strategies p_+ and q_+ as well as the auxiliary variables u and v are uniquely determined by (22), (23), (24) and (25). Moreover, p_+, q_+, u and v are continous functions of p and q.

<u>Proof</u>: Let $\underline{c}$ be the minimum of all a_{ij} and let $\bar{c}$ be the maximum of all a_{ij}. Let $\underline{u}$ be $\alpha\underline{c}$ and let $\bar{u}$ be $\alpha\bar{c}$. For all mixed strategy pairs (p,q) the left-hand side of (24) is at least as great as 1 for $u = \underline{u}$ and at most as great as 1 for $u = \bar{u}$. Moreover, the left-hand side of (24) is non-increasing in u. Therefore, for given (p,q) Eq. (24) has a unique solution u which lies in the closed interval $\underline{u} \leq u \leq \bar{u}$. Consider a sequence $(p^1,q^1),(p^2,q^2),\ldots$ of mixed strategy pairs which converges to (p,q). For $k = 1,2,\ldots$ let u^k be the value of u determined

by (p^k, q^k). Since the interval $\underline{u} \leq u \leq \bar{u}$ is closed and bounded, the sequence $u^1, u^2, \ldots$ has an accumulation point u. It follows by the continuity of H that all accumulation points u of $u^1, u^2, \ldots$ satisfy (24) together with (p,q). Since there is only one such u, the sequence $u^1, u^2, \ldots$ converges to this u. Consequently, u is a continuous function of p and q. An analogous argument shows that v is a continuous function of p and q. It follows by (22) and (23) together with the continuity of H and K that p_+ and q_+ are uniquely determined by p and q and depend continuously on p and q.

3.5 Stationarity

We say that a mixed strategy pair (p,q) is a *stationary point* of the preliminary model if we have $p_+ = p$ and $q_+ = q$ for the strategies p_+ and q_+ determined by (22) to (25). Lemma 2 will show that in the preliminary model equilibrium is necessary and sufficient for stationarity.

<u>Theorem 1</u>: A mixed strategy pair (p,q) is a stationary point of the preliminary model if and only if (p,q) is an equilibrium point. This is true for every parameter pair (α, β) with $\alpha > 0$ and $\beta > 0$ and every game (A,B).

<u>Proof</u>: Let (I,J) be the support of (p,q). (For the definition of "support" see 2.1.) Assume that (p,q) is an equilibrium point. Then, $H(i,q) = H(p,q)$ holds for all $i \in I$ and $K(p,j) = K(p,q)$ holds for all $j \in J$. Therefore,

$$u = \alpha H(p,q) \qquad (26)$$

and

$$v = \beta K(p,q) \qquad (27)$$

solve (24) and (25). The stationarity of (p,q) follows by (22) and (23).

Now, assume that (p,q) is stationary. Then, it follows by (22) that $\alpha H(i,q) = u$ holds for all $i \in I$. Analogously, $\beta K(p,j) = v$ holds for all $j \in J$. Consequently, we must have (26) and (27). It follows by (22) and (23) together with (26) and (27) that we have:

$$H(i,q) = H(p,q) \qquad \text{for } i \in I \qquad (28)$$

$$K(p,j) = K(p,q) \qquad \text{for } j \in J \qquad (29)$$

$$H(i,q) \leq H(p,q) \qquad \text{for } i \in N \backslash I \qquad (30)$$

$$K(p,j) \leq K(p,q) \qquad \text{for } j \in M \backslash J. \qquad (31)$$

This shows that (p,q) is an equilibrium point.

3.6 Local Stability

In this paper, a strong notion of local stability will be applied which includes both attractivity and Liapunov stability. Crawford seems to work with the same kind of local stability even if this is not explicitly spelled out in his paper (Crawford, 1985). It is difficult to exclude the possibility of a snap-back repellor (see Gabisch and Lorenz, 1987:184). A small perturbance may first take the process far away, but eventually it may move to a pure strategy pair from which the stationary point is reached in one step. Clearly, this kind of behavior does not conform to the intuitive notion of local stability.

For every $\varepsilon > 0$, the ε-neighborhood $U_\varepsilon(p,q)$ of (p,q) is defined as the set of all mixed strategy pairs (r,s) whose distance from (p,q) is smaller than ε in the following sense:

$$\sum_{i=1}^{n} (p_i - r_i)^2 + \sum_{j=1}^{m} (q_j - s_j)^2 < \varepsilon^2. \tag{32}$$

Whenever we speak of an *open* neighborhood of (p,q), we mean a neighborhood which is relatively open in the set $P \times Q$ of all mixed strategy pairs. In this sense $U_\varepsilon(p,q)$ is open for every $\varepsilon > 0$.

We say that a stationary point (p,q) is *locally (asymptotically) stable* if for every $\varepsilon > 0$ an open neighborhood V of (p,q) with $V \subseteq U_\varepsilon(p,q)$ can be found such that for

$$(p(0),q(0)) \in V \tag{33}$$

we always have

$$(p(t),q(t)) \in U_\varepsilon(p,q) \quad \text{for } t = 1,2,\ldots \tag{34}$$

and

$$\lim_{t \to \infty} (p(t),q(t)) = (p,q). \tag{35}$$

3.7 The Linear System Connected to a Regular Equilibrium Point

Let (r,s) be a regular equilibrium point and let (I,J) be the support of (r,s). As has been pointed out in 2.4, regularity implies $|I| = |J|$. Define $h = |I|$.

If $h < n$, every ε-neighborhood of (r,s) contains pairs of mixed strategies whose support is different from (I,J). Therefore, the preliminary model is not linear at (r,s). However, if one only considers pairs of mixed strategies (p,q) with the property that (p,q) and (p_+,q_+) have the support (I,J), one obtains a linear relationship between (p_+,q_+) and (p,q). In the following, this linear relationship will be examined in more detail.

Let S be the set of all pairs of mixed strategies (p,q) with support (I,J) for which (22) - (25) yield pairs (p_+,q_+) whose support is (I,J), too. It follows by (22) - (25) that for $(p,q) \in S$ we have:

$$p_{i+} = p_i + \alpha H(i,q) - u \quad \text{for } i \in I \tag{36}$$

$$q_{j+} = q_j + \beta K(p,j) - v \quad \text{for } j \in J. \tag{37}$$

Moreover, p and p_+ have the property that the components with indices in I sum up to 1. An analogous statement holds for q and q_+. Summation over $i \in I$ in (36) and over $j \in J$ in (37) and division by $h = |I| = |J|$ yields:

$$u = \frac{\alpha}{h} \sum_{i \in I} H(i,q) \tag{38}$$

$$v = \frac{\beta}{h} \sum_{j \in J} K(p,j). \tag{39}$$

It is our aim to express the linear relationship between (p_+,q_+) and (p,q) in matrix notation. This is done with the help of the payoff matrices $\underline{A}$ and $\underline{B}$ of the restricted game $(\underline{A},\underline{B})$ of (r,s). For this purpose, we introduce the following *notational conventions:* For every mixed strategy p of player 1 with support I, let $\underline{p}$ be that mixed strategy of player 1 for the restricted game which results from p if first all components with indices not in I are removed and then the remaining components are renumbered from $1,\ldots,h$ in the order of the previous numbering. For every mixed strategy q of player 2 with support J a mixed strategy $\underline{q}$ of player 2 in $(\underline{A},\underline{B})$ is defined analogously. We call $\underline{p}$ and $\underline{q}$ the *restrictions* of p and q, respectively. The symbols $\underline{p}_+$, $\underline{q}_+$, $\underline{r}$ and $\underline{s}$ stand for the restrictions of p_+, q_+, r and s, respectively. The restriction of a pair (p,q) is the pair $(\underline{p},\underline{q})$ of its restrictions. Thus, $(\underline{r},\underline{s})$ is the restriction of (r,s). As we have seen in 2.5, the pair $(\underline{r},\underline{s})$ is a regular equilibrium point of the restricted game $(\underline{A},\underline{B})$.

Let $\underline{D}$ be the $h \times h$-unit matrix with 1's on the diagonal and zeros everywhere else:

$$\underline{D} = \begin{bmatrix} 1 & \cdots & 0 \\ \cdot & \cdot & \cdot \\ \cdot & \cdot & \cdot \\ \cdot & \cdot & \cdot \\ 0 & \cdots & 1 \end{bmatrix} \qquad (40)$$

and let $\underline{z}$ be the h-dimensional column vector whose components are all equal to 1:

$$\underline{z} = \begin{bmatrix} 1 \\ \cdot \\ \cdot \\ 1 \end{bmatrix} . \qquad (41)$$

Eqs. (38) and (39) can now be expressed in matrix notation

$$u = \frac{\alpha}{h} \underline{z}^T \underline{A} \underline{q} \qquad (42)$$

$$v = \frac{\beta}{h} \underline{z}^T \underline{B}^T \underline{p} . \qquad (43)$$

Eqs. (36) and (37) can be rewritten as follows:

$$\underline{p}_+ = \underline{p} + \alpha \underline{A} \underline{q} - \underline{z} u \qquad (44)$$

$$\underline{q}_+ = \underline{q} + \beta \underline{B}^T \underline{p} - \underline{z} v . \qquad (45)$$

Define

$$Z = \underline{D} - \frac{1}{h} \underline{z} \underline{z}^T \qquad (46)$$

and

$$\tilde{A} = Z \underline{A} \qquad (47)$$

$$\tilde{B}^T = Z \underline{B}^T . \qquad (48)$$

We substitute the right-hand side of (42) and (43) for u and v in (44) and (45) and then make use of (46), (47) and (48). This yields

$$\underline{p}_+ = \underline{p} + \alpha \tilde{A} \underline{q} \qquad (49)$$

$$\underline{q}_+ = \underline{q} + \beta \tilde{B}^T \underline{p} . \qquad (50)$$

Obviously, $\underline{z}\underline{z}^T$ is the h×h-matrix whose elements are all equal to 1. Therefore, the elements z_{ij} of Z are as follows:

$$z_{ij} = \begin{cases} \frac{h-1}{h} & \text{for } i = j \\ -\frac{1}{h} & \text{for } i \neq j. \end{cases} \quad (51)$$

In view of the structure of Z, it is clear that left multiplication by Z amounts to the subtraction of the column average from each element. Therefore, the elements $\tilde{a}_{ij}$ of $\tilde{A}$ depend as follows on those of $\underline{A}$:

$$\tilde{a}_{ij} = \underline{a}_{ij} - \frac{1}{h} \sum_{k=1}^{h} \underline{a}_{kj}. \quad (52)$$

Similarly, the elements $\tilde{b}_{ij}$ of $\tilde{B}$ are described by (53):

$$\tilde{b}_{ij} = \underline{b}_{ij} - \frac{1}{h} \sum_{k=1}^{h} \underline{b}_{ik}. \quad (53)$$

The number $\tilde{a}_{ij}$ can be interpreted as a measure of success of player 1's pure strategy i in $(\underline{A},\underline{B})$ against player 2's pure strategy j in $(\underline{A},\underline{B})$. Measuring success in this way involves a comparison with the average payoff for all pure strategies of player 1 in $(\underline{A},\underline{B})$ against player 2's strategy j in $(\underline{A},\underline{B})$. The $\tilde{b}_{ij}$ have an analogous interpretation. Accordingly, we call the $\tilde{a}_{ij}$ and $\tilde{b}_{ij}$ *comparative payoffs*. The matrices $\tilde{A}$ and $\tilde{B}$ are the *comparative payoff matrices* of players 1 and 2, respectively.

The game $(\tilde{A},\tilde{B})$ is called the *comparative payoff game* for (I,J). The definition of the comparative payoff game $(\tilde{A},\tilde{B})$ by (47) and (48) will also be applied to cases where (I,J) is not necessarily the support of a regular equilibrium point; of course, $|I| = |J| > 0$ must hold. In the case considered here, where (I,J) is the support of a regular equilibrium point (r,s), we also say that $(\tilde{A},\tilde{B})$ is the *comparative payoff game* of (r,s).

Since (r,s) is an equilibrium point with support (I,J), all components of $\underline{A}\,\underline{s}$ are equal. The same is true for $\underline{B}^T\underline{r}$. Therefore, it follows by (48) and (49) that we have:

$$\tilde{A}\,\underline{s} = \begin{bmatrix} 0 \\ \vdots \\ 0 \end{bmatrix} \quad (54)$$

$$\tilde{B}^T \underline{r} = \begin{bmatrix} 0 \\ \vdots \\ \vdots \\ 0 \end{bmatrix}. \tag{55}$$

It follows that $(\underline{r},\underline{s})$ is an equilibrium point of the comparative payoff game of (r,s) with zero equilibrium payoffs for both players. It can also be seen without difficulty that the equation system formed by (54) and (55), together with the conditions that for $\underline{r}$ and $\underline{s}$ components sum up to 1, has a unique solution if and only if the system formed by (7), (8), (9) and (10) has a unique solution. Therefore, $(\underline{s},\underline{r})$ is a regular equilibrium point of the comparative payoff game $(\tilde{A},\tilde{B})$.

With the usual notational conventions for composed vectors and matrices, the system formed by (49) and (50) can now be represented as follows:

$$\begin{bmatrix} \underline{p}_+ \\ \underline{q}_+ \end{bmatrix} = \begin{bmatrix} \underline{p} \\ \underline{q} \end{bmatrix} + \begin{bmatrix} 0 & \alpha\tilde{A} \\ \beta\tilde{B}^T & 0 \end{bmatrix} \begin{bmatrix} \underline{p} \\ \underline{q} \end{bmatrix} \tag{56}$$

where the zeros stand for zero-submatrices. The following lemma summarizes the results obtained above.

Lemma 2: Let (r,s) be a regular equilibrium point and let S be the set of all mixed strategy pairs (p,q) with the property that (p,q) and (p_+,q_+) both have the same support as (r,s). Then, for $(p,q) \in S$ the linear system (56) describes the relationship between $(\underline{p}_+,\underline{q}_+)$ amd $(\underline{p},\underline{q})$. Moreover, $(\underline{r},\underline{s})$ is a stationary point of (56) and a regular equilibrium point of the comparative payoff game $(\tilde{A},\tilde{B})$ of (r,s). At this equilibrium point of $(\tilde{A},\tilde{B})$ both players have zero payoffs.

3.8 The Reduced System

We continue to work with the assumptions of Lemma 2. We shall examine the stability properties of the linear system (56). This is a necessary step in the investigation of local stability in the preliminary model.

The system (56) is subject to the constraints that for both vectors $\underline{p}$ and $\underline{q}$ the sum of all h-components must be 1. Let L be the set of all $2h$-dimensional vectors whose first h-components sum up to 1 and whose last h-components sum up to 1. We are interested in *stability within* L in the following sense:

$$\lim_{t \to \infty} (\underline{p}(t),\underline{q}(t)) = (\underline{r},\underline{s}) \tag{57}$$

for every initial value

$$(\underline{p}(0), \underline{q}(0)) \in L . \tag{58}$$

Since (56) is linear, stability in this sense is equivalent to *local stability within* L defined as in 3.6, but with the restriction to initial values in L.

In the following, we shall assume $h = |I| > 1$ or, in other words, that (r,s) is genuinely mixed. It will be convenient to replace (56) by an unconstrained system. For this purpose we eliminate p_h and q_h in (56). We substitute the right-hand side of (59) for p_h and the right-hand side of (60) for q_h:

$$p_h = 1 - \sum_{i=1}^{h-1} p_i \tag{59}$$

$$q_h = 1 - \sum_{j=1}^{h-1} q_j . \tag{60}$$

After the elimination of p_h and q_h, we remove the equations for p_{h+} and q_{h+}. We call the system obtained in this way the *reduced system*.

In order to be able to describe the reduced system in more detail, we introduce the following *notational conventions*: $\bar{p}$, $\bar{q}$, $\bar{p}_+$, $\bar{q}_+$, $\bar{r}$ and $\bar{s}$ denote the $(h-1)$-dimensional vectors obtained by the removal of the h-th component from $\underline{p}$, $\underline{q}$, $\underline{p}_+$, $\underline{q}_+$, $\underline{r}$ and $\underline{s}$, respectively. Moreover, let $\bar{A}$ be the $(h-1) \times (h-1)$-matrix whose elements $\bar{a}_{ij}$ are as follows:

$$\bar{a}_{ij} = \bar{a}_{ij} - \bar{a}_{ih} \tag{61}$$

for $i = 1, \ldots h-1$ and $j = 1, \ldots h-1$. Similarly, let $\bar{B}$ be the $(h-1) \times (h-1)$-matrix whose elements $\bar{b}_{ij}$ are as follows:

$$\bar{b}_{ij} = \bar{b}_{ij} - \bar{b}_{hj} . \tag{62}$$

Let $\bar{a}$ be the column vector formed by the h-1 first components of the last column of $\bar{A}$ and let $\bar{b}$ be the column vector formed by the h-1 first components of the last column of $\bar{B}^T$. It can be seen without difficulty that the reduced system can now be described by the following equation:

$$\begin{bmatrix} \bar{p}_+ \\ \bar{q}_+ \end{bmatrix} = \begin{bmatrix} \alpha \bar{a} \\ \beta \bar{b} \end{bmatrix} + \begin{bmatrix} \bar{p} \\ \bar{q} \end{bmatrix} + \begin{bmatrix} 0 & \alpha \bar{A} \\ \beta \bar{B}^T & 0 \end{bmatrix} \begin{bmatrix} \bar{p} \\ \bar{q} \end{bmatrix}. \qquad (63)$$

Define

$$R = \begin{bmatrix} 0 & \alpha \bar{A} \\ \beta \bar{B}^T & 0 \end{bmatrix} \qquad (64)$$

and

$$c = \begin{bmatrix} \alpha \bar{a} \\ \beta \bar{b} \end{bmatrix}. \qquad (65)$$

Moreover, let D be the $(2h-2) \times (2h-2)$-unit matrix:

$$D = \begin{bmatrix} 1 & \cdots & 0 \\ \vdots & \ddots & \vdots \\ 0 & \cdots & 1 \end{bmatrix}. \qquad (66)$$

With this notation, (63) can be expressed as follows:

$$\begin{bmatrix} \bar{p}_+ \\ \bar{q}_+ \end{bmatrix} = c + (D+R) \begin{bmatrix} \bar{p} \\ \bar{q} \end{bmatrix}. \qquad (67)$$

Obviously, $(\bar{r},\bar{s})$ is a stationary point of the system (67). It is clear that (r,s) is stable within L if and only if $(\bar{r},\bar{s})$ is stable with respect to (67).

<u>Lemma 3</u>: Let (r,s) be a regular equilibrium point. If (r,s) is a pure strategy combination, then $(\underline{r},\underline{s})$ is stable within L with respect to (56). If (r,s) is genuinely mixed, then $(\underline{r},\underline{s})$ is unstable within L with respect to (56).

<u>Proof</u>: The case of pure strategies r and s has been discussed. Assume $h = |I| > 1$. It is sufficient to examine the stability of $(\bar{r},\bar{s})$ with respect to (67).

It is well-known that a system of linear difference equations with constant coefficients is stable if and only if all the eigenvalues of its matrix lie in the interior of the unit circle around the origin in the complex plane. It is also well-known that the sum of all eigenvalues of a quadratic matrix are equal to its *trace*, defined as the sum of all elements on the main diagonal.

The matrix D+R of (66) has 2h-2 eigenvalues. The trace is equal to 2h-2, since all elements on the main diagonal are equal to 1. Therefore, at least one eigenvalue has a real part not smaller than 1. This eigenvalue does not lie in the interior of the unit circle around the origin in the complex plane. $(\bar{r},\bar{s})$ is unstable with respect to (67).

3.9 Movement into the Support of a Regular Equilibrium Point

It will be important for the investigation of local stability properties of the preliminary model that for mixed strategy pairs (p,q) sufficiently near to a regular equilibrium point (r,s) the pair (p_+,q_+) has the same support as (r,s). Loosely speaking, we may say that a solution which converges to (r,s) must move into the support of (r,s).

Lemma 4: Let (r,s) be a regular equilibrium point. Then, for fixed $\alpha > 0$ and $\beta > 0$, a number $\varepsilon > 0$ can be found such that for every pair (p,q) in the ε-neighborhood $U_\varepsilon(r,s)$ of (r,s) the pair (p_+,q_+) determined by (22) to (25) has the same support as (r,s).

Proof: (r,s) is stationary by Theorem 1. Therefore, it follows by (22) and (23) that u and v assume the values H(r,s) and K(r,s) respectively at p = r and q = s. Regularity requires

$$H(i,s) - u < 0 \quad \text{for } r_i = 0 \tag{68}$$

$$K(r,j) - v < 0 \quad \text{for } s_j = 0. \tag{69}$$

It follows by the continuity properties of u and v mentioned in Lemma 1 and the continuity of H and K that for sufficiently small ε inequalities (68) and (69) continue to hold in the ε-neighborhood of (r,s). Therefore, for (p,q) in this ε-neighborhood, we have $p_{i+} = 0$ and $q_{j+} = 0$ outside the support of (r,s). By Lemma 1, the strategies p_+ and q_+ depend continuously on p and q. Therefore, for sufficiently small ε, we have $p_{i+} > 0$ and $q_{j+} > 0$ inside the support of (r,s) for $(p,q) \in U_\varepsilon(r,s)$. Consequently, the assertion of the lemma holds.

3.10 Local Stability Properties of the Preliminary Model

With the help of Lemma 3 and Lemma 4 it is now possible to settle the question of local stability or instability in the preliminary model.

Theorem 2: Let (r,s) be a regular equilibrium point. If r and s are pure strategies, then (r,s) is locally stable in the preliminary model described by (22) to (25). If (r,s) is genuinely mixed, then (r,s) is not locally stable in the preliminary model. This is true for all parameter pairs (α,β) with $\alpha > 0$ and $\beta > 0$.

Proof: Let ε be a number with the property spelled out by Lemma 4. Consider first the case where r and s are pure strategies. Then, a solution $(p(t),q(t))$ with $(p(0),q(0))$ in $U_\varepsilon(r,s)$ immediately moves to (r,s) and stays there forever. Therefore, in this case (r,s) is locally stable.

Now assume that (r,s) is genuinely mixed. Suppose that (r,s) is locally stable. Then, an open neighborhood V of (r,s) with $V \subseteq U_\varepsilon(r,s)$ can be found such that every solution starting there converges to (r,s) and stays in $U_\varepsilon(r,s)$. In particular this is true for solutions which start in $V \cap S$, where S is the set of all mixed strategy pairs with the same support as (r,s). These solutions are described by the linear system (57) which is unstable by Lemma 3. It follows that (r,s) fails to be locally stable with respect to the preliminary model.

Remark: The usual local stability criteria are based on differentiability at the stationary point and therefore cannot be applied to the preliminary model. Lemma 4 overcomes this difficulty.

4. The Anticipatory Learning Process

The preliminary model will now be modified by the introduction of anticipation. The resulting modified model is the *anticipatory learning process*. In this model, the individuals do not react directly to the observed strategies p and q. They first form anticipated strategies $p_>$ and $q_>$. Anticipated strategies depend on p and q in the same way as p_+ and q_+ depend on p and q in the preliminary model. We may say that anticipations follow the preliminary model.

The reactions follow the preliminary model, too. p_+ is computed in the same way as in the preliminary model, but on the basis of p and $q_>$ instead of p and q. Analogously, q_+ is computed on the basis of $p_>$ and q.

Anticipations on the other side are not anticipated. Nevertheless, anticipations are approximately correct if α and β are small.

4.1 Definition of the Anticipatory Learning Process

We continue to use the notation introduced in 3.2. In addition to p, q, p_+ and q_+, the anticipated strategies $p_>$ and $q_>$ appear in the definition of the anticipatory learning process. We also use the notation $p_>(t)$ and $q_>(t)$ where time dependence needs to be emphasized. As in the case of "+", lower indices placed before ">" indicate components. The following equations describe the *anticipatory learning process*:

$$p_{i>} = \max\left[0, p_i + \alpha H(i,q) - u_>\right] \quad \text{for } i = 1,\ldots,n \tag{70}$$

$$q_{j>} = \max\left[0, q_j + \beta K(p,j) - v_>\right] \quad \text{for } j = 1,\ldots,m \tag{71}$$

$$\sum_{i=1}^{n} \max\left[0, p_i + \alpha H(i,q) - u_>\right] = 1 \tag{72}$$

$$\sum_{j=1}^{m} \max\left[0, q_j + \beta K(p,j) - v_>\right] = 1 \tag{73}$$

$$p_{i+} = \max\left[0, p_i + \alpha H(i,q_>) - u\right] \quad \text{for } i = 1,\ldots,n \tag{74}$$

$$q_{j+} = \max\left[0, q_j + \beta K(p_>,j) - v\right] \quad \text{for } j = 1,\ldots,m \tag{75}$$

$$\sum_{i=1}^{n} \max\left[0, p_i + \alpha H(i,q_>) - u\right] = 1 \tag{76}$$

$$\sum_{j=1}^{m} \max\left[0, q_j + \beta K(p_>,j) - v\right] = 1. \tag{77}$$

As before, α and β are positive constants. The column vector $p_>$ of the $p_{i>}$ is player 1's *anticipated strategy* and the column vector of the $q_{j>}$ is player 2's *anticipated strategy*.

Lemma 5: Let α and β be positive constants. Then, for given p and q the anticipated strategies $p_>$ and $q_>$ and next period's strategies p_+ and q_+ as well as the auxiliary variables u, v, $u_>$ and $v_>$ are uniquely determined by (70) to (77). Moreover, $p_>$, $q_>$, p_+, q_+, u, v, $u_>$ and $v_>$ depend continuously on p and q.

Proof: The assertion is an immediate consequence of Lemma 1.

4.2 A Best Reply Property of Stationary Points

In the following, we shall obtain a first result on stationary points of the anticipatory learning process. The investigation of stationarity in the anticipatory learning process is more difficult than in the preliminary model. Finally, we shall show that for sufficiently small α and β equilibrium is necessary and sufficient for stationarity in regular games. However, several auxiliary results must be obtained before this can be done. A numerical example will show that the restriction to sufficiently small α and β cannot be avoided.

Lemma 6: Let (p,q) be a stationary point of the anticipatory learning process. Then, p is a best reply to $q_>$ and q is a best reply to $p_>$. This is true for all $\alpha > 0$ and $\beta > 0$ and for all games (A,B).

Proof: The proof makes use of (74) to (77) only. The argument is essentially the same as in the proof of Lemma 2 and will not be repeated here.

4.3 Equilibrium as a Sufficient Condition of Stationarity

It is easy to answer this question.

Lemma 7: Let (p,q) be an equilibrium point. Then, (p,q) is a stationary point of the anticipatory learning process. Moreover, we have $p_> = p$ and $q_> = q$. This is true for all $\alpha > 0$ and $\beta > 0$ and all games (A,B).

Proof: It follows by Lemma 2 that (p,q) is stationary in the preliminary model. Therefore, we can conclude by (70) to (73) that we have $p_> = p$ and $q_> = q$. It follows by Lemma 6 and the definition of an equilibrium point that (p,q) is stationary.

4.4 A Counterexample

In the following, we shall exhibit an example of stationary points which are not equilibrium points. Figure 1 shows a 2×2-game. Assume $\alpha = \beta = .5$. Let (p,q) be a mixed strategy pair with

$$p_1 = q_1 = x \tag{78}$$

where x is a number with

$$0 \leq x \leq 1. \tag{79}$$

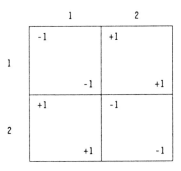

Figure 1: *A counterexample*: Player 1's strategies correspond to rows and player 2's strategies correspond to columns. Payoffs of player 1 are shown in the upper left corner and those of player 2 in the lower right corner of a field corresponding to a pure strategy combination.

We have

$$H(1,q) = K(p,1) = 1-2x \tag{80}$$

$$H(2,q) = K(p,2) = 2x-1. \tag{81}$$

Eqs. (38) and (39) yield $u_> = v_> = 0$. We obtain:

$$p_{1>} = q_{1>} = x + \frac{1}{2} - x = \frac{1}{2} \tag{82}$$

$$p_{2>} = q_{2>} = 1 - x + x - \frac{1}{2} = \frac{1}{2}. \tag{83}$$

Both pure strategies of player 1 yield the same payoff against $q_>$. Therefore, p is a best reply to $q_>$. Similarly, q is a best reply to $p_>$. It follows by Lemma 6 that (p,q) is stationary with respect to the anticipatory learning process with $\alpha = \beta = .5$. This is true for all x with $0 \leq x \leq 1$, but only for $x = 1/2$ the mixed strategy pair (p,q) is an equilibrium point.

The game of Figure 1 has three equilibrium points, two in pure strategies and one in mixed strategies. Each of these equilibrium points is regular. The game of Figure 1 is regular.

Of course, the example depends on the special value of α and β. However, the pure strategy pairs (1,1) and (2,2), which are connected to the extreme points of the interval $0 \leq x \leq 1$, are stationary for every pair (α,β) with $\alpha \geq .5$ and $\beta \geq .5$. We shall show this for the example of (1,1). One obtains:

$$p_{1>} = \max\left[0, 1 - \alpha - u_>\right] \tag{84}$$

$$p_{2>} = \min\left[1, \max(0, \alpha - u_>)\right]. \tag{85}$$

This yields:

$$p_{1>} = \begin{cases} 1-\alpha & \text{for } \alpha \leq 1 \\ 0 & \text{for } \alpha \geq 1 \end{cases} \tag{86}$$

$$p_{2>} = \begin{cases} \alpha & \text{for } \alpha \leq 1 \\ 1 & \text{for } \alpha \geq 1. \end{cases} \tag{87}$$

In both cases we have $p_{2>} \geq p_{1>}$. Similarly, $\beta \geq .5$ yields $q_{2>} \geq q_{1>}$. Therefore, player 1's pure strategy 1 is a best reply to $q_>$ and player 2's pure strategy 1 is a best reply to $p_>$. It follows by Lemma 7 that the pure strategy pair (1,1) is stationary.

The example shows that assertions on equilibrium as a necessary condition for stationarity must be restricted to sufficiently small α and β.

4.5 A Property of Regular Equilibrium Points

In the following, we shall derive an auxiliary result on regular equilibrium points. In order to do this, it is necessary to classify stationary points (p,q) according to the support of (p,q) and $(p_>,q_>)$.

Let (p,q) be a mixed strategy pair and let $(p_>,q_>)$ be the pair of anticipated strategies associated to (p,q). Moreover, let (I,J) be the support of (p,q) and let $(I_>,J_>)$ be the support of $(p_>,q_>)$. We call $(I,J,I_>,J_>)$ the *type* of (p,q). The symbol T is used for the *type function* which assigns the type of (p,q) to every (p,q):

$$(I,J,I_>,J_>) = T(p,q). \tag{88}$$

The following lemma shows that for sufficiently small α and β a regular equilibrium point is the only stationary point of its type.

Lemma 8: Let (r,s) be a regular equilibrium point. Then, numbers $\alpha_0 > 0$ and $\beta_0 > 0$ can be found such that for $0 < \alpha < \alpha_0$ and $0 < \beta < \beta_0$ the following is true: (r,s) is the only stationary point of the anticipatory learning process, whose type is $T(r,s)$.

Proof: Let $(I,J,I_>,J_>)$ be the type of (r,s). Since (r,s) is regular, we have $|I| = |J|$. In view of Lemma 7, the anticipated strategies associated with (r,s) are $r_> = r$ and $s_> = s$. Therefore, we have $I_> = I$ and $J_> = J$. Without loss of generality, we assume $I = \{1,...,h\}$ and $J = \{1,...,h\}$.

Assume that (p,q) is a stationary point of the anticipatory learning process with $T(p,q) = T(r,s)$. We use the notational conventions of 3.7. In view of (49) and (50) together with (70) to (73) we have

$$\underline{p}_> = \underline{p} + \alpha \tilde{A}\underline{q} \tag{89}$$

$$\underline{q}_> = \underline{q} + \beta \tilde{B}^T \underline{p} . \tag{90}$$

Since (p,q) is stationary, (74) to (77) yield:

$$\tilde{A}\underline{q} + \beta \tilde{A}\tilde{B}^T\underline{p} = 0 \tag{91}$$

$$\tilde{B}^T\underline{p} + \alpha \tilde{B}^T \tilde{A}\underline{q} = 0 . \tag{92}$$

In addition to (70) and (71) we have

$$\sum_{i=1}^{h} p_i = 1 \qquad (93)$$

$$\sum_{j=1}^{n} q_j = 1. \qquad (94)$$

Consider the equation system formed by (70), (71), (72) and (73). In the limiting case $\alpha = \beta = 0$ this system has a unique solution, namely $\underline{p} = \underline{r}$ and $\underline{q} = \underline{s}$. This is a consequence of the fact that $(\underline{r},\underline{s})$ is a regular equilibrium point of $(\tilde{A},\tilde{B})$.

The condition for unique solvability of the system (91) to (94) requires that certain determinants do not vanish. Since all these determinants depend continuously on α and β, we can find numbers α_0 and β_0 such that the system continues to have a unique solution for $0 < \alpha \leq \alpha_0$ and $0 < \beta \leq \beta_0$. Therefore, for α and β in these intervals $\underline{p} = \underline{r}$ and $\underline{q} = \underline{s}$ is the only solution of the system. We can conclude that the assertion of the lemma is true.

4.6 Equilibrium as a Necessary Condition

It will be shown that in regular games for sufficiently small α and β a stationary point of the anticipatory learning process must be an equilibrium point. In order to do this, we have to exclude the possibility that for arbitrarily small α and β stationary points can be found which fail to be equilibrium points. Since the number of possible types is finite, it is sufficient to investigate this problem for a fixed type. Lemma 8 already answers this question for types of equilibrium points. We have to show that for other types no stationary point exists for sufficiently small α and β.

<u>Lemma 9</u>: Let (A,B) be a regular game. Let $(I,J,I_>,J_>)$ be a possible type of a mixed strategy pair such that no equilibrium point is of this type. Then, numbers $\alpha_0 > 0$ and $\beta_0 > 0$ can be found such that for $0 < \alpha \leq \alpha_0$ and $0 < \beta \leq \beta_0$ the anticipatory learning process has no stationary point of the type $(I,J,I_>,J_>)$.

<u>Proof</u>: Assume that the assertion does not hold. Let $(I,J,I_>,J_>)$ be a type for which this is the case. Then, we can find monotonically decreasing sequences $\alpha_1, \alpha_2, \ldots$ and $\beta_1, \beta_2, \ldots$ converging to zero such that for $k = 1,2,\ldots$ the anticipatory learning process with $\alpha = \alpha_k$ and $\beta = \beta_k$ has a stationary point (p^k, q^k) of the type $(I,J,I_>,J_>)$. In view of the compactness of $P \times Q$ and the possibility of selecting a convergent subsequence of the (p^k, q^k), it can be assumed that the sequence $(p^1,q^1),(p^2,q^2),\ldots$ converges to a pair (p,q) of mixed strategies:

$$\lim_{k\to\infty} (p^k,q^k) = (p,q). \tag{95}$$

Assume that the sequences of the a_k, β_k and (p^k,q^k) have the properties described above.

Let $(I',J',I'_>,J'_>)$ be the type of (p,q). Since a sequence of positive numbers may converge to zero, the type of (p,q) may be different from $(I,J,I_>,J_>)$, but the sets $I',J',I'_>,J'_>$ must be subsets of $I,J,I_>,J_>$, respectively.

For $k = 1,2,\ldots$ let $p_{k>}$ and $q_{k>}$ be the anticipated strategies and let $u^k,v^k,u^k_>,v^k_>$ be the values of the auxiliary variables connected to (p^k,q^k) by (70) to (77) with $\alpha = a_k$ and $\beta = \beta_k$. It follows by (72) and (73) that we have

$$u^k_> = \frac{\alpha}{|I_>|} \sum_{i \in I_>} H(i,q^k) \tag{96}$$

$$v^k_> = \frac{\beta}{|J_>|} \sum_{j \in J_>} K(p^k,j). \tag{97}$$

In view of (95) and the continuity of H and K, the sequences $u^1_>, u^2_>, \ldots$ and $v^1_>, v^2_>, \ldots$ converge. We have:

$$u_> = \lim_{k\to\infty} u^k_> = \frac{\alpha}{|I_>|} \sum_{i \in I_>} H(i,q) \tag{98}$$

$$v_> = \lim_{k\to\infty} v^k_> = \frac{\beta}{|J_>|} \sum_{j \in J_>} K(p,j). \tag{99}$$

Since the sequences $a_1, a_2, \ldots$ and $\beta_1, \beta_2, \ldots$ converge to zero, it follows by (70) and (71) that the sequence of the $(p^k_>, q^k_>)$ converges to (p,q):

$$\lim_{k\to\infty} (p^k_>, q^k_>) = (p,q). \tag{100}$$

In view of Lemma 6, the stationarity of (p,q) has the consequence that p^k is a best reply to $q^k_>$ and q^k is a best reply to $p^k_>$. Since the property of being a best reply can be expressed by weak inequalities, it is preserved in the limit. It follows by (100) that (p,q) is an equilibrium point. In view of the regularity of (A,B), it is clear that (p,q) is a regular equilibrium point.

The regularity of (p,q) implies the following inequalities:

$$H(i,q) < H(p,q) \quad \text{for} \quad i \in N\setminus I' \tag{101}$$

$$K(p,j) < K(p,q) \quad \text{for} \quad j \in M\setminus J' . \tag{102}$$

In view of the continuity of H and K, inequalities (101) and (102) permit the conclusion that a k_0 exists such that similar inequalities hold for $k > k_0$:

$$H(i,q_>^k) < H(p^k,q_>^k) \quad \text{for} \quad i \in N\setminus I' \tag{103}$$

$$K(p_>^k,j) < K(p_>^k,q) \quad \text{for} \quad j \in N\setminus J' . \tag{104}$$

Since p^k is a best reply to $q_>^k$, inequality (103) cannot hold for $i \in I$. In view of $I' \subseteq I$, this has the consequence $I' = I$. Similarly, we can conclude $J' = J$.

It follows by (100) that for sufficiently great k components $p_{i>}^k$ of $p_>^k$ with $i \in I$ must be positive. This yields $I_> \supseteq I$. Similarly, we obtain $J_> \supseteq J$.

We shall show $I = I_>$ and $J = J_>$. Consider a pure strategy $i \in I_>$. In view of $p_i = 0$ and $p_{i>} > 0$, it follows by (70) that we must have:

$$H(i,q^k) - u_>^k > 0 \quad \text{for} \quad i \in I_> \tag{105}$$

for $k = 1,2,\ldots$. Since (p,q) is an equilibrium point, it follows by (98) that the following is true:

$$u_> \leq H(p,q). \tag{106}$$

In view of the continuity of H, inequalities (105) and (106) yield:

$$H(i,q) \geq H(p,q) \quad \text{for} \quad i \in I_>. \tag{107}$$

It follows by the regularity of (p,q) that every $i \in I_>$ belongs to I. In view of $I_> \supseteq I$, we have $I = I_>$. Similarly, one obtains $J = J_>$.

We have shown that all (p^k,q^k) are of the same type $(I,J,I_>,J_>)$ as (p,q). This contradicts the assumption that no equilibrium point is of the same type as the (p^k,q^k). Consequently, the assertion of the lemma holds.

<u>Theorem 3</u>: Let (A,B) be a regular game. Then, numbers $\alpha_0 > 0$ and $\beta_0 > 0$ can be found such that for $0 < \alpha \leq \alpha_0$ and $0 < \beta \leq \beta_0$ the following is true: A mixed strategy pair is stationary with respect to the anticipatory learning process described by (70) to (77) if and only if (p,q) is an equilibrium point.

<u>Proof</u>: The stationarity of equilibrium points follows by Lemma 7. By Lemma 9, a

stationary point must have the same type as some equilibrium point. Since the number of possible types is finite, the theorem follows by Lemma 8.

4.7 The Linear System Connected to a Regular Equilibrium Point

Let (r,s) be a regular equilibrium point. We shall proceed in a similar fashion as in 3.7. Let (I,J) be the support of (r,s) and define $h = |I|$. Since (r,s) is regular, we also have $|J| = h$.

The set S will from now on be defined as the set of all mixed strategy pairs (p,q) such that (p,q) as well as $(p_>, q_>)$ and (p_+, q_+) determined by (70) to (77) have the support (I,J) of (r,s). We continue to use the notational conventions of 3.7 and extend them to $p_>$ and $q_>$. In the anticipatory learning process, the connection between $(p_>, q_>)$ and (p,q) is the same one as between (p_+, p_+) and (p,q) in the preliminary model. Moreover, in the anticipatory learning process p_+ depends on $(p, q_>)$ in the same way as on (p,q) in the preliminary model. An analagous statement holds for q_+. Therefore, for $(p,q) \in S$, Eq. (56) permits the following conclusions:

$$\begin{bmatrix} p_> \\ q_> \end{bmatrix} = \begin{bmatrix} p \\ q \end{bmatrix} + \begin{bmatrix} 0 & \alpha \tilde{A} \\ \beta \tilde{B}^T & 0 \end{bmatrix} \begin{bmatrix} p \\ q \end{bmatrix} \tag{108}$$

where the zeros stand for $h \times h$-zero-matrices.

$$\begin{bmatrix} p_+ \\ q_+ \end{bmatrix} = \begin{bmatrix} p \\ q \end{bmatrix} + \begin{bmatrix} 0 & \alpha \tilde{A} \\ \beta \tilde{B}^T & 0 \end{bmatrix} \begin{bmatrix} p_> \\ q_> \end{bmatrix}. \tag{109}$$

Eqs. (108) and (109) together yield:

$$\begin{bmatrix} p_+ \\ q_+ \end{bmatrix} = \begin{bmatrix} p \\ q \end{bmatrix} + \begin{bmatrix} 0 & \alpha \tilde{A} \\ \beta \tilde{B}^T & 0 \end{bmatrix} \begin{bmatrix} p \\ q \end{bmatrix} + \begin{bmatrix} 0 & \alpha \tilde{A} \\ \beta \tilde{B}^T & 0 \end{bmatrix}^2 \begin{bmatrix} p \\ q \end{bmatrix}. \tag{110}$$

This system of difference equations has the same significance for the anticipatory learning process as (56) has for the preliminary model. Also with respect to (110) we are interested in stability within L, defined in the same way as in 3.7.

4.8 The Reduced System

We proceed in the same way as in 3.8. If (r,s) is a pair of pure strategies, then $(\underline{r},\underline{s})$ is trivially stable within L, since L consists of a single point. In the following, we assume $h = |I| = |J| > 1$.

In the same way as in 3.8 we eliminate p_h and q_h and remove the equations for $p_{h>}$ and $q_{h>}$ in (108). The $(h-1)$-dimensional vectors resulting from $p_>$ and $q_>$ by taking away the h-th component are denoted by $\bar{p}_>$ and $\bar{q}_>$, respectively. One obtains:

$$\begin{bmatrix} \bar{p}_> \\ \bar{q}_> \end{bmatrix} = c + (D+R) \begin{bmatrix} \bar{p} \\ \bar{q} \end{bmatrix} \tag{111}$$

with D and R defined as in 3.8. Similarly, we eliminate $p_{h>}$ and $q_{h>}$ in (109) and remove the equations for p_{h+} and q_{h+}. This yields

$$\begin{bmatrix} \bar{p}_+ \\ \bar{q}_+ \end{bmatrix} = c + \begin{bmatrix} \bar{p} \\ \bar{q} \end{bmatrix} + R \begin{bmatrix} \bar{p}_> \\ \bar{q}_> \end{bmatrix}. \tag{112}$$

Define

$$g = (D+R)c. \tag{113}$$

In view of (111) and (112), the relationship between $(\bar{p}_+, \bar{q}_+)$ and $(\bar{p}, \bar{q})$ can be expressed as follows:

$$\begin{bmatrix} \bar{p}_+ \\ \bar{q}_+ \end{bmatrix} = g + (D+R+R^2) \begin{bmatrix} \bar{p} \\ \bar{q} \end{bmatrix}. \tag{114}$$

We call this system of difference equations the *reduced system* of (r,s) in the anticipatory learning process. The system (110) is stable within L, if and only if the reduced system (114) is stable.

4.9 The Interaction Covariance Matrix

The stability of the reduced system depends on the eigenvalue of the matrix R. As we shall see, these eigenvalues are closely connected to those of another matrix, which is interesting in view of its interpretation. In order to explain the meaning of this "interaction covariance matrix", we first introduce a decomposition of a payoff matrix into a "biadditive" part and an "interaction" part. The same kind of decomposition underlies statistical procedures of two-way analysis of variance (see, e.g., Darlington, 1975). There the word "interaction" refers to effects which are not attributed to rows or columns, but to the interaction of both. If payoff matrices are decomposed in this way, we can also speak of a separation of non-interactive and interactive effects of pure strategy choices.

Let (r,s) be a genuinely mixed regular equilibrium point and let $(\underline{A},\underline{B})$ be the restricted game of (r,s). In view of the regularity of (r,s), both players have the same number of pure strategies. Let h be this number. $\underline{A}$ and $\underline{B}$ are $h \times h$-matrices.

An $h \times h$-matrix X is called *biadditive* if constants $u_1, \ldots, u_h$ and $v_1, \ldots, v_h$ can be found such that the elements x_{ij} of X satisfy the condition

$$x_{ij} = u_i + v_j \tag{115}$$

for $i = 1, \ldots, h$ and $j = 1, \ldots, h$. The u_i are called *row effects* and the v_i are *column effects*. The payoff matrices $\underline{A}$ and $\underline{B}$ are not biadditive, but we can still try to attribute as much payoff variance as possible to additively composed row and column effects; the remaining payoff parts are then defined as "interaction effects". We decompose $\underline{A}$ accordingly:

$$\underline{a}_{ij} = u_i + v_j + \hat{a}_{ij} \tag{116}$$

for $i = 1, \ldots, h$ and $j = 1, \ldots, h$. We minimize interaction effects in the sense of the least square criterion. Define:

$$Q = \sum_{i=1}^{h} \sum_{j=1}^{h} (\underline{a}_{ij} - u_i - v_j)^2 . \tag{117}$$

The row and column effects u_i and v_i are determined in such a way that Q is minimized. At the minimum, the partial derivatives of Q with respect to the u_i and v_j must be zero. This yields:

$$u_i = \frac{1}{h} \sum_{j=1}^{h} (\underline{a}_{ij} - v_j) \quad \text{for } i = 1, \ldots, h \tag{118}$$

$$v_j = \frac{1}{h} \sum_{i=1}^{h} (\underline{a}_{ij} - u_i) \quad \text{for } j = 1, \ldots, h. \tag{119}$$

These equations do not uniquely determine the u_i and the v_j. However, as we shall see, they uniquely determine the *interaction effects*:

$$\hat{a}_{ij} = \underline{a}_{ij} - u_i - v_j . \quad (120)$$

Define

$$\underline{a}_i = \frac{1}{h} \sum_{j=1}^{h} \underline{a}_{ij} \quad (121)$$

$$\underline{a}'_j = \frac{1}{h} \sum_{i=1}^{h} \underline{a}_{ij} \quad (122)$$

$$\underline{a} = \frac{1}{h^2} \sum_{i=1}^{h} \sum_{j=1}^{h} \underline{a}_{ij} . \quad (123)$$

Obviously, $\underline{a}_i$ is the i-th *row average*, $\underline{a}'_j$ is the j-th *column average* and $\underline{a}$ is the *overall average* of $\underline{A}$. It follows by (118) and (119) that we have:

$$u_i = \underline{a}_i - \frac{1}{h} \sum_{j=1}^{h} v_j \quad \text{for } i = 1,\ldots,h \quad (124)$$

$$v_j = \underline{a}'_j - \frac{1}{h} \sum_{i=1}^{h} u_i \quad \text{for } j = 1,\ldots,h. \quad (125)$$

Summation over i in (118) or over j in (119) yields:

$$\sum_{i=1}^{h} u_i + \sum_{j=1}^{h} v_j = h\underline{a}. \quad (126)$$

Therefore, (124) and (125) permit the following conclusion:

$$u_i + v_j = \underline{a}_i + \underline{a}'_j - \underline{a}. \quad (127)$$

Eq. (127) describes the elements $u_i + v_j$ of the *biadditive part* of $\underline{A}$. We are more interested in the *interaction part* $\hat{A}$ with the elements $\hat{a}_{ij}$. In view of (120) and (127) we have

$$\hat{a}_{ij} = \underline{a}_{ij} - \underline{a}_i - \underline{a}'_j + \underline{a} \quad (128)$$

for $i = 1,\ldots,h$ and $j = 1,\ldots,h$. Player 2's payoff matrix $\underline{B}$ can be decomposed analogously. Let $\underline{b}_i$ the i-th row average, $\underline{b}'_j$ the j-th column average and $\underline{b}$ the overall average of $\underline{B}$. The minimization of interaction effects in the sense of the least square criterion yields:

$$\hat{b}_{ij} = \underline{b}_{ij} - \underline{b}_i - \underline{b}'_j + \underline{b} \tag{129}$$

for $i = 1,\ldots,h$, and $j = 1,\ldots,h$. The h×h-matrix of the $\hat{b}_{ij}$ is denoted by $\hat{B}$. We call $\hat{A}$ and $\hat{B}$ the *payoff interaction matrices* of (r,s) for players 1 and 2, respectively.

It will be analytically advantageous to describe the matrices $\hat{A}$ and $\hat{B}$ as transformations of the payoff matrices $\underline{A}$ and $\underline{B}$ of the restricted game. Consider the matrix Z defined by (46) in 3.7. As we shall see, we have:

$$\hat{A} = Z\underline{A}Z \tag{130}$$

$$\hat{B} = Z\underline{B}Z. \tag{131}$$

Left multiplication by Z amounts to the subtraction of the column average from each element. Right multiplication by Z amounts to the subtraction of the row average from each element. It can be seen immediately that the consecutive performance of both operations yields the results described by (128) and (129).

It is useful to point out two properties of Z. It is clear by (46) that Z is symmetric:

$$Z^T = Z. \tag{132}$$

We have:

$$Z^2 = \underline{D} - \frac{2}{h} zz^T + \frac{1}{h^2} zz^T zz^T. \tag{133}$$

Since the scalar product $z^T z$ is equal to h, the last term on the right-hand side is equal to zz^T/h. This yields:

$$Z^2 = Z. \tag{134}$$

Eq. (134) is expressed by saying that the matrix Z is *idempotent*.

Consider the matrix

$$C = \hat{A}\hat{B}^T. \tag{135}$$

The element c_{ij} of C is the covariance of player 1's payoff interaction effects

$\hat{a}_{ik}$ in row i with player 2's payoff interaction effects $\hat{b}_{jk}$ in row j. Accordingly, we call C the *interaction covariance matrix* of (r,s).

In view of (130), (131) and (134), we have:

$$C = Z\underline{A}Z\underline{B}^T Z . \qquad (136)$$

The definition of the comparative payoff matrices $\bar{A}$ and $\bar{B}$ by (47) and (48) shows that instead of (136) we can write:

$$C = \bar{A}\bar{B}^T Z . \qquad (137)$$

It can be seen with the help of (128) and (129) that column sums and row sums in $\hat{A}$ and $\hat{B}$ are zero. This conclusion can also be reached by (130) and (131) in view of the effects of left and right multiplication by Z. In the same way, it follows by (136) that the row sums and column sums of C are equal to zero.

The interaction covariance matrix C can also be connected to the matrices $\bar{A}$ and $\bar{B}$ defined by (61) and (62) in 3.8. In order to reveal this connection, we first derive relationships of $\bar{A}$ and $\bar{B}$ with $\hat{A}$ and $\hat{B}$, respectively. Define:

$$V = \begin{bmatrix} 1 & \cdots & 0 & 0 \\ \cdot & & \cdot & \\ \cdot & & \cdot & \\ \cdot & & \cdot & \\ 0 & \cdots & 1 & 0 \end{bmatrix} . \qquad (138)$$

This (h-1)×h-matrix V is obtained from (h-1)×(h-1)-unit matrix $\underline{D}$ by the addition of an h-th column, whose elements are all zero. Define:

$$W = \begin{bmatrix} 1 & \cdots & 0 \\ \cdot & & \cdot \\ \cdot & & \cdot \\ \cdot & & \cdot \\ 0 & \cdots & 1 \\ -1 & \cdots & -1 \end{bmatrix} \qquad (139)$$

This (h-1)×h-matrix W results from $\underline{D}$ by the addition of an h-th row, whose elements are all equal to -1. With the help of (61) and (62), it can be seen without difficulty that we have:

$$\bar{A} = V\tilde{A}W \qquad (140)$$

$$\bar{B}^T = V\tilde{B}^T W . \qquad (141)$$

Since left multiplication by Z amounts to the subtraction of column averages, we have:

$$ZW = W. \tag{142}$$

In view of the definitions of $\bar{A}$ and $\bar{B}$ by (47) and (48), it follows by (130) and (131) that $\bar{A}W$ and $\bar{B}^TW$ can be replaced by $\hat{A}W$ and $\hat{B}^TW$, respectively. We obtain:

$$\bar{A} = V\hat{A}W \tag{143}$$

$$\bar{B}^T = V\hat{B}^TW. \tag{144}$$

The reduced system (115) depends on R only, and the definition of R by (64) shows that R is determined by $\bar{A}$, $\bar{B}$, α and β. Therefore, (143) and (144) reveal an interesting property of the reduced system. The reduced system is fully determined by the payoff interaction matrices $\hat{A}$ and $\hat{B}$. Row effects and column effects in the payoff matrices $\underline{A}$ and $\underline{B}$ of the restricted game do not influence the dynamics of the reduced system. Only interaction effects matter.

Eqs. (143) and (144) permit the derivation of a relationship between $\bar{A}\bar{B}^T$ and the interaction covariance matrix C. We have:

$$\bar{A}\bar{B}^T = V\hat{A}WV\hat{B}^TW . \tag{145}$$

It can be seen easily that WV has the following structure:

$$WV = \begin{bmatrix} 1 & \cdots & 0 & 0 \\ \vdots & & \vdots & \vdots \\ 0 & \cdots & 1 & 0 \\ -1 & \cdots & -1 & 0 \end{bmatrix} . \tag{146}$$

The h×h-matrix WV can be described as obtained from W by adding an h-th column of zeros. In view of the fact that right multiplication by Z amounts to the subtraction of row averages, it can be seen easily that the following is true:

$$WVZ = Z . \tag{147}$$

It follows by (131) that we have:

$$WV\hat{B}^T = WVZ\underline{B}^TZ = \hat{B}^T . \tag{148}$$

This together with the definition of C by (135) permits the following conclusion from (145):

$$\bar{A}\bar{B}^T = VCW \tag{149}$$

4.10 Eigenvalue Properties

Lemma 10 reveals the connection between the eigenvalues of the matrix R which appears in the reduced system and those of the interaction covariance matrix C. Later it will be shown that the stability of the reduced system requires that all eigenvalues of C are negative.

<u>Lemma 10</u>: Let (r,s) be a genuinely mixed regular equilibrium point. Let $\tilde{A}, \tilde{B}, \bar{A}, \bar{B}$ and R be defined as in 3.7 and 3.8 and let C be the interaction covariance matrix of (r,s). The following statements (i) to (v) hold:

(i) If λ is an eigenvalue of R, then $-\lambda$ is an eigenvalue of R.

(ii) No eigenvalue of R is equal to zero.

(iii) λ is an eigenvalue of R if and only if for some eigenvalue η of $\bar{A}\bar{B}^T$ the equation

$$\lambda^2 = \alpha\beta\eta \qquad (150)$$

holds.

(iv) η is an eigenvalue of $\bar{A}\bar{B}^T$ if and only if η is a non-vanishing eigenvalue of C.

(v) If $\underline{y}$ is an eigenvector of C, connected to a non-vanishing eigenvalue η of C, then the sum of all components of $\underline{y}$ is zero.

<u>Proof</u>: Assume that λ is an eigenvalue of R. Consider an eigenvector of R associated with λ, let v be the vector of the h-1 first components and let w be the vector of the h-1 last components of this eigenvector. In view of the definition of R we have:

$$\begin{bmatrix} 0 & \alpha\tilde{A} \\ \beta\bar{B}^T & 0 \end{bmatrix} \begin{bmatrix} v \\ w \end{bmatrix} = \begin{bmatrix} \alpha\tilde{A}w \\ \beta\bar{B}^T v \end{bmatrix} = \lambda \begin{bmatrix} v \\ w \end{bmatrix} \qquad (151)$$

with

$$v \neq 0 \quad \text{or} \quad w \neq 0 \qquad (152)$$

since at least one component of an eigenvector must be different from zero. (151) yields

$$\lambda v = \alpha \bar{A} w \qquad (153)$$

$$\lambda w = \beta \bar{B}^T v . \qquad (154)$$

Eqs. (153) and (154) permit the following conclusion

$$\begin{bmatrix} 0 & \alpha\bar{A} \\ \beta\bar{B}^T & 0 \end{bmatrix} \begin{bmatrix} v \\ -w \end{bmatrix} = \begin{bmatrix} -\alpha\bar{A}w \\ \beta\bar{B}^T v \end{bmatrix} = -\lambda \begin{bmatrix} v \\ -w \end{bmatrix} . \qquad (155)$$

This shows that $-\lambda$ is an eigenvalue of R. We have proved (i).

Assume $\lambda = 0$. In view of (153) and (154) we have:

$$\bar{A}w = 0 \qquad (156)$$

$$\bar{B}^T v = 0. \qquad (157)$$

Define

$$\underline{v} = Wv \qquad (158)$$

$$\underline{w} = Ww \qquad (159)$$

where W is the matrix defined by (139). Obviously, $\underline{v}$ and $\underline{w}$ are h-dimensional vectors, whose first h-1 components agree with those of v and w, respectively. For both $\underline{v}$ and $\underline{w}$ the sum of all components is zero. In view of (140) and (141), Eqs. (156) and (157) yield:

$$V\tilde{A}\underline{w} = 0 \qquad (160)$$

$$V\tilde{B}^T \underline{v} = 0. \qquad (161)$$

Since column sums in $\tilde{A}$ and $\tilde{B}$ are zero, the vectors $\tilde{A}\underline{w}$ and $\tilde{B}^T\underline{v}$ have the property that the sum of all components is zero. It follows by the definition of (138) of V that (160) and (161) cannot hold, unless we have:

$$\tilde{A}\underline{w} = 0 \qquad (162)$$

$$\tilde{B}^T \underline{v} = 0. \qquad (163)$$

It follows by (162) and (163) that for sufficiently small ε the mixed strategy pairs (p,q) with

$$\underline{p} = \underline{r} + \varepsilon \underline{v} \tag{164}$$

$$\underline{q} = \underline{s} + \varepsilon \underline{w} \tag{165}$$

is an equilibrium point of the comparative payoff game $(\bar{A},\bar{B})$. (152) permits the conclusion that, contrary to Lemma 2, the equilibrium point $(\underline{r},\underline{s})$ fails to be isolated. Consequently, zero cannot be an eigenvalue of R. We have proved (ii).

We continue to assume that λ is an eigenvalue of R. We want to show that (150) holds for some eigenvalue η of $\bar{A}\bar{B}^T$, or, in other words, that λ^2 is an eigenvalue of $\alpha\beta\bar{A}\bar{B}^T$. It follows by (153) and (154) that we have

$$\lambda^2 v = \alpha\bar{A}\lambda w = \alpha\beta\bar{A}\bar{B}^T v \tag{166}$$

$$\lambda^2 w = \beta\bar{B}^T\lambda v = \alpha\beta\bar{B}^T\bar{A}w. \tag{167}$$

In view of (152), it follows by (166) and (167) that λ^2 is an eigenvalue of $\alpha\beta\bar{A}\bar{B}^T$ or of $\alpha\beta\bar{B}^T\bar{A}$. It is well-known that a change of the order of factors in a product of two matrices does not change the eigenvalue of the matrix product. We can conclude that λ^2 is an eigenvalue of $\alpha\beta\bar{A}\bar{B}^T$ and that (150) is satisfied for some eigenvalue η of $\bar{A}\bar{B}^T$.

Before we turn our attention to the reverse direction of (iii), we show that $\eta = 0$ cannot be an eigenvalue of $\bar{A}\bar{B}^T$. Assume that $\bar{A}\bar{B}^T$ has an eigenvalue zero. Then, either $\bar{A}$ or $\bar{B}^T$ must have a vanishing determinant. This has the consequence that either (156) or (157) has a non-trivial solution. As we have seen above, the existence of such a non-trivial solution permits that $(\underline{r},\underline{s})$ fails to be a regular equilibrium point of the comparative payoff game $(\bar{A},\bar{B})$. It follows that zero cannot be an eigenvalue of $\bar{A}\bar{B}^T$.

We now assume that λ is not necessarily an eigenvalue of R and that (150) holds for some eigenvalue η of $\bar{A}\bar{B}^T$. Let v be an eigenvalue of $\bar{A}\bar{B}^T$ connected to η. In view of (150), we have

$$\lambda^2 v = \alpha\beta\bar{A}\bar{B}^T v. \tag{168}$$

In view of (133) and (135), we have

$$w = \frac{\beta}{\lambda} B^T v . \tag{169}$$

Eqs. (168) and (169) yield (153) and (154). It follows that (151) holds with $v \neq 0$. We can conclude that λ is an eigenvalue of R.

We now turn our attention to (v). Let $\underline{v}$ be an eigenvector of C, connected to a non-vanishing eigenvalue η of C. In view of (136) and (132), we have

$$\eta \underline{v} = ZC\underline{v}. \qquad (170)$$

In view of the effect of left multiplication by Z, the sum of all components of $\underline{v}$ must be zero. We have proved (v).

It remains to show (iv). We continue to assume that η is a non-vanishing eigenvalue of C and that $\underline{v}$ is an associated eigenvector of C. Define

$$v = V\underline{v}. \qquad (171)$$

In view of (138), the (h-1)-dimensional vector v is the vector of the h-1 first components of $\underline{v}$. Since by (v) the component sum of $\underline{v}$ is zero, (139) has the following consequence:

$$\underline{v} = Wv. \qquad (172)$$

In view of (172), we have:

$$\eta \underline{v} = C\underline{v} = CWv. \qquad (173)$$

With the help of (171), we obtain

$$\eta v = \eta V\underline{v} = VCWv. \qquad (174)$$

In view of (149) we can conclude

$$\eta v = \overline{A}\overline{B}^T v . \qquad (175)$$

Since the component sum of $\underline{v}$ is zero, $v \neq 0$ follows by $\underline{v} \neq 0$. Therefore, (175) shows that η is an eigenvalue of $\overline{A}\overline{B}^T$.

Now, assume that η is an eigenvalue of $\overline{A}\overline{B}^T$ and that v is an associated eigenvector of $\overline{A}\overline{B}^T$. Let $\underline{v}$ be defined by (172). With the help of (171) and (172), we can derive (174) from (175) and (173) from (174). This shows that η is an eigenvalue of C. We have proved the lemma.

4.11 A Stability Criterion for the Reduced System

In the following, we shall derive a result on the stability properties of the reduced system (115). It will be shown that for sufficiently small α and β the reduced system is stable if and only if all non-vanishing eigenvalues of the interaction covariance matrix are negative.

A condition which requires negative eigenvalues is much stronger than a codition which requires negative real parts. However, it is not true that the negativity condition on the non-vanishing eigenvalues of the interaction covariance matrix is satisfied only in degenerate cases. This question will be discussed in Section 5.

<u>Lemma 11</u>: Let (r,s) be a genuinely mixed regular equilibrium point. Then, numbers $\alpha_0 > 0$ and $\beta_0 > 0$ can be found such that for $0 < \alpha < \alpha_0$ and $0 < \beta < \beta_0$ the reduced system (115) of (r,s) in the anticipatory learning process is stable if and only if all eigenvalues of the interaction covariance matrix C of (r,s) are negative.

<u>Proof</u>: It follows by a well-known theorem on matrix polynomials that μ is an eigenvalue of $D+R+R^2$ if and only if for some eigenvalue λ of R we have:

$$\mu = 1 + \lambda + \lambda^2 . \tag{176}$$

Therefore, the system (115) is stable if and only if all eigenvalues λ of R satisfy the following condition

$$|1 + \lambda + \lambda^2| < 1 . \tag{177}$$

In view of (iii) in Lemma 10, we can look at the eigenvalues λ of R as functions of α and β. Let λ be an eigenvalue of R and let η be the eigenvalue of $\bar{A}\bar{B}^T$ connected to λ by (150). Let $\rho > 0$ and ϕ with $0 \leq \phi < 2\pi$ be determined by the following equation

$$\lambda = \rho\sqrt{\alpha\beta}\,(\cos \phi + i \sin \phi). \tag{178}$$

In view of (150), Moivre's Theorem yields

$$\eta = \rho^2(\cos 2\phi + i \sin 2\phi). \tag{179}$$

This shows that ρ and ϕ depend only on η and not on α and β. For every η, we obtain two functions of the form (178) of α and β. We use the symbol "Re" in order to indicate the real part of the expression in brackets following after "Re". (178) yields:

$$\text{Re}(\lambda + \lambda^2) = \rho\sqrt{\alpha\beta}\,(\cos \phi + \rho\sqrt{\alpha\beta}\cos 2\phi). \tag{180}$$

This shows that for sufficiently small $\alpha\beta$ the real part of $\lambda+\lambda^2$ has the same sign as the real part of λ. Consequently, for sufficiently small $\alpha\beta$ the real part of $1+\lambda+\lambda^2$ is greater than 1 if λ has a positive real part. It follows that for sufficiently small $\alpha\beta$ inequality (177) cannot be satisfied if at least one eigenvalue λ of R has a positive real part. In view of (i) in Lemma 10, this is the case if at least one eigenvalue of R has a non-vanishing real part. It follows by (iii) and (iv) in Lemma 10 that for sufficiently small α and β the negativity of all non-vanishing eigenvalues η of the interaction covariance matrix C of (r,s) is a necessary condition for the stability of the reduced system (114).

Now, assue that all non-vanishing eigenvalues η of C are negative. In view of Lemma 10, this has the consequence that (178) assumes the following form:

$$\lambda = \gamma \sqrt{\alpha\beta} \; i \tag{181}$$

where γ is a positive or negative constant. Equation (181) yields:

$$|1 + \lambda + \lambda^2| = \sqrt{(1-\gamma^2\alpha\beta)^2 + \gamma^2\alpha\beta} \tag{182}$$

$$|1 + \lambda + \lambda^2| = \sqrt{1-\gamma^2\alpha\beta(1-\gamma^2\alpha\beta)} \; . \tag{183}$$

This shows that (177) is satisfied for $\gamma^2 < 1/\alpha\beta$. It follows that for sufficiently small α and β the negativity of all non-vanishing eigenvalues η of the interaction covariance matrix C of (r,s) is a sufficient condition for the stability of the reduced system.

Remark: From what has been said at the end of 4.8, it follows under the assumptions of Lemma 10 that for sufficiently small α and β the linear system (110) of (r,s) in the anticipatory learning process is stable within L if and only if all eigenvalues of the interaction covariance matrix C of (r,s) are negative.

4.12 Local Stability Under Sufficiently Slow Anticipatory Learning

Local stability conditions for the anticipatory learning process can be obtained only for sufficiently small α and β. In the case of small values of α and β, the learning process is slow in the sense that the changes from period to period are small. This interpretation underlies a convenient way of speaking which will facilitate the statement of stability results: We say that an equilibrium point (r,s) is *locally stable under sufficiently slow anticipatory learning* if numbers $\alpha_0 > 0$ and $\beta_0 > 0$ can be found such that (r,s) is locally stable with respect to the anticipatory learning process for all parameter pairs (α,β) with $0 < \alpha < \alpha_0$ and $0 < \beta < \beta_0$.

Theorem 4: Let (r,s) be a regular equilibrium point of (A,B). If (r,s) is an equilibrium point in pure strategies, then (r,s) is always locally stable under sufficiently slow anticipatory learning. If (r,s) is a genuinely mixed equilibrium point, then (r,s) is locally stable under sufficiently slow anticipatory learning if and only if all eigenvalues of the interaction covariance matrix C of (r,s) are negative.

Proof: As a first step in the proof of the theorem, we show that for fixed $\alpha > 0$ and $\beta > 0$ a number $\varepsilon > 0$ can be found such that for every pair (p,q) in the ε-neighborhood $U_\varepsilon(r,s)$ of (r,s) the pairs $(p_>,q_>)$ and (p_+,q_+) determined by (70) to (77) have the same support as (r,s).

By Lemma 4 we can find a number $\delta > 0$ such that for $(p,q) \in U_\delta(r,s)$ the pair of next period's strategies in the preliminary model has the support of (r,s). Let δ be a number of this kind. In view of the continuity statements of Lemma 5, we can find an $\varepsilon > 0$ with $\varepsilon \le \delta$ such that for (p,q) in $U_\varepsilon(r,s)$ the pairs $(p,q_>)$ and $(p_>,q)$ are in $U_\delta(r,s)$. Let ε be a number of this kind. Since p_+ depends in the same way on $(p,q_>)$ as it depends on (p,q) in the preliminary model and since an analogous statement holds for q_+, it follows that for $(p,q) \in U_\varepsilon(r,s)$ the pair (p_+,q_+) has the same support as (r,s). In view of the fact that $(p_>,q_>)$ depends in the same way on (p,q) as (p_+,q_+) does in the preliminary model, it follows by $\varepsilon \le \delta$ that ε meets the requirements of the statement to be proved.

Consider the case $h = 1$. Here, for $(p,q) \in U_\varepsilon(r,s)$ the pair (p_+,q_+) is nothing else than (r,s) since no other pair has the same support as (r,s). This proves the assertion of the theorem for the case $h = 1$.

From now on assume $h > 1$. It is clear that (r,s) cananot be locally stable unless $(\underline{r},\underline{s})$ is stable within L with respect to the linear system (110) described in 4.7. In view of Lemma 10 and the remark after Lemma 11, it follows that (r,s) cannot be locally stable unless all non-vanishing eigenvalues of C are negative. It remains to show that this condition is sufficient for local stability if α and β are sufficiently small.

Assume that non-vanishing eigenvalues of C are negative. It follows by Lemma 11 that for sufficiently small $\alpha > 0$ and $\beta > 0$ the pair $(\underline{r},\underline{s})$ is stable within L with respect to (110). Let α and β be sufficiently small in this sense. Since $(\underline{r},\underline{s})$ is stable within L, we can find a number $\underline{\delta} > 0$ with $\underline{\delta} \le \varepsilon$ such that the following is true: Let $(p(0),q(0))$ be a pair in $U_{\underline{\delta}}(r,s)$ with the same support as (r,s); then every $(p(t),q(t))$ in the solution starting with $(p(0),q(0))$ is in $U_\varepsilon(r,s)$ and has the same support as (r,s). Here, ε is the number which has been chosen above. The choice of δ guarantees that all $(p(t),q(t))$ remain in $U_\varepsilon(r,s)$ as long as they have the same support as (r,s). The choice of δ guarantees that by induction all $(p(t),q(t))$ have indeed the same support as (r,s). Let $\underline{\delta}$ be a number which meets the requirement.

In view of the continuity statements of Lemma 5, it is now clear that for initial values $(p(0),q(0))$ in a sufficiently small neighborhood of (r,s) the solution starting there will remain in $U_\varepsilon(r,s)$ and converge to (r,s). Moreover, ε can be chosen arbitrarily small. This shows that (r,s) is locally stable. We have proved the theorem.

5. Special Stability Conditions

In games with numerically specified payoffs, the eigenvalue of the interaction covariance matrix of a regular equilibrium point can be computed numerically in order to decide the question of local stability under sufficiently slow anticipatory learning. However, the criterion of Theorem 4 cannot be easily applied to game models whose payoffs depend on unknown parameters. Therefore, we are interested in special stability conditions which permit conclusions based on the structure of numerically unspecified payoff functions.

5.1 Biadditive Equivalence

Let A and F be two matrices with the same number of rows and the same number of columns. We say that A and F are *biadditively equivalent* if we have

$$A = \gamma F + X \qquad (184)$$

with $\gamma > 0$ where X is a biadditive matrix. It is clear that the relationship of "biadditive equivalence" is symmetric and transitive.

Let (A,B) and (F,G) be two games such that each of both players has the same number of pure strategies in both games. We say that (A,B) is *biadditively equivalent* to (F,G) if A is biadditively equivalent to F, and B is biadditively equivalent to G.

Let $(\underline{A},\underline{B})$ be the restricted game of a regular equilibrium point (r,s) of a game (A,B). Assume that $(\underline{A},\underline{B})$ is biadditively equivalent to (F,G):

$$\underline{A} = \gamma_1 F + X \qquad (185)$$

$$\underline{B} = \gamma_2 G + Y \qquad (186)$$

with $\gamma_1 > 0$ and $\gamma_2 > 0$ where X and Y are biadditive matrices. Since right multiplication by Z amounts to the subtraction of row averages and left multiplication to the subtraction of column averages, it follows that ZXZ and ZYZ vanish. Therefore, we have:

$$\hat{A} = \gamma_1 ZFZ \qquad (187)$$

$$\hat{B} = \gamma_2 ZGZ \qquad (188)$$

$$C = \gamma_1 \gamma_2 ZFG^T Z. \qquad (189)$$

This shows that up to a positive factor the interaction covariance matrix of (F,G) agrees with that of (A,B). Consequently, the criterion of Theorem 4 can be applied to (F,G) instead of (A,B). We have obtained the following result:

Theorem 5: Let (r,s) be a genuinely mixed regular equilibrium point and let ($\underline{A},\underline{B}$) be the restricted game of (r,s). Moreover, let (F,G) be biadditively equivalent to ($\underline{A},\underline{B}$). Then, (r,s) is locally stable under sufficiently slow anticipatory learning if and only if all non-vanishing eigenvalues of the interaction covariance matrix

$$C = Z F G^T Z \tag{190}$$

of (F,g) are negative.

5.2 Special Cases of Local Stability

We shall prove a theorem which lists several conditions on the restricted game which yield local stability under sufficiently slow anticipatory learning. One of these conditions requires that the restricted game is biadditively equivalent to a zero-sum game. A *zero-sum game* is a game (A,B) with

$$A = -B. \tag{191}$$

We shall now introduce two other classes of games which also give rise to special stability conditions. Let (A,B) be a quadratic game or, in other words, a game where both players have the same number h of pure strategies. In a game of this kind, a player is called a *pursuer* if his payoff matrix is $\underline{D}$ and he is called an *evader* if his payoff matrix is $-\underline{D}$. A *pursuit game* is a quadratic game with a pursuer and an *evasion game* is a quadratic game with an evader.

The names "pursuer" and "evader" are suggested by the interpretation of pure strategies as locations to be chosen by the players. A pursuer is exclusively motivated to catch the other player and an evader has no other goal than to avoid his opponent.

A pursuit game or an evasion game (A,B) is called *negative definite* if AB is symmetric and negative definite. In the case of a pursuit game (A,B) with $A = \underline{D}$, this means that B is negative definite, whereas in an evasion game (A,B) with $A = -\underline{D}$ negative definiteness requires that B is positive definite.

A *2×2-game* is a game in which each of both players has two pure strategies. One of the special stability conditions requires that the restricted game is a 2×2-game without pure equilibrium points.

Theorem 6: Let (r,s) be a regular equilibrium point and let ($\underline{A},\underline{B}$) be the restricted game of (r,s). The equilibrium point is locally stable under sufficiently slow anticipatory learning if one of the following five conditions (i) to (v) is satisfied:

(i) (r,s) is an equilibrium point in pure strategies.

(ii) ($\underline{A},\underline{B}$) is a 2×2-game without equilibrium points in pure strategies.

(iii) $(\underline{A},\underline{B})$ is biadditively equivalent to a zero-sum game.

(iv) $(\underline{A},\underline{B})$ is biadditively equivalent to a negative definite pursuit game.

(v) $(\underline{A},\underline{B})$ is biadditively equivalent to a negative definite evasion game.

Proof: In view of Theorem 4, the assertion holds in the case of (i). Assume (ii). We explore the consequences of the absence of pure equilibrium points. No pure strategy can be dominating in $(\underline{A},\underline{B})$ since otherwise $(\underline{A},\underline{B})$ would have an equilibrium point in pure strategies. Therefore,

$$\underline{a}_{11} \geq \underline{a}_{21} \tag{192}$$

implies

$$\underline{a}_{22} > \underline{a}_{12} \tag{193}$$

and consequently

$$\underline{a}_{11} + \underline{a}_{22} - \underline{a}_{12} - \underline{a}_{21} > 0. \tag{194}$$

Moreover, (192) implies

$$\underline{b}_{11} < \underline{b}_{12} . \tag{195}$$

Otherwise, (1,1) would be an equilibrium point. In view of the absence of dominance, (195) yields:

$$\underline{b}_{21} > \underline{b}_{22} . \tag{196}$$

It follows by (195) that we have:

$$\underline{b}_{11} + \underline{b}_{22} - \underline{b}_{12} - \underline{b}_{21} < 0 . \tag{197}$$

Eqs. (194) and (197) yield

$$(\underline{a}_{11} + \underline{a}_{22} - \underline{a}_{12} - \underline{a}_{21})(\underline{b}_{11} + \underline{b}_{22} - \underline{b}_{12} - \underline{b}_{21}) < 0. \tag{198}$$

In the case $\underline{a}_{11} < \underline{a}_{21}$, the same inequality can be derived in a similar way. As we shall see, (198) has the consequence that the non-vanishing eigenvalues of C is negative. (C has only one non-vanishing eigenvalue.) With the help of (128) in Section 4.9 we obtain

$$\hat{a}_{11} = \underline{a}_{11} - \frac{\underline{a}_{11}+\underline{a}_{12}}{2} - \frac{\underline{a}_{11}+\underline{a}_{21}}{2} + \frac{\underline{a}_{11}+\underline{a}_{12}+\underline{a}_{21}+\underline{a}_{22}}{4} \qquad (199)$$

$$\hat{a}_{11} = \frac{\underline{a}_{11}+\underline{a}_{22}-\underline{a}_{12}-\underline{a}_{21}}{4} \qquad (200)$$

$$\hat{a}_{22} = \hat{a}_{11} \qquad (201)$$

$$\hat{a}_{12} = \hat{a}_{21} = -\hat{a}_{11} . \qquad (202)$$

Analogous formulas hold for the elements of $\hat{B}$. It can be seen easily that C has the following structure

$$C = \begin{bmatrix} c & -c \\ -c & c \end{bmatrix} \qquad (203)$$

with

$$c = 2\hat{a}_{11}\hat{b}_{11} . \qquad (204)$$

It follows by (198) that c is negative. It can be seen immediately that C has two eigenvalues, namely 0 and $2c$. This shows that the non-vanishing eigenvalue of C is negative.

We now turn our attention to (iii). In view of Theorem 5, it is sufficient to look at the eigenvalues of C in the zero-sum case. Assume $\underline{A} = -\underline{B}$. we have:

$$\lambda v^T v = -v^T \underline{A} Z \underline{A}^T Z v. \qquad (205)$$

Obviously, C is symmetric. Therefore, all eigenvalues of C are real. Let λ be a non-vanishing eigenvalue of C and let v be an associated eigenvector. We have:

$$w = \underline{A}^T Z v . \qquad (206)$$

Left multiplication by v^T yields:

$$\lambda v v^T = -v^T \underline{A} Z \underline{A}^T Z v. \qquad (207)$$

Define

$$w = \underline{A}^T Z v. \qquad (208)$$

Since Z is symmetric and idempotent, we obtain

$$\lambda v^T v = -w^T w. \qquad (209)$$

In view of the fact that the components of v are real, it follows by $v \neq 0$ that vv^T is positive. Moreover, w^Tw is non-negative. It follows by $\lambda \neq 0$ that λ is negative.

We now turn our attention to (iv). It is sufficient to consider the case $\underline{A} = \underline{D}$. The other case in which player 2 is the pursuer is analogous. since Z is idempotent, we have

$$C = Z\underline{B}^T Z. \tag{210}$$

Let λ be a non-vanishing eigenvalue of C and let v be an eigenvector associated with λ. Since C is a matrix product beginning with Z, the sum of all components of v is zero and we have:

$$v = Zv. \tag{211}$$

Since $\underline{B}$ is symmetric, C is symmetric, too. Therefore, all eigenvalues of C are real and v has real components. With the help of

$$\lambda v = Z\underline{B}v \tag{212}$$

we obtain:

$$\lambda v^T v = v^T \underline{B} v. \tag{213}$$

In view of $v \neq 0$ and the negative definiteness of $\underline{B}$, it is clear that $v^T v$ is positive and the right hand side of (213) is negative. It follows by $\lambda \neq 0$ that λ is negative.

5.3 Bailiff and Poacher

In the following, Theorem 6 will be applied to a simple example. Imagine a situation where a bailiff (player 1) is responsible for three fish ponds 1, 2 and 3. A poacher (player 2) wants to steal fish in one of the ponds. The bailiff can watch only one pond and the poacher can steal at only one pond. This means that each of both players has three pure strategies, namely 1, 2 and 3. the bailiff has no other interest than to catch the poacher. If both go to the same pond, the poacher is caught by the bailiff, the bailiff gets a payoff of +1 and the poacher gets a payoff of -1. If both choose different ponds, the bailiff gets a payoff of zero and the poacher gets a payoff of u_1, u_2, u_3, depending on the pond; the ponds may be different with respect to the value of a successful act of poaching. The game is shown in Figure 2.

230 Game Theory and Economic Behaviour I

		poacher		
		1	2	3
bailiff	1	1 ⋱ -1	0 ⋱ u_2	0 ⋱ u_3
	2	0 ⋱ u_1	1 ⋱ -1	0 ⋱ u_3
	3	0 ⋱ u_1	0 ⋱ u_2	1 ⋱ -1

$u_i > 0$ for $i = 1, 2, 3$

<u>Figure 2</u>: Bailiff and poacher. In every field, player 1's payoff is indicated in the upper left corner and player 2's payoff is shown in the lower right corner.

The parameters u_1, u_2 and u_3 are assumed to be positive. Under appropriate additional conditions, the game has a regular completely mixed equilibrium point (p^*,q^*) whose probabilities for pure strategies are as follows:

$$p_i^* = 1 - \frac{\frac{2}{1+u_i}}{\frac{1}{1+u_1} + \frac{1}{1+u_2} + \frac{1}{1+u_3}} \qquad (214)$$

$$q_i^* = \frac{1}{3} \qquad (215)$$

for $i = 1, 2, 3$. It can be seen easily that (p^*,q^*) is a regular equilibrium point, if p_1^*, p_2^* and p_3^* are positive. Assume that this is the case. We are going to show that (p^*,q^*) is locally stable under sufficiently slow anticipatory learning.

If in each column i of Figure 2, the constant u_i is subtracted from player 2's payoff, one receives the biadditively equivalent game shown in Figure 3. Obviously, this game is a negative definite pursuit game. It follows by Theorem 6 that the completely mixed equilibrium point (p^*,q^*) of the game of Figure 2 is locally stable under sufficiently slow anticipatory learning.

Of course, (p^*,q^*) is not necessarily an equilibrium point of the game of Figure 3. It is not important for the application of Theorem 6 whether this is the case or not.

"Bailiff and poacher" is not a zero-sum game. However, the interests of both players are opposed in the sense that the bailiff wants to catch the poacher and the poacher wants to avoid being caught. Game situations with a similar flavor seem to offer a good chance for local stability under sufficiently small anticipatory learning.

	1	2	3
1	1 $-1-u_1$	0 0	0 0
2	0 0	1 $-1-u_2$	0 0
3	0 0	0 0	1 $-1-u_3$

Figure 3: A game which is biadditively equivalent to the game of Figure 2.

5.4 The 3×3-case

It is interesting to look at the special case of a restricted game with three pure strategies for each of both players. Consider a 3×3-interaction covariance matrix C. Since all row sums and column sums of C are zero, it is possible to characterize a 3×3-interaction matrix by four parameters, namely the three diagonal elements, in the following denoted by c_1, c_2 and c_3, and by a "skewness parameter" η:

$$C = \begin{bmatrix} c_1 & c_3 - \frac{s}{2} + \eta & c_2 - \frac{s}{2} - \eta \\ c_3 - \frac{s}{2} - \eta & c_2 & c_1 - \frac{s}{2} + \eta \\ c_2 - \frac{s}{2} + \eta & c_1 - \frac{s}{2} - \eta & c_3 \end{bmatrix} \quad (216)$$

where

$$s = c_1 + c_2 + c_3 \quad (217)$$

is the trace of C. It can be seen easily that every 3×3-interaction covariance matrix can be described in this way. Let C_{ii} be the adjunct of the diagonal element of c_i. We have:

$$C_{11} = c_2 c_3 - \left[c_1 - \frac{s}{2}\right]^2 + \eta^2 \qquad (218)$$

$$C_{22} = c_1 c_3 - \left[c_2 - \frac{s}{2}\right]^2 + \eta^2 \qquad (219)$$

$$C_{33} = c_1 c_2 - \left[c_3 - \frac{s}{2}\right]^2 + \eta^2 . \qquad (220)$$

The non-vanishing eigenvalues λ of C are roots of the following quadratic equation:

$$\lambda^2 - s\lambda + (C_{11} + C_{22} + C_{33}) = 0. \qquad (221)$$

Equations (218), (219), and (220) yield:

$$C_{11} + C_{22} + C_{33} = c_1 c_2 + c_1 c_3 + c_2 c_3 - c_1^2 - c_2^2 - c_3^2 + s^2 - \frac{3}{4} s^2 + 3\eta^2 \qquad (222)$$

$$C_{11} + C_{22} + C_{33} = -\frac{1}{2}\left[(c_1 - c_2)^2 + (c_1 - c_3)^2 + (c_2 - c_3)^2\right] + \frac{1}{4} s^2 + 3\eta^2. \qquad (223)$$

Both roots of (221) are negative if and only if the following conditions are satisfied:

$$s < 0 \qquad (224)$$

$$\frac{s^2}{4} \geq C_{11} + C_{22} + C_{33} > 0. \qquad (225)$$

With the help of (223) and (225) it can be seen that the non-vanishing eigenvalues of C are negative if and only if the trace of C is negative and the following condition is satisfied:

$$\eta^2 \leq \frac{1}{6}\left[(c_1 - c_2)^2 + (c_1 - c_3)^2 + (c_2 - c_3)^2\right] < \eta^2 + \frac{1}{12} s^2. \qquad (226)$$

If C is symmetric and all diagonal elements of C are equal, (226) is satisfied. Sufficiently small deviations from symmetry can be compensated by differences between the diagonal elements which are sufficiently big, but not too big. Suppose, for example, that the differences among the diagonal elements are not too big in the sense that

$$(c_i - c_j)^2 < \frac{s^2}{6} \quad \text{for} \quad i,j = 1,2,3 \tag{227}$$

holds. If this is the case, the middle term in (226) is smaller than the right-hand term. Thus, (226) will be satisfied if at least two diagonal elements differ from one another and if the deviations from symmetry are small enough.

Obviously, the non-negativity of both non-vanishing eigenvalues of C is a restrictive condition; however, the part of the parameter space where the condition is satisfied is not a set of lower dimension.

6. Anticipated Anticipations

The anticipatory learning process assumes that anticipations in the opponent population are not anticipated. This is quite natural in a model of limited rationality. Nevertheless, it is of interest to anwer the question whether a learning process with anticipated anticipations would have different stability properties. We shall restrict our attention to the local stability of completely mixed regular equilibrium points. Since the problem is not of central importance for the theory presented here, no attempt will be made to explore the more general case, which probably will not yield additional insights.

In the anticipatory learning process the change of strategies from one period to the next may be decomposed into a *first order effect* due to the observed strategy on the other side and a *second order effect* due to the anticipated change of the opponent strategy. To these effects one may add a *third order effect* due to the anticipated effect of the opponents anticipation. Of course, this is not yet the end of the story; for any k we may add a *k-th order effect* due to the opponents anticipation of the (k-1)-th order effect. Finally, one may wish to consider the limiting case of full anticipation, where effects of any order k are taken into account.

The preliminary model relies on first order effects. The first order effects do not stabilize genuinely mixed equilibrium points, but in favorable cases they may also fail to amplify deviations from equilibrium. Under these conditions the second order effects have a chance to exert a stabilizing influence in the anticipatory learning process. The higher order effects turn out to be much weaker for sufficiently slow learning and therefore cannot reverse the influence of the second order effect. Of course, these remarks do not intend to provide more than an intuitive interpretation of the results to be obtained below.

6.1 The Learning Process of Order k

Let (A,B) be a quadratic game with h pure strategies $1,\ldots,h$ for each of both players and let (r,s) be a regular completely mixed equilibrium point of (A,B).

Define

$$\tilde{R} = \begin{pmatrix} 0 & \alpha\tilde{A} \\ \beta\tilde{B}^T & 0 \end{pmatrix}. \qquad (228)$$

The *learning process of order* k is defined as follows:

$$\begin{bmatrix} p_+ \\ q_+ \end{bmatrix} = (\underline{D} + \tilde{R} + \tilde{R}^2 + \ldots \tilde{R}^k) \begin{bmatrix} p \\ q \end{bmatrix}. \qquad (229)$$

Here $\underline{D}$ is the h×h-unit matrix:

$$\underline{D} = \begin{bmatrix} 1 & & 0 \\ & \ddots & \\ 0 & & 1 \end{bmatrix}. \qquad (230)$$

In view of (56) and (110) the learning process of order 1 is the linear system connected to (r,s) in the preliminary model and the learning process of order 2 is the linear system connected to (r,s) in the anticipatory learning process. Define:

$$\begin{bmatrix} p_>^k \\ q_>^k \end{bmatrix} = (\underline{D} + \tilde{R} + \ldots + \tilde{R}^{k-1}) \begin{bmatrix} p \\ q \end{bmatrix}. \qquad (231)$$

We call $p_>^k$ and $q_>^k$ the *anticipated strategies of order* k. In the case k = 1 the matrix $\underline{D} + \ldots + \tilde{R}^{k-1}$ is interpreted as $\underline{D}$. With the help of (231) the learning process of order k can be described as follows:

$$\begin{bmatrix} p_+ \\ q_+ \end{bmatrix} = \begin{bmatrix} p \\ q \end{bmatrix} + \tilde{R} \begin{bmatrix} p_>^k \\ q_>^k \end{bmatrix}. \qquad (232)$$

Players adapt to anticipated strategies which would be next period's opponent strategies if the learning process were of order k-1 instead of order k.

6.2 Full Anticipations

Let β be the maximum of the absolute values of elements of $\tilde{R}$. For sufficiently small $\alpha\beta$ we have

$$\mu < \frac{1}{h}. \qquad (233)$$

Assume that $\alpha\beta$ is sufficiently small in this sense. It can be seen immediately that the maximum of the absolute values of elements of $\tilde{R}^k$ is at most μ^k. Consequently the infinite sequence

$$Y = \underline{D} + \tilde{R} + \tilde{R}^2 + \ldots \tag{234}$$

converges. Moreover,

$$\underline{D} + \tilde{R}Y = Y \tag{235}$$

yields

$$(\underline{D} - \tilde{R})Y = D \tag{236}$$

$$Y = (\underline{D} - \tilde{R})^{-1}. \tag{237}$$

The *learning process with full anticipations* is defined as follows:

$$\begin{bmatrix} p_+ \\ q_+ \end{bmatrix} = (\underline{D} - \tilde{R})^{-1} \begin{bmatrix} p \\ q \end{bmatrix}. \tag{238}$$

Full anticipations are the limit of anticipations of order k for $k \to \infty$. Within a theory of bounded rationality, it is not reasonable to assume that full anticipations are formed. Nevertheless, it is interesting to look at this limiting case.

6.3 Local Stability Condition

In the same way as in 3.8 and 4.8, the learning process of order k can be replaced by a *reduced system* (see (67) and (114)). One receives:

$$\begin{bmatrix} \bar{p}_+ \\ \bar{q}_+ \end{bmatrix} = (\underline{D} + R + \ldots + R^k) \left[c + \begin{bmatrix} \bar{p} \\ \bar{q} \end{bmatrix} \right]. \tag{239}$$

The notational conventions are the same as in (3.8). The *reduced system* of the learning process with full anticipations assumes the following form:

$$\begin{bmatrix} \bar{p}_+ \\ \bar{q}_+ \end{bmatrix} = (\underline{D} - R)^{-1} \left[c + \begin{bmatrix} \bar{p} \\ \bar{q} \end{bmatrix} \right]. \tag{240}$$

It can be seen immediately that the learning process of order k is locally stable if and only if the following condition holds for every eigenvalue λ of R:

$$|1 + \lambda + \ldots + \lambda^k| < 1. \tag{241}$$

The eigenvalues of (240) have the form $1/(1-\lambda)$ where λ is an eigenvalue of R. Therefore, the learning process with full anticipation is locally stable if and only if we have:

$$1 < |1 - \lambda|. \tag{242}$$

The learning process of order k and the learning process with full anticipations are defined as linear systems. There is no difference between local and global stability in linear systems. However, the definition of these processes does not make sense unless one restricts it to a sufficiently small neighborhood of (r,s). Therefore, it seems to be better to speak of local stability in order to emphasize the local character of the analysis.

<u>Theorem 7:</u> Let (A,B) be a quadratic game with h pure strategies 1,...,h for each of both players and let (r,s) be a completely mixed regular equilibrium point of (A,B). Then numbers $\alpha_0 > 0$ and $\beta_0 > 0$ can be found such that for $0 < \alpha < \alpha_0$ and $0 < \beta < \beta_0$ the following is true for the learning process of order k with k at least equal to 2 and for the learning process with full anticipations: (r,s) is (locally) stable if and only if all non-vanishing eigenvalues of the interaction covariance matrix C of (r,s) are negative.

<u>Proof:</u> The theorem is a generalization of Lemma 11, which covers the case k = 2. Let λ be an eigenvalue of R. We use the notational conventions of the proof Lemma 11. In particular, we express λ as a function of α and β as in (178).

In view of

$$1 + \lambda + \ldots + \lambda^k = \frac{1-\lambda^{k-1}}{1-\lambda} \tag{243}$$

the stability condition (241) can be expressed as follows:

$$|1 - \lambda^{k+1}| < |1 - \lambda|. \tag{244}$$

Let η be the absolute value of λ:

$$\eta = |\lambda| = \rho\sqrt{\alpha\beta} . \tag{245}$$

152

In view of (178) we have

$$|1 - \lambda| = \sqrt{(1-\eta\cos\phi)^2+\eta^2\sin^2\phi} \tag{246}$$

$$|1 - \lambda| = \sqrt{1-2\eta\cos\phi+\eta^2} . \tag{247}$$

Similarly, we obtain

$$|1 - \lambda^{k+1}| = \sqrt{1-2\eta^{k+1}\cos(k+1)\phi+\eta^{2(k+1)}} . \tag{248}$$

It follows that (244) holds if and only if we have:

$$-2\eta^{k+1}\cos(k+1)\phi + \eta^{2(k+1)} < -2\eta\cos\phi + \eta^2 \tag{249}$$

or equivalently

$$-2\eta^{k}\cos(k+1)\phi + \eta^{2k+1} < -2\cos\phi + \eta. \tag{250}$$

Suppose that at least one eigenvalue λ of R has a non-vanishing real part. In this case it follows by (i) in Lemma 10 that at least one eigenvalue of R has a positive real part. Assume that λ has a positive real part. Then for $\alpha\beta \to 0$ the right-hand side of (250) converges to the negative constant $-2\cos\phi$ whereas the left-hand side converges to zero. It follows that (250) does not hold for sufficiently small α and β. In view of (iii) and (iv) in Lemma 10, we can conclude that the negativity of all non-vanishing eigenvalues of C is a necessary condition for the local stability with respect to the learning process of order k.

In the full anticipation case local stability requires (242). With the help of (247), this condition can be expressed as follows:

$$0 < -2\eta\cos\phi + \eta^2 \tag{251}$$

or, equivalently

$$0 < -2\cos\phi + \eta. \tag{252}$$

This shows that the negativity of all non-vanishing eigenvalues of C is also a necessary condition for local stability with respect to the learning process with full anticipation.

Now assume that all non-vanishing eigenvalues of C are negative. Then all eigenvalues of R are imaginary and we have $\cos\phi = 0$. Condition (250) assumes the following form

$$\eta^{2k+1} < \eta. \tag{253}$$

Clearly this is the case for sufficiently small α and β. It is also clear that (252) is satisfied: It follows that for the learning process of order k with k at least equal to 2 and for the learning process with full anticipation the non-negativity of all non-vanishing eigenvalues of C is not only necessary, but also sufficient for the local stability of (r,s) if α and β are sufficiently small.

6.4 Comment

Modifications of the anticipatory learning process which consider higher order anticipations do not change the conclusions on the local stability of completely mixed regular equilibrium points. Even if we avoided the full specification of the modified processes for the whole space of mixed strategy pairs including its boundaries, the result suggests that the introduction of higher order anticipations does not yield a fundamentally different theory.

7. Concluding Remark

The theory presented here exhibits anticipation as a potentially stabilizing force in game equilibrium learning. The anticipatory learning process can serve as a point of departure in the analysis of laboratory experiments. Of course, adjustments have to be made for the finiteness of laboratory populations. Frequency distributions over pure strategies cannot be identified with mixed strategies. Moreover, one must expect that the parameters α and β vary from subject to subject. In principle, this does not pose insurmountable obstacles for the theoretical analysis of specific data sets, even if the direct application of the results of this paper may not be possible.

It would be naive to expect that the picture of individual behavior portrayed by the anticipatory learning process is an exact description of reality. Probably modifications will be necessary in the light of experimental evidence. Nevertheless, the author hopes that this paper will not be completely without significance for the development of empirically validated game learning theories.

References

Bomze, I.M. (1986). Non-Cooperative Two-Person Games in Biology: A Classification. International Journal of Game Theory **15**: 31-57.

Brown, G. (1951). Iterative Solution of Games by Fictitious Play. In: T. Koopmanns (Ed.): "Activity Analysis of Production and Allocation". pp. 374-376. New York: Wiley.

Bush, R.R., and F. Mosteller (1955). Stochastic Models for Learning. New York: Wiley.

Crawford, V.P, (1985). Learning Behavior and Mixed Strategy Nash Equilibria. Journal of Economic Behavior and Organization **6**: 69-78.

Cross, J.G. (1983). A Theory of Adaptive Economic Behavior. Cambridge: Cambridge University Press.

Darlington, R.B. (1975). Radicals and Squares. New York: Logan Hill Press.

Gabisch, G., and H.W. Lorenz (1987). Business Cycle Theory. Lecture Notes in Economics and Mathematical Systems **283**. Berlin: Springer-Verlag.

Harley, C.B. (1981). Learning the Evolutionarily Stable Strategy. Journal of Theoretical Biology **89**: 611-633.

Harsanyi, J.G. (1973). Games with Randomly Disturbed Payoffs. A New Rationale for Mixed-Strategy Equilibrium Points. Internat. Journal of Game Theory **2**: 235-250.

Maynard Smith, J. (1982). Evolution and the Theory of Games. Cambridge: Cambridge University Press.

Maynard Smith, J., and G.A. Price (1973). The Logic of Animal Conflict. Nature **246**: 15-18.

Milinski, M. (1979). An Evolutionarily Stable Feeding Strategy in Sticklebacks. Zeitschrift für Tierpsychologie **51**: 36-40.

Miyasawa, K. (1961). On the Convergence of the Learning Process in a 2×2 Non-Zero-Sum Two-Person Game. Economic Research Program, Research Memorandum No. 33. Princeton, N.J.: Princeton University.

Nash, J. (1951). Non-Cooperative Games. Annals of Mathematics **54**: 286-295.

O'Neill, B. (1987). Nonmetric Test of the Minimax Theory of Two-Person Zerosum Games. Proceedings of the National Academy of Sciences, U.S.A., Vol. **84**: 2106-2109.

Robinson, J. (1951). An Iterative Method of Solving a Game. Annals of Mathematics **54**: 296-301.

Rosenmüller, J. (1971). Über Periodizitätseigenschaften spieltheoretischer Lernprozesse. Zeitschr. Wahrsch. Verw. Gebiete **17**: 259-308.

Selten, R. (1980). A Note on Evolutionarily Stable Strategies in Asymmetric Animal Conflicts. Journal of Theoretical Biology **83**: 93-101.

Selten, R., and P. Hammerstein (1984). Gaps in Harley's Argument on Evolutionarily Stable Learning Rules and in the Logic of "Tit for Tat". The Behavioral and Brain Sciences **7**: 115-116.

Selten, R., and R. Stoecker (1986). End Behavior in Sequences of Finite Prisoner's Dilemma Supergames. Journal of Economic Behavior and Organization **7**: 47-70.

Shapley, L.S. (1964). Some Topics in Two-Person Games. In: M. Dresher, L. Shapley, and A. Tucker, "Advances in Game Theory." Annals of Mathem. Studies, No. **52**. pp. 1-28. Princeton, N.J.: Princeton University Press.

Taylor, P.D., and L.B. Jonker (1978). Evolutionarily Stable Strategies and Game Dynamics. Mathem. Biosci. **40**: 145-156.

END BEHAVIOR IN SEQUENCES OF FINITE PRISONER'S DILEMMA SUPERGAMES

A Learning Theory Approach

Reinhard SELTEN and Rolf STOECKER

University of Bonn, D-5300 Bonn 1, FRG

Received 7 June 1983, final version received 15 January 1985

A learning theory is proposed which models the influence of experience on end behavior in finite Prisoner's Dilemma supergames. The theory is compared with experimental results. In the experiment 35 subjects participated in 25 Prisoner's Dilemma supergames of ten periods each against anonymous opponents, changing from supergame to supergame. The typical behavior of experienced subjects involves cooperation until shortly before the end of the supergame. The theory explains shifts in the intended deviation period. On the basis of parameter estimates for each subject derived from the first 20 supergames, successful predictions could be obtained for the last five supergames.

1. Introduction

In a finite Prisoner's Dilemma supergame the same game is repeated for a fixed number of times known to both players in advance. It is well known that such games have a definite game theoretical solution which prescribes non-cooperative behavior in all periods of the supergame. However, experimental behavior does not conform to this theoretical prediction. Early experiments with finite Prisoner's Dilemma supergames [Rapoport–Dale (1966), Morehous (1966), Lave (1965)] already have shown that subjects sufficiently often choose the cooperative alternative. At first glance, the situation seems to be similar to comparable experiments where the number of periods is not known to players in advance. However, such games are more akin to the infinite Prisoner's Dilemma supergame which permits equilibrium points resulting in cooperative behavior.

More recently, experiments have been performed where subjects played the same finite Bertrand Duopoly or Prisoner's Dilemma supergames many times against changing anonymous opponents: the subjects played against this opponent within one supergame, but could not expect to meet the same opponent again in a later supergame [Stoecker (1980, 1983)]. The results show that subjects develop a pattern of behavior which may be described as tacit cooperation until shortly before the end of the supergame followed by non-cooperative choices until the end. As soon as one of the players deviates to non-cooperative behavior the other reacts with non-cooperative choices and cooperation is not established any more. This pattern of cooperation followed by an end-effect is observed in almost all supergames between experienced players. Obviously, straightforward game theoretical reasoning cannot explain experienced behavior in finite Prisoner's Dilemma supergames. One could try to account for this by the assumption that the players' utility is different from the monetary rewards. Players may for example value cooperation as such and therefore refrain from non-cooperative behavior in spite of monetary incentives. Such explanations fail to be convincing in view of the end-effect which indicates that monetary incentives are stronger than the desire to be cooperative for those who deviate to non-cooperative behavior. A more detailed discussion of this point can be found elsewhere [Selten (1978)].

An attempt to explain cooperation followed by an end-effect as the result of fully rational behavior may be based on the idea of slightly incomplete information on the other player's payoff [Kreps, Milgrom, Roberts and Wilson (1982)]. However, such theories predict the mature pattern of behavior already for inexperienced subjects. This does not agree with experimental observations. Subjects first have to learn cooperation and only afterwards do they discover the end effect. Descriptive theories cannot ignore the limited rationality of human subjects.

In this paper we shall present a learning theory approach to the explanation of end behavior in finite Prisoner's Dilemma supergames. We assume that players are motivated by monetary rewards. However, we do not assume optimizing behavior. Our theory is based on a Markov learning model where subjects change their intention to deviate from cooperation in a certain period with transition probabilities depending on experience in the last supergame.

There are obvious analogies between learning and evolution. The evolution of cooperative behavior in the infinite Prisoner's Dilemma supergame has been discussed in Axelrod's (1984) stimulating book. We shall not comment on this work in detail, since here we are concerned with the end-effect, a phenomenon which is excluded by the nature of the infinite supergame.

We shall also present the result of an experiment where each of 35 subjects participated in 25 Prisoner's Dilemma supergames of ten periods each. The data exhibit remarkable individual differences between subjects. Therefore, the parameters of the learning model are fitted separately for each subject.

If one allows for random perturbances which occasionally result in reactions which are excluded by the model, the learning theory could be viewed as roughly in agreement with the behavior of 34 of 35 subjects (one subject behaves in a rather chaotic way). The intention to deviate from cooperation can be moved forward or backwards in time or remain constant from one supergame to the next. The learning model always excludes either the forward shift or the backwards shift. In only 21 out of 585 cases could a reaction in the excluded direction be observed in the data.

A careful look at the data suggests a distinction between different groups of subjects which differ with respect to the degree of conformance between the learning model and observed behavior. In the last 13 supergames where all of the 34 subjects already had some experience with the end effect, 18 subjects never showed any response excluded by the model. However, four of these subjects had constant intentions to deviate in these supergames and therefore could be explained in a simpler way. A slightly more general model would be compatible with all responses of nine further subjects in the last 13 supergames. Each of the remaining seven subjects exhibits only one response in the direction excluded by the model in these supergames.

Statistical computations support the impression that the general ideas underlying our learning model provide a reasonable picture of observed behavior. Computer simulation based on individually estimated parameters produces results which tend to agree with the experimental observations.

2. Experimental procedure

The experiments are based on the Prisoner's Dilemma game shown in fig. 1. The payoffs shown are in German Pfennigs (one German Mark equals 100 Pfennig). In each supergame the game of fig. 1 was repeated ten times. Each subject played 25 supergames. They were told that they played against the same opponent within one supergame but against different opponents in different supergames.

Subjects were placed in separate rooms. They did not communicate with each other. The experimenters asked for each decision by intercom and announced the opponent's decision at the end of each period. Subjects kept records of previous decisions and gains.

		Player 2 HP	Player 2 NP
Player 1	HP	60 / 60	−50 / 145
Player 1	NP	145 / −50	10 / 10

Fig. 1. The game used in the experiment – payoffs for player 1 are shown in the upper left corner and payoffs for player 2 are shown in the lower right corner. The strategies were introduced as high price (HP, hoher Preis) and low price (NP, niedriger Preis).

The experimenters did not only ask for the subjects' decisions but also for expectations on opponents' decisions. Moreover, the subjects were required to write down reasons for each period-decision.

Subjects came to the laboratory for two afternoon sessions of four hours each. Part of the time was used for introductory explanations and for tests on altruism and risk-taking. These tests are not described here since their results will not be used in the evaluation of the experiments. The actual playing of the 25 supergames took about four hours. After some experience one period took less than a minute. It is important to point out that payoff incentives are quite high relative to such a short time span (see fig. 1).

The experimenters tried to create the impression that 26 subjects participated in each session and that they never would meet the same opponent again in a later supergame. Actually, in each session there were only 12 subjects. Unknown to the subjects the experimental design separated the 12 subjects into two groups of six. Each subject played only against changing opponents among the other five subjects in his group (see appendix A).

It was intended to have six groups of six subjects. However, in one of the second sessions one of the subjects did not come and was substituted by a fixed strategy administrated by the experimenter. This strategy prescribes cooperation until a non-cooperative choice of the opponent is observed and non-cooperative behavior from then on. This means cooperation up to the end if the opponent does not deviate in the first nine periods. There were actually three subjects who followed this policy and explicitly explained it in their written reasons.

The subjects were male and female economics and business administration students of the University of Bielefeld in their first year.

3. Experimental results

In the course of the experiment subjects learned a pattern of behavior involving cooperation followed by a non-cooperative end-effect. In order to make this statement precise we introduce the following definitions.

Definition 1. The play of a supergame is called *cooperative* if the following three conditions are satisfied:

(a) In the first m periods, where m is at least four, both players choose the cooperative alternative HP.
(b) In period $m+1$ (for $m<10$) at least one player chooses the non-cooperative alternative NP.
(c) In all periods $m+2,\ldots,10$ (if there are any) both players choose the non-cooperative alternative.

Note that this definition does not exclude the case $m=10$ where both players cooperate from the beginning to the end. Admittedly, the requirement $m \geq 4$ is to some extent arbitrary. However, it is necessary to have some criterion in order to distinguish plays with an end-effect from plays where no cooperation has been reached at all. Moreover, in the experiment no additional case would have to be classified as cooperative if in (a) the condition $m \geq 4$ is weakened to $m \geq 1$.

An end-effect may also occur in plays where cooperation has been reached only after initial non-cooperation. In order to capture this possibility we adopt the following definition of an *end-effect play:*

Definition 2. An end-effect play is characterized by three conditions, (a'), (b) and (c).

(a') Both players choose the cooperative alternative in at least four consecutive periods $k, \ldots, m$.

The conditions (b) and (c) are the same as in the definition of a cooperative play. By definition, a cooperative play is also an end-effect play.

We say that a supergame belongs to round n if it was played as the nth supergame by the subjects. Since there were 36 players [including one simulated player for rounds (9) to (25)] each round has 18 plays. Table 1 shows for every round how many plays were end-effect plays and how many of those were cooperative ones. This is indicated for each of the six groups of interacting subjects separately.

Table 1 shows that for experienced subjects most plays tend to be cooperative; however, there are some subjects who sometimes tried to gain an advantage by choosing the non-cooperative alternative in the first period hoping that the other player would not retaliate. Such behavior results mostly in end-effect plays which fail to be cooperative in the sense of the definition given above. Group 1 contains one subject who seemed to have great difficulty understanding the situation until round (21). In the first 20 rounds his behavior was highly irregular. In the last five rounds 99 percent of the plays are end-effect plays and 96 percent are cooperative plays. Appendix B gives a detailed account of the observed end-effect behavior for all subjects separately.

The learning model to be explained later contains an intended period of deviation as an internal state of the subject. In all cases where a subject deviated before the opponent or simultaneously with the opponent the intended deviation period is nothing else than the observed deviation period. However, if the opponent deviated before the subject the intended deviation period is not uniquely determined by the decisions observed in the play. This situation occurs in 198 out of 621 cases. In 84 of these cases the reasons written down by the subjects indicated the intended deviation period. In the

Table 1
Number of end-effect plays (EEP) and cooperative plays (CP) by rounds and subject groups.

Round	Group I EEP	I CP	II[a] EEP	II CP	III EEP	III CP	IV EEP	IV CP	V EEP	V CP	VI EEP	VI CP	Total EEP	Total CP
(1)									2	2	1		3	2
(2)									1	1			1	1
(3)	1				1	1	1	1	2	2	1		6	4
(4)					1	1	2	2	3	1	2	2	8	6
(5)					1	1	1	1	2		1	1	5	3
(6)					1	1	2	2	3	1	1	1	7	5
(7)			1		1	1	2	2	2	1	1	1	7	5
(8)	1		1		2	2	2	2	3	1	2	2	11	7
(9)	3	3	3	2	1	1	2	2	2	1	2	2	13	11
(10)	2	2	3	3	1	1	1	1	2	1	2	2	11	10
(11)	2	2	3	3	1	1	2	2	1		2	2	11	10
(12)	2	2	3	2	2	2	2	2	2		2	2	13	10
(13)	2	2	2	2	3	3	3	3	3	3	3	3	16	16
(14)	2	2	2	1	3	3	3	3	3	3	3	3	16	15
(15)	2	2	3	1	3	3	3	3	3	3	3	3	17	15
(16)	2	2	3	2	3	3	3	3	3	3	3	3	17	16
(17)	2	2	3	2	3	3	3	3	3	3	3	3	17	16
(18)	2	2	3	2	3	3	3	3	3	3	3	3	17	16
(19)	2	2	3	2	3	3	3	3	3	3	3	3	17	16
(20)	2	2	3	2	3	3	3	3	3	3	3	3	17	16
(21)	3	3	3	1	3	3	3	3	3	3	3	3	18	16
(22)	3	3	3	2	3	3	3	3	3	3	3	3	18	17
(23)	3	3	3	3	3	3	3	3	3	3	3	3	18	18
(24)	3	3	3	3	3	3	3	3	3	3	3	3	18	18
(25)	3	3	2	2	3	3	3	3	3	3	3	3	17	17

[a] This group contains the simulated player for rounds (9) to (25).

remaining 114 cases an estimate of the intended deviation period was based on reported expectations together with observed behavior and reasons from previous rounds.

Table 2 shows the means and standard deviation of intended deviation periods in end-effect plays for all 35 subjects who participated in rounds (13) to (25) for rounds and groups separately. In the last 12 rounds all subjects can be described as experienced in the sense that each of them had been in at least one end-effect play in an earlier round. In the computations deviation period 11 was assigned to those cases where the subject did not intend to deviate at all.

It can be seen that the end-effect has a clear tendency to shift to earlier periods in the last 13 supergames. For each of the six groups the Spearman rank correlation coefficient between the mean of the intended deviation period and the number of the supergame is negative and significant at the 0.1 percent level (two-sided) for the last 13 supergames.

Table 2

Means and standard deviation of intended deviation period in end-effect plays for rounds and groups, separately.

	Round													Spearman rank correlation coefficient[a]
	(13)	(14)	(15)	(16)	(17)	(18)	(19)	(20)	(21)	(22)	(23)	(24)	(25)	
Group I														
Mean	7.8	7.8	7.5	7.0	6.8	7.0	6.5	6.5	6.2	6.0	5.7	5.7	5.5	−0.99
Standard deviation	0.96	0.5	0.58	0.0	0.5	0.82	0.58	0.58	0.41	0.0	0.52	0.82	0.55	
Group II[b]														
Mean	9.2	8.7	8.5	8.5	8.3	8.5	8.2	8.0	7.7	7.8	8.0	7.7	7.5	−0.95
Standard deviation	1.3	1.5	1.2	1.2	1.4	1.2	1.5	1.6	1.6	1.6	1.6	1.9	2.4	
Group III														
Mean	10.2	10.0	10.0	9.8	10.0	9.8	9.8	9.8	9.7	9.7	9.7	9.7	9.0	−0.95
Standard deviation	0.98	1.3	1.3	1.8	1.6	1.8	1.8	1.8	2.1	2.1	2.1	2.1	2.3	
Group IV														
Mean	7.3	7.8	7.7	7.3	7.7	6.8	6.8	7.0	6.7	6.5	6.2	5.8	6.0	−0.93
Standard deviation		2.0	1.0	1.0	1.9	0.8	0.8	0.9	0.5	0.6	0.8	1.0	0.6	
Group V														
Mean	10.0	10.0	10.0	9.3	9.5	9.2	9.3	9.3	9.0	8.8	9.0	8.5	8.3	−0.95
Standard deviation	1.3	1.3	1.3	1.4	1.2	1.2	1.0	1.0	1.1	1.2	0.6	0.6	0.5	
Group VI														
Mean	10.3	10.2	10.2	10.2	9.8	9.8	9.7	9.0	8.8	8.8	8.8	8.3	8.2	−0.99
Standard deviation	0.8	0.8	1.0	1.0	1.0	0.8	0.8	0.6	0.4	0.4	0.8	0.5	0.8	
Total														
Mean	9.2	9.1	9.0	8.7	8.7	8.7	8.5	8.3	7.9	7.9	7.8	7.5	7.4	−1.00
Standard deviation	1.5	1.6	1.5	1.6	1.7	1.6	1.7	1.6	1.7	1.7	1.8	1.8	1.8	

[a]Values are rounded but rank correlations are computed for exact means.
[b]This group contains the simulated player.

Even if it is very clear from the data that there is a tendency of the end-effect to shift to earlier periods, it is not clear whether in a much longer sequence of supergames this trend would continue until finally cooperation is completely eliminated. It is also possible that the mean of the intended deviation period would have a tendency to decrease in such a way that it finally converges to a stable limit. It is interesting to note that some groups show a very strong shift to earlier periods. In round (25) the means of groups I and IV are at 5.5 and 6.0, respectively, whereas this mean is at 9.0 for group III.

4. A learning theory of end-effect behavior

Our learning model contains the intended deviation period k as the internal state of the subject. A subject is assumed to change his internal state from round t to round $t+1$ according to constant transition probabilities. Each subject is characterized by three parameters α, β and γ.

If the subject observes that in round t his opponent deviated earlier than he intended to deviate, then with probability α he will shift his intended deviation period k to $k-1$. The probability that the subject's internal state remains k in this case is $1-\alpha$.

If the subject observes that in round t his opponent deviated at the same period k as he did, then with probability β the subject's intended deviation period will be shifted to $k-1$ and with probability $1-\beta$ it will stay where it is.

If the subject observes that in round t he deviated before the opponent, then with probability γ he will shift the intended deviation period to $k+1$ provided we have $k<10$. With probability $1-\gamma$ he will not change his internal state. There is no change for $k=10$. The assumptions of the learning model are summarized by table 3.

In the explanations given above it was assumed that in round t the subject experienced an end-effect play. It is assumed that no change of intention

Table 3
Transition probabilities from round to round for the intended deviation period of a subject. k is the intended deviation period in round t.

Subject's intended deviation period in round t	Intended deviation period in round $t+1$		
	One period sooner	Unchanged	One period later
Later than his opponent	α	$1-\alpha$	
Together with his opponent	β	$1-\beta$	
Sooner than his opponent		$1-\gamma$ for $k<10$, 1 for $k=10$	γ for $k<10$, 0 for $k=10$

takes place after a round which did not result in an end-effect play. This convention is unimportant for our theoretical derivations and simulations but it has some minor significance for the interpretation of our data.

We now proceed to discuss our motivations behind the assumption of transition probabilities. A subject who has observed that his opponent deviated earlier than he himself intended to do will think that it might have been better to deviate earlier. The same is true to a lesser degree if the opponent deviated in the same period as he did. Therefore, it is reasonable to assume $\alpha \geq \beta > 0$. In both cases there is no reason to shift the intention to deviate to later periods.

Now consider a subject who in round t deviated in a period $k < 10$ and observed that his opponent did not deviate from cooperation up to period k. He does not know exactly in which period the opponent intended to deviate. Therefore, it could have been better to deviate in a later period. We may for example look at $k = 8$. The subject does not know whether the opponent intended to deviate in period 9, 10 or not at all. In the latter two cases a deviation in period 9 would have been more advantageous. It is plausible to assume that this kind of uncertainty produces a tendency to shift the deviation periods towards the end of the supergame. Of course, for $k = 10$ there is no such uncertainty and the subject must conclude that it was right to deviate in the last period if he observed that the opponent cooperated up to the end.

In the mathematical learning models considered in the literature [see for example Restle–Greeno (1970), Bush–Mosteller (1955)] it is generally clear whether reinforcement of behavior has taken place or not. However, in a situation where a subject deviated earlier than his opponent in a period $k < 10$ he does not know whether his decision was right or wrong.

Unobserved features of the opponent's behavior prevent him from having a clear experience of success or failure. However, he knows that here is a possibility that his decision was wrong.

Our specification of the general ideas explained above contain certain simplifying assumptions. We exclude the possibility that the intended deviation period shifts by more than one period. It is, of course, easy to construct a more general learning model where shifts of two or three periods are permitted. However, the scarcity of data forces us to restrict our attention to models with as few parameters as possible.

In a situation where a subject deviated earlier than his opponent his uncertainty on the nature of his experience is the greater the earlier his deviation was. The more periods there are after the deviation until the end of the supergame, the more chances there are that the deviation was too early. Therefore, one could think of making γ dependent on k in such a way that γ increases with decreasing k. This would be a theoretically attractive modification of the model but also here the necessary increase of the number of parameters prevents us from comparing such models with the data.

5. Theoretical considerations

In the following we shall look at the consequences of our theory in an idealized situation which is not that of the experiment.

Consider a very large population of subjects where the parameters α, β and γ are the same for all subjects. With this population we imagine a fictitious experiment over a very long sequence of supergames. At each round the subjects are paired randomly.

We may ask the question how in this system the probabilities of intended deviation periods evolve. In order to describe the process which governs the evolution of these probabilities we introduce the following notations:

$$p_k^t$$

is the probability that in round t a randomly chosen subject has the intention to deviate in period k, where $k=11$ stands for the intention not to deviate at all ($k=1,\ldots,11$),

$$S_k^t = \sum_{m=1}^{k} p_m^t$$

is the probability that in round t a randomly chosen subject has the intention to deviate in periods $1,\ldots,k$, and

α, β and γ

are the parameters of table 3.

It is useful to look at the situation in a way which is similar to that of a Markov chain. We may ask the following question: what are the probabilities that a subject will intend to deviate in period $k-1$, k or $k+1$ in round $t+1$ if he intended to deviate in period k in round t? These 'transition probabilities' can be arranged in a matrix where columns correspond to intended deviation periods in round t and rows correspond to intended deviation periods in round $t+1$. A part of this matrix is shown in table 4.

With the help of table 3 it can be seen easily that the transition probabilities are in fact those shown in table 4. From what has been said up to now it is clear that the probabilities p_k^{t+1} are determined by the following equation system:

$$p_{11}^{t+1} = [(1-\alpha)S_{10}^t + (1-\beta)p_{11}^t]p_{11}^t,$$

$$p_{10}^{t+1} = [\alpha S_{10}^t + \beta p_{11}^t]p_{11}^t + [(1-\alpha)S_9^t + (1-\beta)p_{10}^t]$$

$$+ (1-\gamma)(1-S_{10}^t)]p_{10}^t + \gamma(1-S_9^t)p_9^t,$$

$$p_9^{t+1} = [\alpha S_9^t + \beta p_{10}^t]p_{10}^t + [(1-\alpha)S_8^t + (1-\beta)p_9^t$$
$$+ (1-\gamma)(1-S_9^t)]p_9^t + \gamma(1-S_8^t)p_8^t,$$

$$\vdots \qquad\qquad\qquad \vdots$$

$$p_2^{t+1} = [\alpha S_2^t + \beta p_3^t]p_3^t + [(1-\alpha)p_1^t + (1-\beta)p_2^t$$
$$+ (1-\gamma)(1-S_2^t)]p_2^t + \gamma(1-p_1^t)p_1^t,$$

$$p_1^{t+1} = [\alpha p_1^t + \beta p_2^t]p_2^t + [p_1^t + (1-\gamma)(1-p_1^t)]p_1^t.$$

Table 4
Transition probabilities for a subject between rounds t and $t+1$ (explanation in the text).

Round $t+1$	Round t				
	(11)	(10)	(9)	(8)	(7)
(11)	$(1-\alpha)S_{10}^t +$ $(1-\beta)p_{11}^t$	0	0	0	0
(10)	$\alpha S_{10}^t + \beta p_{11}^t$	$(1-\alpha)S_9^t +$ $(1-\beta)p_{10}^t +$ $(1-\gamma)(1-S_{10}^t)$	$\gamma(1-S_9^t)$	0	0
(9)	0	$\alpha S_9^t + \beta p_{10}^t$	$(1-\alpha)S_8^t +$ $(1-\beta)p_9^t +$ $(1-\gamma)(1-S_9^t)$	$\gamma(1-S_8^t)$	0
(8)	0	0	$\alpha S_8^t + \beta p_9^t$	$(1-\alpha)S_7^t +$ $(1-\beta)p_8^t +$ $(1-\gamma)(1-S_8^t)$	$\gamma(1-S_7^t)$
(7)	0	0	0	$\alpha S_7^t + \beta p_8^t$	$(1-\alpha)S_6^t +$ $(1-\beta)p_7^t +$ $(1-\gamma)(1-S_7^t)$

Starting from an initial distribution $(p_1^1, \ldots, p_{11}^1)$ the probability vector $(p_1^t, \ldots, p_{11}^t)$ can be computed for every round t. We may ask the question whether this probability vector converges to a stable equilibrium distribution.

We shall not try to give a rigorous theoretical answer to the question of convergence. However, we have run a large number of numerical computations whose results show a definite pattern which will be described in the following. It must first be pointed out that the difference equations have the following property: If $p_k^t = 0$ holds for $k = m, \ldots, 11$ for some $t = t_0$ then the same conditions will be satisfied for every $t > t_0$. For this reason alone, the result of the simulation cannot be completely independent of the initial conditions.

However, if p^1_{10} and p^1_{11} are sufficiently high, the results of our computations do not depend on the exact initial conditions. The results obtained for $p^1_{11}=1$ do not change as long as the initial conditions remain in a neighbourhood of this extreme case. The size of this neighbourhood depends on the parameters but for most cases it seems to be quite large.

Table 5 shows numerical results for selected parameter combinations. All these computations have been run starting from the initial condition $p^1_{11}=1$. Our experimental results suggest that subjects learn to cooperate before they learn to show any end-effect. Therefore, the assumption $p^1_{11}=1$ is quite reasonable.

All the computations with $p^1_{11}=1$ converged to a stationary distribution which was always mostly concentrated either at the end or at the beginning of the supergame. In table 5 either the first three or the last three periods obtained at least 97 percent of the total mass of the probability.

The parameter combinations of table 5 are arranged in groups with constant β and γ and increasing α. If α is small in comparison to $\gamma-\beta$ the distribution is mostly concentrated near the end of the supergame. With increasing α this concentration becomes less pronounced until a critical value of α is reached beyond which the stationary distribution is mostly concentrated at the beginning of the supergame. As can be seen in table 5 the critical value for α is a little below $\gamma-\beta$. It can be checked analytically without much difficulty that for $\alpha+\beta=\gamma$ the distribution $p_1=p_2=0.5$ is stationary. In fact, in cases with $\alpha+\beta=\gamma$ the process converges to this distribution.

The results of these computations suggest an abrupt change of the stationary distribution at the critical value of α. In table 5 the critical values of α are enclosed by intervals of the length of 10^{-3}. It can be seen that within this small interval the stationary distribution reached by the process changes drastically. The change is somewhat less pronounced if the interval is narrowed down to the length of 10^{-7} but even there p_5 and p_6 are practically 0 before and after the change from a concentration at the end to a concentration at the beginning.

In the experiments a group of interacting subjects had only six members and the parameter values varied considerably from subject to subject. Moreover, the experimental pairings of subjects are not random, but follow a repetitive scheme (see appendix A). Nevertheless, the model applied to the experimental situation can be looked upon as a Markov chain with a suitably defined state space. The highest among the intended deviation periods of the subjects cannot increase from one supergame to the next, but if α and β are positive and λ is smaller than one for all subjects, then there always is a positive probability that the highest intended deviation time will

Table 5

Stable probability distributions over intended deviation periods for selected parameter combinations.

α	β	γ	p_1	p_2	p_3	...	p_8	p_9	p_{10}
0.100	0.1	0.1	1.000	—	—		—	—	—
0.100	0.1	0.4	—	—	—		0.017	0.250	0.733
0.200	0.1	0.4	—	—	—		0.038	0.346	0.615
0.264	0.1	0.4	—	—	—		0.135	0.520	0.340
0.265	0.1	0.4	0.119	0.509	0.371		—	—	—
0.300	0.1	0.4	0.500	0.500	—		—	—	—
0.300	0.1	0.5	—	—	—		0.034	0.356	0.610
0.355	0.1	0.5	—	—	—		0.105	0.520	0.373
0.356	0.1	0.5	0.098	0.512	0.390		—	—	—
0.400	0.1	0.5	0.500	0.500	—		—	—	—
0.500	0.1	0.5	0.990	0.010	—		—	—	—
0.200	0.2	0.5	—	—	—		0.086	0.400	0.511
0.300	0.2	0.5	0.500	0.500	—		—	—	—
0.400	0.2	0.5	0.667	0.333	—		—	—	—
0.100	0.1	0.6	—	—	—		0.005	0.167	0.829
0.200	0.1	0.6	—	—	—		0.007	0.201	0.791
0.300	0.1	0.6	—	—	—		0.013	0.256	0.731
0.400	0.1	0.6	—	—	—		0.030	0.364	0.606
0.449	0.1	0.6	—	—	—		0.095	0.533	0.370
0.450	0.1	0.6	0.087	0.522	0.391		—	—	—
0.200	0.2	0.6	—	—	—		0.044	0.333	0.622
0.300	0.2	0.6	—	—	—		0.083	0.417	0.498
0.353	0.2	0.6	—	—	—		0.188	0.516	0.282
0.354	0.2	0.6	0.151	0.500	0.349		—	—	—
0.400	0.2	0.6	0.500	0.500	—		—	—	—
0.500	0.2	0.6	0.667	0.333	—		—	—	—
0.600	0.2	0.6	0.993	0.007	—		—	—	—
0.100	0.1	0.7	—	—	—		0.003	0.143	0.854
0.200	0.1	0.7	—	—	—		0.004	0.167	0.828
0.300	0.1	0.7	—	—	—		0.006	0.203	0.791
0.400	0.1	0.7	—	—	—		0.011	0.259	0.730
0.500	0.1	0.7	—	—	—		0.028	0.370	0.602
0.543	0.1	0.7	—	—	—		0.073	0.517	0.409
0.544	0.1	0.7	0.072	0.515	0.413		—	—	—
0.600	0.1	0.7	0.500	0.500	—		—	—	—
0.700	0.1	0.7	0.990	0.010	—		—	—	—
0.440	0.2	0.7	—	—	—		0.164	0.525	0.302
0.441	0.2	0.7	0.133	0.506	0.361		—	—	—
0.529	0.2	0.8	—	—	—		0.144	0.530	0.320
0.530	0.2	0.8	0.119	0.509	0.371		—	—	—
0.445	0.3	0.8	—	—	—		0.213	0.510	0.257
0.446	0.3	0.8	0.165	0.495	0.340		—	—	—
0.538	0.4	0.9	—	—	—		0.227	0.504	0.244
0.539	0.4	0.9	0.173	0.492	0.335		—	—	—
0.100	0.1	1.0	—	—	—		—	0.100	0.900

decrease by one. Therefore, with probability one the highest deviation time will finally decrease to one in an infinite sequence of supergames. This means that in the long run behavior converges to complete non-cooperation.[1]

Even if the possibility of convergence to a stationary distribution exhibiting a stable end-effect is excluded by our model, if it is applied to a finite population of subjects, the computations for the idealized situation with an infinite population are not without interest. They suggest that in the finite situation convergence to non-cooperation may be very slow if the parameter values for α and β are relatively small and those for λ are relatively large. Moreover, in the light of the computations for the infinite case one must consider the possibility that the conclusion on convergence to non-cooperation is not robust with respect to slight misspecifications of the learning model. Suppose that the probabilities for the excluded transitions are not really zero, but only relatively small. It is reasonable to expect that under this condition a stationary distribution exhibiting a stable end-effect might be obtained for suitable parameter combinations in the finite case.

However, it can be expected that the results obtained for the large group case with equal parameters are indicative for what can be expected to happen in the experimental situation if the model is correct.

6. Subject differences

After the theoretical considerations of the last section we shall now turn our attention to some important features of our experimental results. The behavioral assumptions of our model do not fit all subjects equally well.

There are several deviations from the theoretical behavior which may occur. Some subjects occasionally change the intended deviation period by more than one step from one round to the next. Even if this is not a deviation from the spirit of our model, it is a deviation from the specification which had to be used in view of the scarcity of observations. A more serious deviation which occurred only rarely is a shift of the intended deviation period in the wrong direction. Some subjects do not show any reaction excluded by the model but they have a constant intended deviation period. An intended deviation period which does not change over time can be explained in simpler and possibly more adequate ways than by our model.

Table 6 distinguishes several groups of subjects according to the conformance of their behavior to the model in the last 13 rounds. We restricted this evaluation to the second half of the experiment since there almost all subjects had learned to cooperate. Only a subject who has learned to cooperate can experience an end-effect play.

It can be seen that only 20 percent of all subjects show a shift in the

---

[1] We are grateful to an anonymous referee who directed our attention to this point.

Table 6

Grouping of subjects by conformance of behavior to the model in the last 13 rounds. Each subject is listed only once.

Subject category	Number of subjects	Number of deviations	Total number of cases
No deviations from the model and varying intended deviation period	14	—	158
Constant intended deviation period[a]	4	—	48
Shifts of more than one step but no other deviations from the model	9	14	108
Shifts in the wrong direction[b]	7	7	81
Failure to learn cooperation[c]	1	—	—

[a]Three of these subjects never intended to deviate as a matter of principle.
[b]Two of these subjects also showed jumps of more than one step.
[c]Unlike all other subjects this subject did not learn to cooperate in the first half of the experiment. He began to experience end-effect plays only in the last five rounds.

wrong direction. For each of these seven subjects such a shift occurs only once.

Three subjects always had the intention to cooperate until the last period. The protocols written by these subjects show that they did this on principle. Obviously, the learning model does not adequately describe the motivations of these subjects even if it formally fits their behavior. One subject always intended to deviate in period (8). He thought that this is the optimal deviation period. Since this opinion was based on experience rather than theoretical reasoning, his behavior may be adequately explained by the model.

Up to occasional deviations, a learning theory approach like that of our model seems to offer a plausible explanation for the behavior of the vast majority of subjects. A fundamentally different theory may be required for those three subjects who never intended to deviate in rounds (13) to (25) as a matter of principle. The learning model cannot be compared with the behavior of the subject who failed to learn to cooperate in rounds (1) to (20). With these exceptions the learning model can be proposed as an idealized picture of observed behavior. The next section will try to throw further light on the extent to which the data agree with the learning model.

7. Parameter estimates

The observations for rounds (1) to (20) have been used in order to obtain parameter estimates $\hat{\alpha}$, $\hat{\beta}$ and $\hat{\gamma}$ of α, β and γ, respectively, for all subjects with the exception of subject 1 who failed to learn to cooperate in rounds (1) to (20). On the basis of these parameter estimates Monte Carlo simulations have been run in order to generate predictions for rounds (21) to (25) which can be compared with the data. The Monte Carlo simulations will be discussed in section 9.

As far as possible relative frequencies of transitions have been taken as parameter estimates. In the determination of relative frequencies shifts of more than one step in the right direction have been counted as if they were shifts of one step. Shifts in the wrong direction have been counted as if they were cases of unchanged deviation periods.

The parameter estimates are shown in table 7. In the three cases indicated by the superscript 'a', relative frequencies were not available due to lack of observations and estimates had to be obtained in another way.

It is plausible to assume $\alpha \geq \beta$ since there is more reason for a shift to an earlier deviation if the opponent has deviated earlier than in the case that he has deviated at the same time. In fact, in 26 of the 31 cases where relative frequencies estimates $\hat{\alpha}$ and $\hat{\beta}$ are available, the inequality $\hat{\alpha} \geq \hat{\beta}$ is satisfied. Therefore, it seems to be reasonable to take the following inequality as a point of departure:

$$0 \leq \hat{\beta} \leq \hat{\alpha} \leq 1.$$

Accordingly, an auxiliary estimate $\hat{\beta} \leq \hat{\alpha}/2$ is formed at the midpoint of the relevant interval delineated by this inequality if a relative frequency estimate is available for α but not for β. Analogously, an auxiliary estimate $\hat{\alpha} = (\hat{\beta}+1)/2$ is formed if a relative frequency estimate is available for β but not for α.

It can be seen that the estimates in table 7 vary considerably from subject to subject. This is also true for the 14 subjects whose behavior completely conforms to the model. For these 14 subjects a second set of parameter estimates has been obtained in the same way on the basis of the data from rounds (1) to (25). These estimates will be used for a comparison of the learning model with a simple alternative hypothesis to be explained in the next section.

8. Comparison with a simple alternative hypothesis

In the following we want to look at the question whether our model provides a better explanation of the data than a simple alternative hypothesis based on the assumption that no learning takes place at all. We compare the learning model with the simplest alternative theory of this kind. In the alternative hypothesis each subject is assumed to have a probability distribution over his intended deviation period which does not vary over time. The intended deviation period of each round is assumed to be stochastically independent from those of other rounds.

Table 7

Parameter estimates based on rounds (1) to (20). Subjects are grouped according to the categories of table 6, in the same order.

Subject	$\hat{\alpha}$	$\hat{\beta}$	$\hat{\gamma}$
2	1.00	0.67	0.50
3	1.00	1.00	0.17
6	0.50	0.30	0.25
7	1.00	0.33	0.20
9	0.60	0.00	0.00
11	0.33	0.40	0.50
12	0.50	0.00	0.00
13	1.00[a]	1.00	0.00
18	0.22	0.00	0.50
22	0.67	0.40	0.50
28	0.25	0.00	0.00
30	0.25	0.00	0.50
31	0.57	0.33	0.57
34	1.00	0.50	0.50
15	0.00	0.00	0.00
17	0.00	0.00	0.00
21	0.00	0.20	0.50
25	1.00	0.00	0.00
16	0.43	0.00	0.67
19	0.00	0.00	0.67
20	0.75	0.50	0.14
24	0.00	0.50	0.09
27	0.00	0.13	0.00
29	0.00	0.14	1.00
32	0.38	0.00	0.00
33	0.14	0.00	0.33
35	1.00	1.00	0.75
4	0.33	0.20	0.00
5	0.50	0.00	0.00
8	1.00	0.75	0.33
14	0.17	0.08[a]	1.00
23	1.00	0.50[a]	0.18
26	0.50	0.50	0.33
36	1.00	0.50	0.50

[a] No relative frequency estimate available; auxiliary estimate according to $\hat{\alpha} = (\hat{\beta}+1)/2$ or $\hat{\beta} = \hat{\alpha}/2$, respectively.

The comparison will be restricted to those 14 subjects which never showed a reaction excluded by the learning model in the last 13 rounds and also had varying intended deviation periods in these rounds. For each of these subjects the probabilities for the actually observed intended deviation periods have been computed under the assumption of the model and under the alternative hypothesis. The parameters α, β and γ have been estimated on the basis of all 25 rounds. In the computation of the probabilities for the learning model the behavior of the other subjects in the same group had been taken as given. The probabilities therefore are conditional on the behavior of the other players. For each of the 14 subjects a conditional likelihood ratio has been formed as the quotient of the probability generated by the model divided by the probability generated by the alternative hypothesis. The conditional likelihood ratios are shown in table 8.

Table 8
Conditional likelihood ratios for the 14 subjects in the first group of table 6.

Subject no.	Ratio	Subject no.	Ratio
2	5.2	13	8830.0
3	131.0	18	540.0
6	23796.0	22	1628.0
7	2421.0	28	4.5
9	34.0	30	41086.0
11	660.0	31	2.0
12	1077.0	34	13.0

It can be seen that all 14 of the conditional likelihood ratios are greater than one; most of them are quite high.

The results of table 8 support the assumption that learning is an important factor in the choice of the intended deviation period. It would not make much sense to extend the method to the other subjects. Of course, a subject with a constant intended deviation period is better explained by the alternative hypothesis. The probabilities generated by the learning model for subjects who do not conform to it, are always 0, even in cases where there is only one isolated deviation. However, it is important to exclude the possibility that even for those subjects whose behavior does not violate the restrictions of the model the simpler alternative hypothesis yields a better explanation.

It would have been desirable to compute likelihood ratios for whole groups of interacting subjects rather than for individuals. Unfortunately, every group had at least one member not among those subjects to which the comparison was restricted.

Even if it must be admitted that our method of comparison is not entirely satisfactory the results confirm the impression that the learning model captures important aspects of the dynamics of end-effect behavior.

9. Monte Carlo simulations

The Monte Carlo simulations which already have been mentioned in the section on theoretical considerations serve the purpose to examine the predictive potential of the learning model. Therefore, in table 7 the parameters α, β and γ have been estimated individually for the subjects on the basis of observed behavior in the first 20 rounds. With these parameters the last five rounds have been simulated starting from the observed values of intended deviation periods in round (20) as initial conditions. The pairing of the subjects followed the schedule of appendix A. The simulations only cover five of the six groups. The first group had one member who did not learn to cooperate before the last five rounds (see table 6). Therefore, for this subject no parameter estimates could be computed on the basis of the first 20 rounds.

The size of the end-effect is best described by the 'intended deviation time' which is defined as 11 minus the intended deviation period.

For each of the five groups fig. 2 shows the means of the intended deviation times over the six subjects for each of the last five rounds. These means are indicated both for the actual experiment and for the eight Monte-Carlo simulations.

It can be seen that the actual observed means are not too dissimilar from those generated by the Monte-Carlo simulations. It must be pointed out, however, that some shifts of more than one period occurred in the last five rounds. There was, for example, one subject in group III who shifted his intended deviation period from 11 to 6 from round (24) to round (25). Of course, the Monte-Carlo simulations cannot reproduce the effects of such jumps. This explains the special features in the drawing for groups III and V.

A meaningful statistical comparison of the simulations and the observations must be based on some features of the simulations which do not vary too much from realization to realization. It is plausible to conjecture that the rank order of the cumulative shifts of intended deviation periods over the last five rounds satisfies this criterion. The cumulative shift is the difference of the intended deviation periods in round (25) and in round (20). For each simulation run we obtain a rank order of these shifts over the six subjects of the group. In this way, the eight simulation runs for each group yield eight rank orders. Kendall's concordance coefficient W has been computed for the eight rankings in each of the five groups, separately. All five concordance coefficients are significant on the 0.01 level. This supports the conjecture that the rank order of cumulative shifts is a variable which can be predicted with some reliability if the model is correct. The predicted mean rank order has been computed by the sum of ranks following Kendall's proposal [Siegel (1957), Kendall (1948)].

For each of the five groups we have correlated the mean rank order of

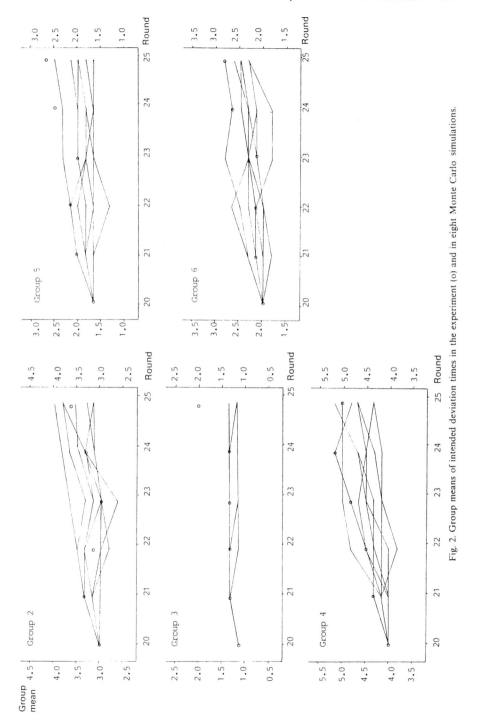

Fig. 2. Group means of intended deviation times in the experiment (o) and in eight Monte Carlo simulations.

cumulative shifts derived from the eight simulation runs with the rank order of cumulative shifts observed in the experiment. The Spearman Rank correlation coefficients are 0.880 ($p<0.05$), 0.548, 0.956 ($p<0.01$), 0.462 and 0.926 ($p<0.05$) for groups II,..., VI, respectively.

The cumulative shift between round (20) and round (25) rather than the intended deviation period of round (25) has been chosen as the basis of the comparison between simulations and observations since the latter variable could reflect the initial conditions of round (20) more than the effects of the parameter values. On the other hand, the cumulative shift is a measure which can be expected to be more closely connected to the dynamics of the learning process.

If the learning model had no predictive value one would expect positive and negative rank correlation coefficients between predicted and observed rank orders of cumulative shifts with equal probability. The binomial test rejects this null hypothesis on the 0.05 level (one-sided). Moreover, three of the five-rank correlation coefficients are significant at the 0.05 level.

The result of the comparison of predicted and observed rank orders of cumulative shifts support the learning model as an idealized picture of end-effect behavior in repeated Prisoners' Dilemma supergames.

Appendix A

The six groups of interacting subjects were composed as shown in table A.1. Within each group of six interacting subjects the pairings were determined according to the scheme in table A.2. The same pattern was repeated in rounds (6) to (10), (11) to (15), (16) to (20) and (21) to (25).

For group II the numbers 1, 3, 5, 7, 9, 11 have to be replaced by 2, 4, 6, 8, 10, 12, in that order. The pairings within the other groups are obtained analogously.

Table A.1
Composition of groups I to VI.

Group	Subjects
I	1, 3, 5, 7, 9, 11
II	2, 4, 6, 8, 10, 12
III	13, 15, 17, 19, 21, 23
IV	14, 16, 18, 20, 22, 24
V	25, 27, 29, 31, 33, 35
VI	26, 28, 30, 32, 34, 36

Table A.2
Pairings in group I for rounds (1) to (5).

Round	Pair 1	Pair 2	Pair 3
(1)	1, 3	5, 7	9, 11
(2)	1, 5	3, 11	7, 9
(3)	1, 7	3, 9	5, 11
(4)	1, 9	3, 5	7, 11
(5)	1, 11	3, 7	5, 9

Appendix B

Table B.1 Subjects' intended deviation period and the observed behavior of the opponent in end-effect plays. Subjects are ordered according to interacting groups (see appendix A).[a]

Subject		(1)	(2)	(3)	(4)	(5)	(6)	(7)	(8)	(9)	(10)	(11)	(12)	(13)	(14)	(15)	(16)	(17)	(18)	(19)	(20)	(21)	(22)	(23)	(24)	(25)
1.	intended									θ													6	6	7	6
	observed									10													6	5	6	5
3.	intended									9	8								6	6	7	7	6	6	6	5
	observed									9	7			7	7	7	7	7	^	^	6	6	6	6	6	5
5.	intended			9					9	9	9	9	7	7	^	7	7	7	8	^	^	^	6	6	5	5
	observed			9					9	9	^	8	^	^	8	8	7	^	7	7	6	6	6	6	5	^
7.	intended									7	7	8	8	9	7	7	7	6	7	6	^	6	6	5	5	5
	observed									^	^	^	8	7	7	8	7	^	6	6	^	6	6	^	^	5
9.	intended									10	10	9	8	8	8	8	7	6	7	7	6	6	6	6	6	6
	observed									θ	9	8	8	7	^	7	7	^	6	6	^	6	6	5	^	5
11.	intended			9								8	^	^	8	7	7	7	6	7	7	6	6	6	^	5
	observed			9								8	8	9	8	8	7	7	7	7	^	7	7	5	5	7
2.	intended								9	^	^	^	^	^	^	8	8	^	8	^	7	7	7	^	7	6
	observed								9	10	10	9	8	9	8	8	8	8	8	7	^	7	7	^	^	^
4.	intended									10	10	9	^	9	8	8	8	8	8	8	8	7	7	^	8	^
	observed									^	9	9	9	9	8	8	8	7	8	7	7	7	7	7	6	^
6.	intended							8	^	^	10	10	^	^	8	^	7	^	^	^	^	7	7	^	^	6
	observed							8	8	9	9	9	9	^	8	8	8	8	8	8	8	7	7	7	7	^
8.	intended									^	^	^	^	^	^	^	^	^	8	^	^	7	8	8	^	6
	observed								9	8	9	9	9	9	^	8	7	8	8	8	^	7	^	7	6	^
10.	intended	Simulated player for rounds (8) to (25)
	observed																									
12.	intended									9	9	9	9	9	8	8	8	8	8	8	8	7	7	7	6	6
	observed									8	^	^	8	9	^	^	^	8	8	8	7	^	7	7	^	^
13.	intended													9	8	8	8	8	8	8	8	^	7	7	7	7
	observed													9	^	^	^	8	8	8	7	7	^	^	^	^
15.	intended				θ				θ	θ			θ	θ	θ	θ	θ	θ	θ	θ	θ	θ	θ	θ	θ	θ
	observed				θ			θ	10	θ			10	θ	θ	10	8	8	θ	8	7	7	7	7	7	6
17.	intended			θ	θ	θ		θ	θ	θ				θ	θ	θ	θ	θ	θ	θ	θ	θ	θ	θ	θ	θ
	observed			9	θ	θ		9	θ	θ				θ	θ	θ	θ	θ	θ	θ	θ	θ	θ	θ	θ	θ
19.	intended									θ	θ			10	9	10	θ	θ	8	7	θ	θ	θ	θ	7	θ
	observed								θ	θ	10			9	9	θ	θ	θ	θ	θ	θ	θ	θ	θ	7	6
21.	intended						10	10	10	θ	10	θ	9	9	^	θ	θ	θ	θ	θ	θ	7	7	7	7	θ
	observed						9	θ	θ	θ	θ	10	θ	θ	θ	7	7	θ	7	8	θ	7	7	7	7	8
23.	intended			9			9	9	9	θ		θ	θ	10	10	9	9	8	7	7	7	7	7	7	7	8
	observed			^			^	^	^	^		θ	θ	θ	9	8	^	^	^	^	^	^	^	^	^	7

262 *Game Theory and Economic Behaviour I*

Table B.1 (continued)

Subject		Round																								
		(1)	(2)	(3)	(4)	(5)	(6)	(7)	(8)	(9)	(10)	(11)	(12)	(13)	(14)	(15)	(16)	(17)	(18)	(19)	(20)	(21)	(22)	(23)	(24)	(25)
14.	intended				9	9	θ	9	8	8				8	9	9	9	θ	θ	θ	8	7	7	7	7	6
	observed				>	7	6	7	8	>				>	7	7	7	8	6	6	7	7	7	6	5	>
16.	intended				9	9	6	9	8	8	8	6	8	8	5	6	7	7	7	7	8	7	6	6	5	6
	observed				>	7	6	7	θ	>	6	6	6	6	>	>	>	8	6	7	6	7	6	6	>	6
18.	intended			5	θ	6	6	7	θ	θ		6	5	8	8	8	8	>	8	8	7	7	7	7	>	6
	observed			>	9	>	6	>	7	8		6	5	7	5	8	6	8	7	7	6	7	7	6	7	6
20.	intended				θ	6	θ	θ	8	7	6	6	5	θ	θ	6	6	6	6	6	6	6	6	6	5	5
	observed				7	>	>	7	8	7		6	5	8	7	8	7	6	6	6	6	6	7	6	6	6
22.	intended			7	7		θ	7	8	7	6	6	5	θ	7	8	7	6	6	6	6	6	>	6	>	6
	observed			5	7		7	7	7	7	>	6	5	8	7	>	7	7	7	7	7	7	7	5	5	5
24.	intended			9	8	>	>	>	>	7			6	7	>	7	7	>	>	>	>	>	6	>	>	>
	observed	9		9	>	>	8	>	8	8			>	>	7	8	7	8	8	8	8	8	8	8	8	7
25.	intended	9		7	>	9	>	9	>	8	8			>	>	>	>	>	>	>	>	>	>	>	>	6
	observed	9		9	9	9	9	>	9	9	8		9	θ	θ	θ	9	9	9	9	9	9	θ	9	9	8
27.	intended	9		>	9	9	8	>	8	9	9	9	8	10	θ	θ	8	9	9	9	9	8	θ	9	8	9
	observed				9	9	8	8	9	9	9	7	8	θ	θ	9	9	8	9	9	9	9	θ	9	8	8
29.	intended				9	9	>	>	9	9	9	7	8	θ	θ	9	9	>	9	9	>	>	8	>	>	9
	observed				9	9	8	8	8	9	9	>	8	8	7	θ	9	8	8	9	9	θ	θ	9	9	8
31.	intended		9	7	9	9	9	>	8	9	>	>	>	θ	10	θ	>	10	10	10	10	>	>	10	9	9
	observed		>	>	8	9	9	9	8	9				θ	9	>	θ	θ	8	>	8	9	9	9	9	8
33.	intended	10	10	10	10	9	9	9	9	9	9	9	8	10	θ	θ	θ	θ	9	>	8	>	9	10	>	9
	observed	9	9	9	8	9	9	9	8	9	7	7	8	θ	8	9	9	9	9	9	9	9	9	9	9	8
35.	intended	9			8		8	10	>	8	>	>	8	θ	9	10	8	9	8	>	10	10	>	9	9	8
	observed	>			>		>	9	9				>	θ	>	8	>	9	>	θ	8	8	9	9	9	8
26.	intended													θ	10	10	10	10	10	10	10	>	>	10	>	9
	observed													θ	10	10	θ	θ	θ	9	8	9	9	9	9	8
28.	intended				9	9	9	9	9	9	9	9	9	10	9	9	>	9	θ	>	9	9	9	9	9	9
	observed				9	9	9	>	7	7	7	7		9	>	>	>	θ	θ	>	9	>	9	8	8	8
30.	intended	9	9	9	9				9	9	9	9	8	10	>	9	>	9	10	10	9	9	9	8	9	9
	observed	>	8	8					8	9	7	7		θ	θ	9	θ	θ	10	9	9	9	9	8	9	9
32.	intended	10			10		9			9	>	>	7	θ	10	9	θ	θ	10	9	9	9	9	9	9	8
	observed	9			9		9			8	7	9	6	θ	10	9	θ	θ	θ	9	9	9	9	9	9	7
34.	intended								6			>	6	9	10	9	θ	9	10	9	9	9	9	>	>	>
	observed								>			9	8	θ	10	θ	10	9	10	9	8	9	9	9	9	8
36.	intended			8	9			10	8	8	8	8	8	θ	10	10	10	10	10	9	8	8	8	8	8	>
	observed			>	>			9	>	>	>	>	>	10	θ	10	9	9	10	>	>	>	>	8	>	8

Note: [a] The entries are as follows: 'intended': intended deviation period; 'observed': opponent's deviation period as observed by the subject; >: the subject deviated before the opponent; θ: in an 'intended'-row: the subject did not intend to deviate, and in an 'observed'-row: the opponent cooperated up to the end.

References

Axelrod, R., 1984, The evolution of cooperation (Basic Books, New York).
Bush, R.R. and F. Mosteller, 1955, Stochastic models for learning (Wiley, New York).
Kendall, M.G., 1948, Rank correlation methods (Griffin, London).
Kreps, D. and R. Wilson, 1982, Reputation and imperfect information, Journal of Economic Theory 27, 253–279.
Kreps, D., P. Milgrom, J. Roberts and R. Wilson, 1982, Rational cooperation in the finitely repeated Prisoner's Dilemma, Journal of Economic Theory 27, 245–252.
Lave, L.B., 1965, Factors affecting cooperation in the Prisoner's Dilemma, Behavioral Science 10, 26–38.
Milgrom, P. and J. Roberts, 1982, Predation, reputation and entry deterrence, Journal of Economic Theory 27, 280–312.
Morehous, L.G., 1967, One-play, two-play, five-play and ten-play runs of Prisoner's Dilemma, Journal of Conflict Resolution 11, 354–362.
Rapoport, A. and Ph.S. Dale, 1967, The 'end' and 'start' effects in interated Prisoner's Dilemma, Journal of Conflict Resolution 11, 354–462.
Restle, F. and J.G. Greeno, 1970, Introduction to mathematical psychology (Addison-Wesley, Reading, MA).
Selten, R., 1978, The chain store paradox, Theory and Decision 9, 127–159.
Siegel, S., 1956, Nonparametric statistics for the behavioral sciences (McGraw-Hill, New York).
Stoecker, R., 1980, Experimentelle Untersuchung des Entscheidungsverhaltens im Bertrand–Oligopol (Pfeffer, Bielefeld).
Stoecker, R., 1983, Das elernte Schlußverhalten — eine experimentelle Untersuchung, Zeitschrift für die gesamte Staatswissenschaft, 100–121.

[9]
Experimental Sealed Bid First Price Auctions with Directly Observed Bid Functions

Reinhard Selten
Joachim Buchta
University of Bonn

In the literature one finds quite a number of experimental papers on first price sealed bid auctions with independent private values (Cox, Roberson, & Smith, 1982; Cox, Smith, & Walker, 1983, 1985a, 1985b, 1988; Kagel & Levin, 1985; Kagel, Harstad, & Levin, 1987; Walker, Smith, & Cox, 1987). In these studies bid functions have been estimated from the data. In some cases it was assumed that all subjects had the same bid functions and in others separate bid functions were estimated for each subject, but in all cases it had been assumed that bid functions do not change in the course of the experiment.

The new feature of the experiment presented here consists in the fact that bid functions were not estimated but directly observed. This makes it possible to check whether bid functions have the properties predicted by game-theoretic analysis. It can also be explored how bid functions are changed by experience.

In our experiment three bidders compete for 50 periods; private values are uniformly and independently distributed over the interval [0,100]. It is well known that under these conditions risk-neutral equilibrium requires that for three competitors the bid is two thirds of the private value. In experiments the bids tend to be higher. In the literature this fact has been explained by the assumption that subjects are risk-averse (Cox, Robertson, & Smith, 1982). This interpretation of the data does not agree well with our findings as we show later.

We look at the data in the light of an approach that we call "learning direction theory." This approach is not yet a full-fledged learning theory because it only explains directions of change without any attempt to predict amounts of change. The same kind of theory has been applied successfully in several other studies (Selten & Stoecker, 1986; Mitzkewitz & Nagel, 1993; Nagel, 1993; Kuon, 1994).

Learning direction theory predicts that subjects tend to lower or not increase their bids at last period's value if they have obtained the object because in this case they might have been able to buy it at a lower price. Similarly it is predicted that subjects tend to increase or not lower their bids at last period's value if it was sold to another competitor below this value. In this case they lost the opportunity to make a profit by a bid slightly higher than the price. Learning direction theory is based on the idea that after an experience subjects think about what might have been a better decision in the last period. They then adjust their behavior accordingly. Of course, this does not conform to the full rationality of economic theory, but it does conform to the bounded rationality of experimental subjects.

THE EXPERIMENTAL DESIGN

The data were gathered in three sessions involving 9, 6, and 12 subjects, respectively, subdivided into independent subject groups of three subjects each that stayed together during all 50 periods. There was no interaction across subject groups.

In each of the nine independent subject groups the three subjects interacted in sealed bid first price auctions for 50 periods. The subjects interacted anonymously via computer terminals. The private value of the object was uniformly distributed over the interval [0,100]. Only one object was sold in every auction.

At the beginning of every period each subject had to specify a piecewise linear bid function shown on the computer screen. The corners of the bid function could be fixed by a graphical input mode or could be entered numerically. The program then automatically computed the linear connections between adjacent corners and showed them on the screen. The input precision was two decimal points but otherwise the number of corners was not limited. The subjects had the option not to change their bid functions from one period to the next. Only in the case of a change, new corners had to be specified. It was also possible to make a direct linear connection between two corners whereby intermediate corners were eliminated. Bids higher than the private value were not permitted. Several changes could be made consecutively and the result of each change was immediately visible on the screen. Finally, the subjects had to confirm that they did not want further changes.

After the subjects had determined their bid functions an object value was drawn randomly and bids were determined accordingly. The object was sold to the highest bidder at the price of the bid. After the auction the price was made public to all three participants. Hidden information about values and bids of the competitors was not revealed. As far as this is possible, common knowledge about the rules was established by the introductory talk.

The subjects had the possibility to look at a table on the screen that showed

Table 5.1
The Last Nonmonotonous Bid Function

Period of Last Nonmonotonous Bid Function	Number of Subjects
50	6
40 – 49	3
30 – 39	3
4 – 5	2
Always monotonous	13

the corners of the current bid function. Another table, which was also available, exhibited the subject's history of past play.

After each auction period the profits of the highest bidders were added to his or her account. At the end of the session money payoffs were determined by an exchange rate of DM 0.30 per point.

All sessions were performed in 1992 at the Bonn Laboratory of Experimental Economics. Subjects were separated from each other and interacted anonymously by formal bids only. Each session lasted for 3 – 4 hours including an introductory talk of about 30 minutes and exercises guided by a computerized learning program by which subjects were made familiar with the technique of fixing piecewise linear bid functions.

ARE BID FUNCTIONS MONOTONOUS?

All theoretical models proposed in the literature (Cox, Roberson, & Smith, 1982) including those involving risk aversion and/or incomplete information yield monotonously increasing bid functions. Surprisingly, in our experiment we observed many bid functions that were decreasing over parts of the range.
We speak of a new bid function if a bid function has been changed from the previous period to the current one. In all sessions, 475 new bid functions were observed. Two hundred and nineteen (46%) of these new bid functions are decreasing over parts of the range. It is quite possible that in the beginning some subjects did not have a sufficient understanding of the strategic situation and others indulged in a certain playfulness, which led them to misuse the graphical program to draw interesting landscapes. However, the phenomenon of nonmonotonicity is not an initial effect. Among the 383 new bid functions in periods 6 to 50 there are 174 (45%) that are nonmonotonous. Here, we use the word *nonmonotonous* in the sense that a bid function is decreasing over parts of

Selten & Buchta

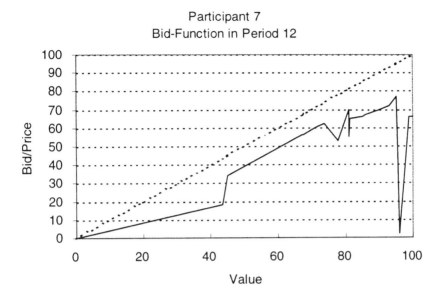

Fig. 5.1. The bid function of Subject 7 in period 12.

---

the range. Similarly, *monotonous* is used in the sense of not decreasing.

Of the 27 subjects, only 15 specified monotonous bid functions in periods 6 to 50. Table 5.1 gives an impression of when subjects had a nonmonotonous bid function for the last time in the course of play.

It can be seen that those 12 subjects who specified nonmonotonous bid functions in periods 6 to 50 did not switch to monotonicity until late in the session. The phenomenon of nonmonotonicity is quite persistent even if there seems to be a weak tendency to more monotonicity with increasing experience. Figure 5.1 shows an example of a nonmonotonous bid function observed in the experiment.

At the moment we can only speculate about the reasons why some subjects specify nonmonotonous bid functions. One often finds sudden drops of the bid function in the high parts of the range that may be due to the desire to have a small chance to make an extraordinarily high profit. This goal can be achieved by specifying very low bids for a small interval of high values.

Especially in the last two periods some subjects substantially decrease their bids in the high part of the range, presumably, in order to provide a last chance for a high profit. In order to exclude such end effects as well as initial effects from the evaluation of the data, we often restrict our attention to periods 5 to 48.

Table 5.2
Linearity in Bid Functions

	Period 5	Period 48
Exactly linear	5	2
Not exactly linear	22	25

Note. Entries show numbers of subjects.

Table 5.3
Measures of Determination of the Ray-Approximations of Observed Bid Functions

	Period 5	Period 48
Minimum	.889	.780
Median	.996	.997
Maximum	1.000	1.000
Number of subjects with $R^2 > .99$	18	22

ARE BID FUNCTIONS LINEAR?

Relatively few of the bid functions are exactly linear. Table 5.2 shows that this does not change with experience.

Even if exact linearity is the exception rather than the rule, approximate linearity is very common. For each observed bid function we determined a ray approximation by computing a linear regression through the origin which, for the sake of simplicity, was based on integer values only in the range [0,100]. The extent to which a bid function is approximately linear can be judged by the measure of determination R^2 connected to this ray approximation. Table 5.3 gives an impression of the distribution of the measures of determination in periods 5 and 48.

Within the range of the measures of determination shown in Table 5.3 considerable deviations from the visual impression of linearity are possible. Thus, the ray approximation of the bid function of Fig. 5.1 has a measure of

Table 5.4
Slopes of the Ray-Approximations of Observed Bid Functions

	Period 5	Period 48
Minimum	.520	.463
Median	.785	.791
Maximum	.949	.980
Average distance from the median	.089	.073

determination of $R^2 = .936$. Only bid functions whose ray approximations have very high measures of determination can really be considered as approximately linear. In Table 5.3 we have drawn an arbitrary border line at $R^2 = .99$. Already in period 5 the measure of determination is higher for most subjects. In this sense, we can say that approximately linear bid functions are very common.

SLOPES OF THE RAY APPROXIMATIONS

Table 5.4 gives an impression of the distributions of the slopes of the ray-approximations in periods 5 and 48.

Table 5.4 conveys the impression that the distribution of the ray approximation slopes does not change much from period 5 to period 48. The medians show that most of the slopes are higher than 2/3, the slope of the risk-neutral equilibrium bid function. An interpretation of this fact in terms of risk aversion would require that some subjects, the more risk-averse ones, have high slopes both in periods 5 and 48, whereas others, the less risk-averse ones, have low slopes in both periods. However, this is not the case. The Spearman rank correlation between the subjects' slopes in periods 5 and 48 is only + .197. Of course, this very low rank correlation is not significant.

It is interesting to see that the slopes of the ray approximations of a subject's bid functions exhibit considerable variation over time. The average of the absolute difference between the slopes of period 5 and period 48 taken over all 27 subjects is .128. This number is much higher than the average distances from the median shown in Table 5.4. The fact that from period 5 to period 48 the slopes travel farther than the distance to the median clearly indicates that an interpretation in terms of degrees of risk aversiveness is not adequate.

The slopes of the ray approximations do not show a clear tendency to decrease or increase over time. For 12 subjects the slope at period 48 is smaller than at period 5, for 13 subjects the slope at period 48 is greater than at period 5,

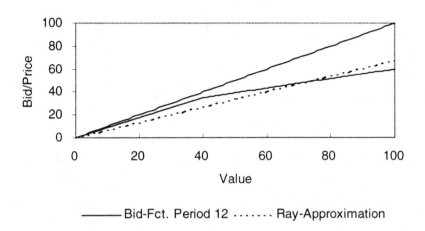

Fig. 5.2. Example of a flat top (Subject 24 period 12).

---

and for 2 subjects the slope is the same in both periods.

FLAT AND STEEP TOPS

Game equilibrium models of the first price sealed bid private value auction involving risk aversion predict bid functions with a *flat top* in the sense that for high values the slope is smaller than for low values (Cox, Roberson, & Smith, 1982). In the opposite case of higher slopes for high values we speak of a *steep top*. In order to examine whether flat or steep tops prevail in our data we have defined a measure called *top deviation*. The top deviation is the bid at 100 minus 100 times the slope of the ray approximation. As Figs. 5.2 and 5.3 illustrate, a negative top deviation corresponds to a flat top and a positive one to a steep top.

In period 48 we observe 15 subjects with positive top deviations, 10 with negative top deviations, and 2 have zero top deviations. (These are the two subjects with exactly linear bid functions in period 48.) The data do not show a clear tendency toward flat or steep tops in the bid functions of experienced individuals. This negative result also speaks against an interpretation of the distribution of slopes in terms of risk aversion.

Selten & Buchta

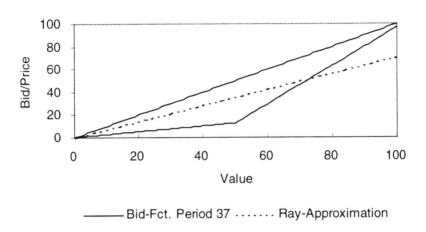

Fig. 5.3. Example of a steep top (Subject 7, period 37).

LEARNING DIRECTION THEORY

This section looks at the data in the light of an approach called learning direction theory. The name indicates that only the direction of learning is explained. The theory is qualitative rather than quantitative. It predicts the direction of change in response to experience but does not specify the amount of change.

As an illustration of the basic idea underlying the approach consider the example of a marksman who tries to shoot an arrow at the trunk of a tree. If he misses the trunk to the right, he will shift the position of the bow to the left and if he misses the trunk to the left, he will shift the position of the bow to the right. The marksman looks at his experience from the last trial and adjusts his behavior according to a simple qualitative picture of the causal relationship between the position of the bow and the path of the arrow.

Learning direction theory generalizes the kind of behavioral adjustment illustrated by the marksman example to repetitive decision tasks of the following kind: A parameter p_t has to be fixed in a sequence of periods $t = 1, 2, \ldots, T$. After each period the decision maker obtains feedback information that permits qualitative conclusions on whether a parameter choice $p < p_{t-1}$ or $p > p_{t-1}$ might have been better in the last period. Of course, such conclusions require some qualitative causal picture of the relationship between the parameter and the degree of success achieved. Learning direction theory predicts $p_t \leq p_{t-1}$, if the feedback information indicates, that p_{t-1} might have been too high and $p_t \geq p_{t-1}$, if there is reason to suppose, that p_{t-1} might have been too low.

The theory does not exclude the case that the decision maker does not respond to his or her experience. However, if one responds, one is expected to respond in the right direction. The behavior predicted by learning direction theory is boundedly rational because it makes rational use of the qualitative conclusions drawn from the feedback information. Of course, this is very different from the full rationality usually assumed in economic theory. Whereas full rationality is ex ante in the sense that it looks to the future, the bounded rationality of learning direction theory is ex post in the sense that it looks at the past. The reasoning based on the feedback information does not really concern the question "What will be the best decision in this period?" but rather "What would have been a better decision in the past period?" Moreover, the reasoning is qualitative rather than quantitative. Qualitative conclusions are drawn on the basis of qualitative features of the feedback information. Accordingly we may say that learning direction theory involves qualitative ex post rationality.

Admittedly, learning direction theory does not provide a general, fully specified model of behavior. In every specific case one has to fill in suitable assumptions about the qualitative causal relationships guiding the interpretation of feedback information. Moreover, in applications to experimental data it will rarely be the case that the influences on behavior embodied in learning direction theory are the only ones that are present in the situation. Therefore, one cannot expect more than conformance to the weak prediction that disregarding cases of unchanged parameters more parameter changes will be in the right direction than in the wrong one.

In research on the end effect in 10 period prisoners dilemma (PD) supergames (Selten & Stoecker, 1986), the learning direction theory approach was applied to the period in which subjects intended to switch from cooperation to noncooperation toward the end of the supergame. It was postulated that this intended deviation period would shift in the direction toward the beginning after an experience in which the opponent deviated to noncooperation earlier or at the same time as the subject, and in the direction toward the end if the subject deviated first. In the first case, the subjects would have earned more by an earlier deviation and in the second case more might have been earned by a later deviation. In the relevant part of the experiment, after both cooperation and end effect had been learned, only 7 of 34 subjects ever changed their intended deviation periods in the wrong direction, each of them only once. Learning direction theory was strongly supported by the data.

In a study of Mitzkewitz and Nagel (1993), the learning direction approach was successfully applied to demands and offers in ultimatum games with incomplete information. Another application can be found in a guessing game study by Nagel (1993), in which each of 15 subjects had to name a number repeatedly with the aim of coming as near as possible to 2/3 (or 1/2) of the average of all numbers named. Further evidence in favor of learning direction theory has been found in the evaluation of alternate move bargaining experiments with incomplete information by Kuon (1994) where the approach was applied to characteristics of bargaining behavior like first demand and concession rate.

APPLICATION OF LEARNING DIRECTION THEORY TO BID FUNCTION CHOICE

It was the intention of our study to generalize learning direction theory beyond decision tasks involving a single parameter to situations in which a whole function has to be fixed by the subjects. Admittedly we succeeded to do this only to a limited extent. It is difficult to say something about the way in which the bid function as a whole will be changed in response to last period's experience but learning direction theory suggests a hypothesis about changes at last period's value.

Let v, b, p be last period's value, bid, and price respectively. We distinguish the following three "experience conditions".

Successful bid: $p = b$
Lost opportunity: $b < p < v$
Outpriced value: $v \leq p$

If we say that a change tends to be a decrease (increase) in an experience condition we mean that in this condition a decrease (increase) is more frequent than an increase (decrease). We now state the hypothesis suggested by learning direction theory.

Bid change hypothesis: A change of the bid for last period's value tends to be a *decrease* in the *successful bid* condition and an *increase* in the *lost opportunity* condition. No systematic tendency is expected in the outpriced value condition.

Interpretation: In the successful bid condition the subject could have obtained the object at a lower price. Of course, it is not clear how high the second bid was but a higher profit could have been made by some lower bid. Therefore, in the successful bid condition the feedback information suggests that a lower bid might have been better. Accordingly, the bid change hypothesis predicts a tendency toward a decrease of the bid.

In the lost opportunity condition the subject could have obtained the object by a bid slightly higher than last period's price. This means that the feedback

Table 5.5
Number of Bid Change Directions in the Three Experience Conditions in Periods 6 to 50

	Decrease	Increase	Unchanged	Row sums
Successful bid	49	8	348	405
	.12	.02	.86	.33
Lost opportunity	9	47	94	150
Outpriced value	53	62	545	660
	.08	.09	.83	.55
Column sums	111	117	987	1215
	.09	.10	.81	1.00

Note. In each field absolute frequencies are shown above and relative frequencies are shown below. In the first three columns relative frequencies relate to the row sums in the last column relative frequencies relate to the column sum.

information suggests that a higher price might have been better. Accordingly, the bid change hypothesis predicts a tendency toward an increase of the bid in the lost opportunity condition.

In the outpriced value condition the subject did not have any opportunity to make a positive profit by another bid in the last period. Therefore, the feedback information does not permit any conclusions on what might have been better. Accordingly, we do not expect any tendency towards a decrease or an increase of the bid in the outpriced value condition.

Experimental results: Table 5.5 gives an overview over the bid changes in the three experience conditions.

Obviously, the numbers are in agreement with the bid change hypothesis. In the successful bid condition we observe 12% decreases and 2% increases. In the lost opportunity condition we find 31% increases and 6% decreases. In the outpriced value condition the table shows 8% decreases and 9% increases. Whereas in the successful bid condition and in the lost opportunity condition the numbers of changes in the right direction are five to six times as great as those in the wrong direction there is not much difference in the outpriced value condition.

A change of last period's bid occurs in 14% of all cases in the successful bid conditions and in 37% of all cases in the lost opportunity condition. We may say that the subjects are more inclined to react to the lost opportunity condition than to the successful bid condition. This is understandable because the feedback information is much clearer in the lost opportunity condition, where a subject can easily see the profit that could have been gained and where it is also obvious which bid should have been made. Of course, clarity arises only by hindsight, but this lies in the nature of ex post rationality.

The observations counted in Table 5.5 are not independent of each other. A meaningful test of the bid change hypothesis must look at aggregated data from each of the nine three-person subject groups that constitute our independent observations. In the successful bid condition there are more changes in the right direction than in the wrong direction in every group. (Cases of no change and periods 1 to 5 are not considered in this comparison and the following one.) In the lost opportunity condition there are more changes in the right direction than in the wrong one in eight of nine groups. In one group both numbers are equal. In both cases the null hypothesis that the probability for a higher number of changes is the same in both directions can be rejected by a binomial test in favor of the alternative predicted by the bid change hypothesis on a significance level of 2% (one-tailed). It is also interesting to notice, that in the lost opportunity condition no subject made more changes in the wrong direction than in the right direction. In the successful bid condition only 1 of 27 subjects had more changes in the wrong direction than in the right direction. (This subject had only one change in this condition.)

The data confirm the bid change hypothesis but this does not mean that learning direction theory captures all influences on observed behavior. The bid for last period's value was changed only in 228 (60%) of the 383 new bid functions in periods 6 to 50. Obviously learning direction theory cannot explain the changes in the remaining 40% of the cases.

TYPICITY AND CONFORMITY

This section examines the question whether conformity to learning direction theory is a feature of typical subject behavior. The data confirm the predictions of learning direction theory but it is at least a theoretical possibility that this is mainly due to the behavior of a minority of subjects that is not really typical.

A measure of typicity has been introduced in research on the evaluation of strategy programs for a finitely repeated asymmetric Cournot duopoly (Selten, Mitzkewitz, & Uhlich, 1989). The strategies were described by a number of characteristics that were either present or absent in each case. The information about the characteristics of strategies can be summarized by an incidence matrix with rows for characteristics and columns for subjects and entries of one in the

Experimental Sealed Bid First Price Auctions

case that the strategy of the subject has the characteristic and zero otherwise. For the sake of brevity we speak of characteristics of a subject instead of characteristics of a subject's strategy. On the basis of the incidence matrix numbers between zero and one called *typicities* were assigned both to characteristics and to subjects according to the following principles:

1. The typicity of a subject is the sum of the typicities of its characteristics.
2. The typicity of a characteristic is proportional to the sum of the typicities of the subjects with this characteristic.
3. The sum of the typicities of the characteristics is 1.
4. The typicities of subjects and characteristics form eigenvectors of AA' and A'A respectively, where A is the incidence matrix and A' its transpose. The eigenvectors are connected to the greatest eigenvalue that is the same one in both cases.

It follows by 1, 2, and 3 that the typicities of subjects and characteristics form right eigenvectors of AA' and A'A respectively. In addition to this it is required that these are eigenvectors connected to the common greatest eigenvalue. This eigenvalue is a positive real number if each of the characteristics is shared by some majority of the subjects (Kuon, 1993).

The method of computing typicities just described cannot be directly applied if important features of the subjects are naturally expressed by variables measured on an interval scale or a ratio scale. For this case Kuon (1993) developed a method that transforms the measurements to characteristics. The method can best be explained with the help of an illustrative example. One of the variables used for the exploration of what is typical in our case is the number of a subject's new bid functions from period 6 onward. Figure 5.4 shows the distribution of this variable.

One first determines a *typical interval* defined as the majority interval with the greatest occupation surplus. A *majority interval* is an interval that contains more than half of all subjects' values. The *occupation surplus* is the difference of the relative occupation (number of subjects with values in the interval divided by the total number of subjects) minus the relative length of the interval (length of the interval divided by length of the range). In our case [2,10] is the typical interval. Its occupation surplus is computed as follows:

occupation:	15
relative occupation:	$15/27 = .556$
relative length:	$9/44 = .204$
occupation surplus:	.352

In our example the variable can assume only integer values. In such cases the length of an interval is the number of integer points in the interval. In the case of

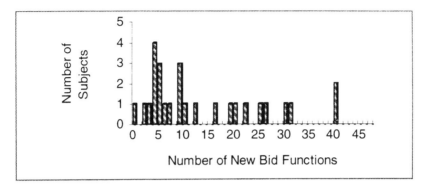

Fig. 5.4. Distribution of the numbers of new bid functions from period 6 on for the 27 subjects.

a continuous variable length is defined in the usual euclidean way. It may happen that the maximal occupation surplus does not uniquely determine one of the majority intervals as the typical one. For this case Kuon defines the typical interval as the union of all majority intervals with maximal occupation surplus. This problem does not arise in the evaluation of our data.

The typicity analysis for the 27 subjects of our experiments is based on nine characteristics derived from the following variables:

1. Number of new bid functions from period 6 onward.
2. Fraction of nonmonotonic new bid functions from period 6 onward.
3. Period of last new bid function.
4. Mean slope of ray approximation (all periods).
5. Mean measures of determination (all periods).
6. Variance of slopes (all periods).
7. Variance of measures of determination (all periods).
8. Mean top deviation (all periods).
9. Number of non negative top deviations (all periods).

Of course, the selection of the variables used for the typicity analysis is to some degree arbitrary. We have tried to cover the most important features of observed bid functions and their change over time.

It is our intention to examine whether subjects tend to conform the more closely to the bid change hypothesis the more typical they are. For this purpose

Experimental Sealed Bid First Price Auctions

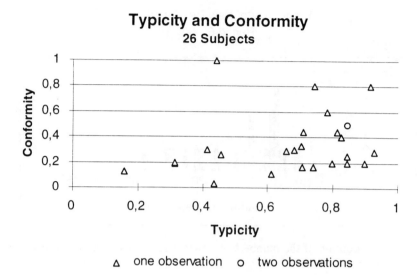

Fig. 5.5. Typicity and conformity.

we define a *conformity index* as the fraction of bid changes at last period's value in the right direction in the successful bid and lost opportunity conditions among all bid function changes from period 6 onward. Here, *right direction* means an increase in the lost opportunity condition and a decrease in the successful bid condition. The number of all bid function changes contains also those cases where the bid function was changed but not at last period's value. The conformity index shows which fraction of all bid function changes from period 6 onward is explainable by the bid change hypothesis. The higher the index is the more the subject conforms to the bid change hypothesis.

For one of the subjects a conformity index is not defined because this subject never changed the bid function. The Spearman rank correlation coefficient between typicity and conformity index for the remaining 26 subjects is .355. This is significant at the 5% level (one-tailed). The result indicates that the more typical a subject is the more it tends to conform to the bid change hypothesis.

Figure 5.5 provides a visual impression of the connection between typicity and conformity.

BID FUNCTIONS OF THE MOST TYPICAL SUBJECT IN PERIODS 6 TO 11

Participant 3
Bid-Functions from Period 6 to Period 7

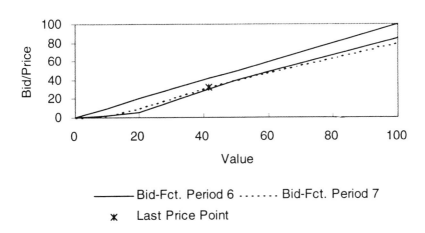

Fig. 5.6. Bid functions of Participant 3 for periods 6 and 7.

Participant 3
Bid-Functions from Period 7 to Period 8

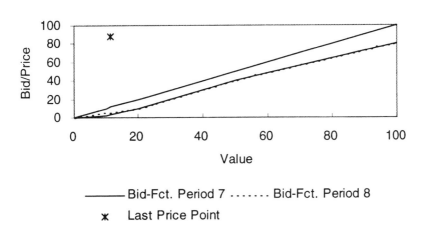

Fig. 5.7. Bid functions of Participant 3 for periods 7 and 8.

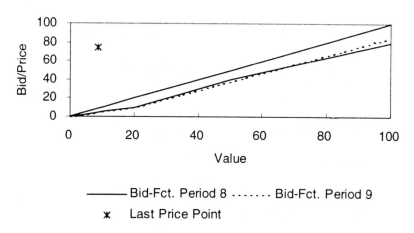

Fig. 5.8. Bid functions of Participant 3 for periods 8 and 9.

This section illustrates behavior by the example of the most typical subject's bid functions in periods 6 to 11. This subject, participant 3, changed the bid function in each of the periods 7 to 11. Figures 5.6 to 5.10 show the old and the new bid function for each of the five cases. The figures also show the last price point, which indicates the subject's value and the price observed in the last period.

In Fig. 5.6, the last price point is above last period's bid function and below the line at which price equals value. This means that Participant 3 experienced the lost opportunity condition in period 7. The change of the bid function is in agreement with the bid change hypothesis. At last period's value the new bid function is above the old one. It is interesting to see that the increase of the bid function in the lower part was accompanied by a decrease in the upper part. Maybe, the subject wanted to behave as indicated by the bid change hypothesis without modifying the overall aggressiveness of the bid function.

As Fig. 5.7 shows, Participant 3 was in the outpriced value condition in period 8. Nevertheless, the participant changed the bid function. The change seems to be motivated by the desire to achieve a gestalt improvement. The new bid function is smoother than the old one. The kink at the value 10 is eliminated; the new kink at value 20 is less sharp. We did not yet analyze bid function changes in the outpriced value conditions systematically but we have the impression that gestalt improvement is an important motive.

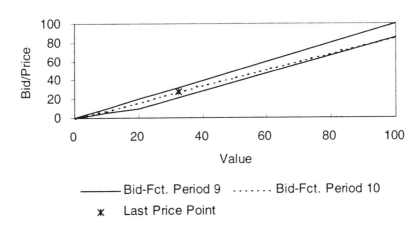

Fig. 5.9. Bid functions of Participant 3 for periods 9 and 10.

Figure 5.8 shows that in period 8, too, Participant 3 experienced the outpriced value condition. Again, the bid function change is a gestalt improvement. The kink at value 50 is eliminated by specifying a straight line for the interval [20,100].

In period 9, Participant 3 experienced the lost opportunity condition (see Fig. 5.9). As predicted by the bid change hypothesis, the new bid function is above the old one at last period's value. The new bid function is a straight line. We may say that a gestalt improvement is combined with the response to last period's experience.

In period 10, Participant 3 received the object (see Fig. 5.10), and lowered the bid at last period's value as predicted by the bid change hypothesis. Interestingly, the new bid function for period 11 is the bid function of period 9.

Admittedly, the behavior of the most typical subject is not representative of the whole sample because typicity is positively correlated to conformity to the bid change hypothesis. The behavior of less typical subjects is often much less easy to understand, particularly in cases of nonmonotonic bid functions.

ARE PAYOFF INCENTIVES SUFFICIENTLY STRONG?

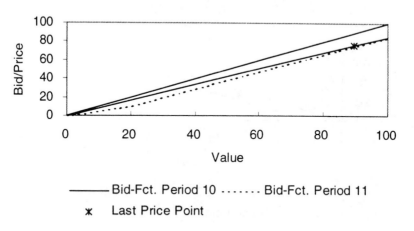

Fig. 5.10. Bid functions of Participant 3 for periods 10 and 11.

Harrison (1989) criticized first price sealed bid auction experiments with independent private values. He maintains that the incentives for optimization are typically very weak such that even sizeable deviations entail only very small losses of fractions of a penny. He argues that a subject will not find it profitable to go to the trouble to determine an optimal bid function under these circumstances. Harrison does not explain how the subject can find out that losses caused by the failure to optimize are very small without first having determined what is optimal. Presumably, a subject would have to compute the equilibrium bid function and the consequences of various deviations from it under the condition that it is used by the other players in order to find out whether it is worthwhile to spend the effort involved in this computation. However, a subject who has already computed what is optimal can behave accordingly even if the cost of deviation is small.

Quite apart from this inherent inconsistency of Harrison's critique we argue that at least in our case the incentive for optimization is not small. Suppose, a subject computes the risk-neutral equilibrium bid function at the beginning of the game before the first decision is made. The subject will then be able to use this function for all 50 periods of the game. The right measure of the incentives is not the gain in a single period but the gain over all 50 periods.

The risk-neutral equilibrium bid function is a straight line through the origin

Table 5.6
Losses by Deviations from the Equilibrium Ray Bid Function to the Average Ray Approximation of Observed Bid Functions

	Ray bid function slopes of opponents		
	a,a	*a,r*	*r,r*
Own slope *a*	125.00	107.14	96.84
Own slope *r*	105.44	92.45	83.33
Loss in DM	19.56	14.69	13.51
Loss in $ (1 DM = $.6)	11.73	8.81	8.11

Note. Entries in the first two columns are payoffs in DM obtained in fifty periods.

with the slope

$$a = \frac{2}{3} = .667.$$

Consider a risk-neutral player who wants to behave optimally. This cannot be done without forming expectations about the behavior of the others. If he or she expects that all players are risk-neutral and fully rational, the player will come to the conclusion to use the risk-neutral equilibrium strategy. Of course, the player also may form different expectations. We consider only the case that linear bid functions through the origin are used. This is not unreasonable because bid functions very often are nearly linear. We refer to linear bid functions through the origin as *ray bid functions*. The average slope of the ray approximations observed in the experiments are roughly equal to

$$r = \frac{7}{9} = .778.$$

In order to examine the consequences of deviations from optimal behavior it is reasonable to look at the advantage of using the ray bid function with the equilibrium slope *a* compared with using the ray bid function with the average slope *r* under the assumption that the opponents, too, use ray bid functions with slopes *a* or *r*. Table 5.6 shows payoffs in DM for the six possibilities that arise in

Experimental Sealed Bid First Price Auctions

this way. The loss is highest in the case of equilibrium strategies used by the opponents. From the point of view of rational theory this is the most relevant case. The loss is smaller but still considerable if one or two opponents use the ray bid function with average slope.

Interestingly, the ray bid function with slope a is not only the equilibrium strategy but also the optimal bid function for a wide range of behaviors on the other side provided that the opponents use ray bid functions. The ray bid function with slope $a = 2/3$ is optimal if the slopes s and t of the opponents' ray bid functions are not smaller than 2/3. This can be seen as follows:

Let v, u, and w be the values of the object for Players 1, 2, and 3, respectively, and assume that players 2 and 3 use the following bid functions:

$$b_2 = su$$
$$b_3 = tw.$$

We are interested in the expected value $E(b,v)$ of Player 1's payoff if the value is v and the bid is b. Player 1 does not obtain any payoff unless b_2 and b_3 are smaller than b. (Cases of equality can be neglected.) We have $b > b_2$ if and only if

$$u < \frac{b}{s} v$$

holds. Because u cannot be greater than 1 this has the consequence that the probability for $b_2 < b$ is as follows:

$$\text{Prob}(b_2 < b) = \min\left[\frac{b}{s}, 1\right].$$

Analogously we have

$$\text{Prob}(b_3 < b) = \min\left[\frac{b}{t}, 1\right] P.$$

This yields

$$E(b, v) = (b - v) \min\left[\frac{b}{s}, 1\right] \min\left[\frac{b}{t}, 1\right].$$

It follows that

$$B = (b-v)\frac{b^2}{st}$$

is an upper bound of $E(b,v)$:

$$E(b,v) \le B.$$

It can be seen easily that the maximum of B with respect to b is obtained at

$$b = \frac{2}{3}v.$$

The maximum of $E(b,v)$ cannot be greater than the maximum of its upper bound B. Therefore, if $E(b,v)$ and B coincide at $b = 2v/3$, then $E(b,v)$ assumes its maximum at this value, too. If s and t are both not smaller than 2/3 then b/s and b/t with $b = 2v/3$ are not greater than 1. This has the consequence that for $t \ge$ 2/3 and $s \ge 2/3$ the bid function $b = 2v/3$ is optimal for Player 1. We can conclude that the ray bid function with the slope $a = 2/3$ is optimal for a risk-neutral player, as long as the opponents use ray bid functions with slopes at least as high.

SUMMARY AND CONCLUSION

Our experiment differs from other experiments on sealed bid first price auctions with independent private values by the direct observation of bid functions. This makes it possible to examine the shape of bid functions without any assumption of constancy over time. Surprisingly, many bid functions are nonmonotonous. Nevertheless, most bid functions have high measures of determination and can be considered to be almost linear in this sense. Therefore, it is reasonable to concentrate attention on the slopes of the ray approximations. The distribution of these slopes does not change much from period 5 to period 48 but the slopes of the individual subjects show much movement over time. This throws doubt on conclusions from earlier experiments based on the assumption that subjects have constant bid functions. In particular, the explanation of bids higher than at risk neutral equilibrium by game theoretical models involving risk aversion is not in agreement with the data. There seems to be no systematic tendency towards flat or steep tops.

The movement of bid functions in time permits a partial explanation by learning direction theory. In earlier studies this theory has been applied to situations in which a one-dimensional decision parameter has to be fixed repeatedly over a number of periods. The theory focuses on the direction of change of the decision parameter. This direction is seen as determined by qualitative ex post reasoning on the basis of the feedback information. In our case the decision task requires the determination of a whole function and not just a single one-dimensional parameter. Nevertheless, learning direction theory can be applied to the question how the bid at last period's value can be expected to change in response to last period's experience. We have distinguished three experience conditions, the successful bid condition, the lost opportunity condition, and the outpriced value condition. The bid change hypothesis derived from learning direction theory predicts that in the case of a change of the bid at last period's value this bid tends to decrease in the successful bid condition and to increase in the lost opportunity condition. No systematic tendency is expected in the outpriced value condition. As we have seen, the bid change hypothesis is confirmed by the data. It has also been shown with the help of a typicity analysis that the more typical a subject is the more it conforms to the bid change hypothesis. The behavior of the most typical subject illustrates the fact that the bid change hypothesis does not fully describe the motivation of subjects. Bid functions may be changed for other reasons, for example, in order to achieve gestalt improvements. Finally we have argued that in our case the incentives for optimization are not small.

Our results show that a behavioral approach based on learning direction theory yields a better explanation of the data than normative game-theoretic models. Admittedly, the bid change hypothesis makes only weak predictions about relative frequencies of directions of change. But these predictions are derived without any parameter estimates. Game equilibrium models involving degrees of risk aversion or distributions over such degrees in games of incomplete information and possibly other parameters, like utilities of winning, fully specify behavior as a function of these parameters, but because these parameters have to be estimated, it is doubtful whether the predictions derived from such models are really stronger than those obtained from the bid change hypothesis. Moreover, a basic hypothesis underlying these models, namely constancy of bid functions over time, is clearly rejected by the data.

Some aspects of our data still need to be explored in more detail, in particular the surprisingly high frequency of nonmonotonous bid functions. At the moment we can only speculate about the reasons for nonmonotonicity. It seems possible that some subjects have a mixed attitude toward risk that induces them to behave risk-aversive in most of the range of possible values and risk-seeking in some subintervals of high values. Of course, this does not agree with classical ideas about the way in which decision makers relate to risk but, nevertheless, such behavior may be quite frequent in practice.

ACKNOWLEDGMENT

1. Support by the Deutsche Forschungsgemeinschaft through SFB 303 is gratefully acknowledged.

REFERENCES

Cox, J. C., Roberson, B., and Smith, V. L. (1982). Theory and behavior of single object auctions. In V. L. Smith (Ed.), *Research in Experimental Economics* (Vol. 2, pp. 1 – 43). Greenwich, CT: JAI Press.

Cox, J. C., Smith, V. L., & Walker, J. M. (1983). Test of a heterogenous bidders theory of first price auctions. *Economic Letters, 12,* 207-212.

Cox, J. C., Smith, V. L., & Walker, J. M. (1985a). Expected revenue in discriminative and uniform price sealed-bid auctions. In V. L. Smith (Ed.), *Research in Experimental Economics* (Vol. 3, pp. 183-232). Greenwich, CT: JAI Press.

Cox, J. C., Smith, V. L., & Walker, J. M. (1985b). Experimental development of sealed-bid auction theory; calibrating controls for risk aversion. *American Economic Review, 75,* 160-165.

Cox, J. C., Smith, V. L., & Walker, J. M. (1988), Theory and individual behavior of rirst-price auctions. *Jounal of Risk and Uncertainty, 1,* 61-99.

Harrison, G. W. (1989). Theory and misbehavior of first-price auctions. *American Economic Review, 79,* 749-762.

Kagel, J. H., Harstad, R. M., & Levin, D. (1987). Information impact and allocation rules in auctions with affiliated private values: A laboratory study. *Econometrica, 55,* 1275-1304.

Kagel, J. H., & Levin, D. (1985). Individual bidder behavior in first-price private value auctions. *Economic Letters, 19,* 125-128.

Kuon, B. (1993). Measuring the typicalness of behavior. *Mathematical Social Sciences, 26,* 35-49.

Kuon, B. (1994). Two-person bargaining experiments with incomplete information. *Springer lecture notes in economics and mathematical systems* (No. 412). Berlin, Heidelberg, New York.

Mitzkewitz, M. and Nagel, R. (1993), Experimental results on ultimatum games with incomplete information. *International Journal of Game Theory, 22,* 171-198

Nagel, R. (1993), Experimental results on interactive competitive guessing. *University of Bonn Discussion Paper No. B-236.*

Selten, R., & Stoecker, R. (1986). End behavior in sequences of finite prisoner's dilemma supergames: A learning theory approach. *Journal of Economic Behavior, 47-70.*

Selten, R., Mitzkewitz, M. & Uhlich, G. R. (1997). Duopoly strategies programmed by experienced players. *Econometrica, 65,* 517-555.

Walker, J. M., Smith, , V. L., & Cox, J. C. (1987). Bidding behavior in first price sealed bid auctions. *Economic Letters, 23,* 239-244.

PART IV

POLITICAL AND SOCIAL INTERACTION

[10]
The scenario bundle method

'Scenario bundles' are game-like models of possible international conflict situations. It is the purpose of this paper to describe a systematic way of using qualitative expert judgements as a basis for the construction and evaluation of scenario bundles for a given geographical area (for example, the Persian Gulf region). This procedure is known as the 'scenario bundle method'.

The approach is semi-formal rather than mathematical, even if it is inspired by game theory. The main emphasis is on modelling rather than analysis. Scenario bundles are very simple game structures with straightforward game theoretical equilibrium point solutions. Once the difficult modelling task has been completed the analysis is easy.

The development of the scenario bundle method is the outcome of a joint effort by Amos Perlmutter, a political scientist, and the author, a game theorist.[1] The shape of the method emerged in a period of intense cooperation in the summer of 1973. Our first experiences with tentative applications to the Middle East convinced us of the necessity to seek the active help of a panel of experts. Finally, the 'Research Conference on Strategic Decision Analysis Focusing on the Persian Gulf', which was held in the first week of October 1976, at Hotel Schwaghof, Bad Salzuflen, Federal Republic of Germany, gave us the opportunity to realize this idea.[2]

The discussions at the conference added further impulses to the development of the method. Comments and suggestions by the participants have contributed to the content of this paper. The conference had an important influence on the elaboration of the scenario bundle method in its present form.[3]

1 Short outline

The social scientist who wants to construct a game model of a specific section of reality has to answer the following questions:

- Who are the players?
- What are the motivating factors which determine the players' preferences?
- What are the strategic possibilities of the players?
- What are the consequences of various combinations of strategic choices?
- What are the players' preferences over these consequences?

The scenario bundle method is a systematic way of submitting such questions to a panel of knowledgeable persons. Group discussions produce qualitative judgements which serve as a basis of model construction. Scenario bundles do not require the specification of numerical parameters. Qualitative judgements are sufficient.

A necessarily incomplete short summary of the method may serve to give a first impression. There are a number of tasks which have to be performed one after the other. These tasks are listed below under subsections 1.1 to 1.11. More detailed explanations will be given later.

1.1   A list of actors
Most of the actors are states but non-states are considered, too.

1.2   A list of goals for every actor
Goals are middle range objectives. The actors are viewed as motivated by goals which have their roots in historical developments.

1.3   A list of fears for every actor
As important as goals are fears, mostly connected to real or imaginary dangers of military attack.

1.4   Ranking of military strength
The ranking of the actors' military strength serves to assist the experts' judgement on expected outcomes of possible military conflicts.

1.5   Structure of protective relationships
Stronger actors may have protective relationships to weaker actors. The structure of these relationships serves to assist the experts' judgements on plausible coalitions.

1.6   List of plausible coalitions
The huge number of combinatorially possible coalitions is narrowed down to a small number of plausible coalitions which suggest themselves in the light of the goals and fears of their members.

1.7   Lists of initial options for every actor
Initial options are plausible actions whose unobstructed completion furthers goals or diminishes fears.

1.8   Lists of initial options for plausible coalitions
Not only actors but also coalitions may have initial options.

1.9   Lists of internal events
Goals and fears may be changed by revolutions, military coups, the death of a ruler or similar internal events.

1.10   Scenario bundles
An initial option or an internal event may have the consequence that some actors must choose between several reactive options, such as support or non-support of an ally after military attack. A reactive option is characterized by an immediate necessity of decision. A reactive option may be followed by other reactive options, and so on until a natural end-point is reached. In this way, a scenario bundle is built up which is

graphically represented by a tree. The elaboration of the scenarios involves expert judgements about the expected outcome of possible military conflicts. In the same way as initial options, reactive options are connected to the goals and fears of the actors. Each initial option or internal event leads to a separate scenario bundle.

11.11 Preference judgements and analysis of scenario bundles
The main questions to be answered by the analysis of scenario bundles are the following:

- Which initial options are likely to be taken?
- Which initial options are not likely to be taken?
- What are the likely consequences of internal events?

In order to answer these questions the scenario bundles must be examined working backwards from the end to the starting point. Judgements about the preferences between various outcomes from the point of view of the actors, in terms of their goals and fears, must be made in the process of analysis. Making these judgements within the process of analysis rather than beforehand drastically reduces the number of judgements to be made.

2 General features
In this section the most important general features of the scenario bundle method are discussed as far as this is possible on the basis of the short outline given above. Further explanations will be given in later sections.

2.1 Emphasis on modelling
The scenario bundle method is a systematic procedure for the construction and analysis of game models. It is not a gaming approach. The experts whose cooperation is needed do not play games. They construct games.

Gaming has proved to be a valuable research tool in the study of international politics. Guetzkow's international simulation is an admirable contribution to the field.[4] The scenario bundle method does not compete with this line of research.

Scenario bundles are only slightly more complicated than extensive games with perfect information. The game theoretical analysis is easy. (The typical gaming model is too complicated to permit analytical treatment.) In the sense that most of the effort is spent in model construction rather than analysis the scenario bundle method can be described as primarily a modelling technique.

2.2 Semi-formal character
The construction of scenario bundles is a systematic procedure whose final output are formal mathematical structures. Apart from this the method is informal rather than mathematical. Such features as the connection between the lists of goals and fears and the structure of protective relationship on one side and the nature of options and preferences on the other side necessarily remain unformalized. Since both formal and informal modes of reasoning are employed, it seems to be appropriate to characterize the method as semi-formal.

2.3 Qualitative judgemental basis

Only qualitative expert judgements are needed for the construction of scenario bundles. It is not necessary to specify quantitative parameters. Preference judgements are strictly ordinal. Probability judgements are non-numerical. The purpose of the procedure is the mobilization of judgemental power rather than rigorous empirical research. Originally unconnected elementary judgements are systematized into a consistent whole. No attempt is made to derive the elementary judgements from more basic facts and figures.

2.4 States as decision centres

The method may be described as state-centric, since states, or more precisely, their governments, are viewed as decision centres. In principle this is not an indispensable feature of the approach since it may be possible to construct scenario bundles where a state is disaggregated into several decision units. Nevertheless, in practice, the state-centric view recommends itself as the appropriate level of aggregation, since the number of actors must be kept down to a manageable level.

The scenario bundle method tends to neglect those aspects of the governmental decision process, which are emphasized by the bureaucratic theory of policy.[5] Only experience can show whether this is a serious deficiency. Excellent textbook discussions of the relative merits of different levels of aggregation are available.[6]

2.5 Rationality assumptions

The actors are assumed to be rational decision makers. This means that they do what they expect to be best in terms of their preferences. Rationality is understood in the narrow sense of purposeful behaviour.

The preferences are determined by goals and fears that need not be rational in any sense. It may be very hard to understand why an actor thinks that it is worth while to pursue one of his goals. His fears may be based on misperceptions.

As a theory of unlimited rationality the Bayesian view of decision making has a firmly established axiomatic basis. Intuitively compelling axioms justify the expected utility maximization hypothesis.[7,8]

Nevertheless, practical rational decision making cannot be expected to obey the pattern of Bayesian theory. Human rationality is limited rather than unlimited. Human decision makers do not satisfy the stringent Bayesian requirement of unbounded computational capabilities. Theories of limited rationality try to do justice to this basic fact.[9,10]

The Bayesian treatment of uncertainty requires numerical probability and utility judgements. In this respect the rationality assumptions of the scenario bundle method are non-Bayesian. Otherwise it would not be sufficient to rely on ordinal preference judgements only.

An important non-Bayesian way of dealing with uncertainty may be described as 'focusing'. Instead of visualizing a probability distribution a decision maker may form a point estimate of a typical outcome. This typical outcome is his focus point. He is aware of the uncertainty but as far as his decision process is concerned the focus point is treated as if it were a certain event. In this non-technical sense the focus point is a certainty equivalent.[11]

An elaborate theory of focusing is due to Shackle.[12] This theory need not concern us here. Shackle's concept of potential surprise may or may not be combined with the idea of focusing.

The scenario bundle method assumes that focusing is used as an instrument of practical rationality, wherever this seems to be reasonable. Typically the uncertainty about the outcome of hypothetical military compaigns is handled in this way.

In some situations it may be necessary to form more than one focus point. If we consider the possibility of a military coup in a specified country we must form a picture of a typical event of this kind, but we also have to consider the possibility that no such event will happen. Crude non-numerical probability judgements are unavoidable here.

Governments cannot always be expected to act rationally even if rationality is understood in the narrow sense of purposeful behaviour. Textbooks on international relations discuss this obvious objection against theories based on the idea of rational decision making.[13, 14]

It is not clear how important the possibility of irrational behaviour really is. There seems to be no strong evidence against the hypothesis that rationality is the rule rather than the exception. As long as this is the case, it seems to be necessary to explore the consequence of rational decision making.

The scenario bundle method can be adapted to a limited amount of theorizing about irrational behaviour. If the experts feel that a certain actor has a considerable probability of taking an irrational action, this possibility can be treated as an 'internal event'. The scenario bundle model is flexible enough to permit this. It is possible to explore the rational consequences of irrational actions.

2.6 Motivational assumptions

Scenario bundles are topical models, in the sense that they relate to a specific geographical area at a specific point of time. General theories of government motivation like that of Hans Morgenthau[15] are very attractive from the abstract point of view, but it seems to be very difficult to apply them to topical models. Single variable explanations like the idea of power maximization do not seem to be adequate here. Therefore, the scenario bundle method takes a different approach to the problem of government motivation.

Government preferences are assumed to be determined by goals and fears which are the result of historical developments. Traditional values and orientations rather than abstract principles are perceived as the source of foreign policy objectives. Different countries may have very different systems of goals and fears.

Karl Deutsch rightly criticizes the assumption of a constant motivational structure.[16] It is a misconception to think that game models cannot avoid to make this assumption. The scenario bundle method treats the motivational structure as variable. Sometimes the most important consequence of an action is its influence on other actors' goals. A rational decision maker must take this into account.

2.7 Time perspective

The scenario bundle method starts with the situation in a specific geographical area at a specific point of time and aims to cover a period of about five years. In principle one might investigate a period in the past, but it is a more challenging task to look at the future. The particular application at the 'Conference on Strategic Decision Analysis Focusing on the Persian Gulf' took the time of the meeting as its starting point (October, 1976).

2.8 Nature of the results

Scenario bundles indicate possible future developments. The conclusions of the analysis can be confirmed or refuted by the history of the next five years. Due to the systematic character of the method the reasoning behind a wrong conjecture has a high degree of transparency. One can trace the causes of failure.

It would be presumptuous to promise predictive reliability. Nobody can claim to have a sure way to look into the future of international relations. If there were such a method, governments could not afford to ignore it.

Journalistic and scientific speculations about future developments on the international scene are not deterred by the lack of predictive reliability. There is an urgent need to form expectations which must be satisfied in some way.

Only experience can show whether a systematic procedure for the integration of expert judgements achieves better results than the unaided intuition of well-informed observers. The obvious advantages of transparency and consistency do not guarantee predictive success.

3 Actors

Here and in later sections it will often be useful to refer to the 'Results of the conference on Strategic Decision Analysis Focusing on the Persian Gulf.[17] For the sake of shortness we shall simply speak of the 'conference results' wherever we do this. For selected results see the appendix to this paper.

The list of actors in the conference results contains 22 actors, categorized as local powers, peripheral powers, superpowers and non-states.

Local powers are states in the geographical area under consideration and peripheral powers are states in the vicinity of the area. Peripheral powers must be included in the list of actors if it seems to be possible that they might become involved in international conflict originating in the area.

Superpowers are put on the list in view of their worldwide political influence. Originally the conference participants considered only three superpowers; the United States, the Soviet Union and China. Europe, or more precisely the European Community, was added later. Generally, one does not think of Europe as a superpower and it is only a matter of convenience that Europe is classified in this way.

It is justified to model an aggregate of states as a single actor, if one can expect a coordinated foreign policy with respect to the area under consideration.

The list in the conference results contains three non-states. These non-states may be characterized as ethnic and/or revolutionary movements who want to achieve governmental control or at least autonomy in some territory which need not be precisely defined. One might think of such actors as potential states. The inclusion of non-states of this kind only slightly modifies the state-centricity of the approach.

4 Goals and fears

The scenario bundle method is based on a theory of government motivation which perceives the goals and fears of governments as the results of historical developments. Traditional attitudes, historical experiences, ideological or religious orientations and sometimes the personal character of the ruler may determine the preferences of a government.

4.1 Goals

We think of each actor as having a list of goals ranked in order of importance. As an example we look at the goals of Iran as they appear in the conference results:

1. economic growth, independence from oil
2. hegemony over the gulf – open gulf
3. ascendancy to military middle power status
4. internal stability.

A goal may be seen as derived from an unsolved problem. The unsatisfactory level of economic development and the dependence of the economy on oil production pose a problem which requires the attention of the Iranian government.

From this example we may proceed to a tentative general definition of a goal:

> A goal may be characterized as the desired solution of an unsolved political problem which attracts a substantial part of the government's attention.

Every government wants to stay in power. Nevertheless, internal stability is not considered a goal if it does not pose a problem. Problem-relatedness is an important feature of the goal concept. The problem must be serious enough to attract considerable government attention.

A goal may be directed towards the achievement of something the actor does not have, but it may also be concerned with something which cannot be maintained without constant effort. We may distinguish both cases by calling a goal of the first kind an 'achievement goal' and a goal of the second kind a 'maintenance goal'.

As far as possible a goal should be *operational* in the very weak sense that after a change of the state of the world it should be possible to judge whether one is nearer to the goal or not. In the process of working out a list of goals one must try to approach this ideal, even if a certain vagueness may be unavoidable.

Some actors, mainly superpowers and peripheral powers, may have goals which are not important for the geographical area under consideration. Such goals are irrelevant for the analysis and need not be included in the lists.

The scenario bundle method looks at a goal as a basic datum of an actor's rational decision process. An actor does not select his goals and fears, he takes them as given.

Governmental goals have been discussed in the literature.[18, 19] The approach of the scenario bundle method is not dissimilar to the point of view of these authors, even if there are some important differences connected to the explanations given above.

4.2 Fears

The scenario bundle method is based on the assumption that the motivational structure of an actor has two kinds of basic elements: goals and fears. In the same way as a goal a fear is a datum of the rational decision process. As an example of a list of fears we look at the list for Iran in the conference results:

1. Soviet attack
2. Iraqi attack against the oil centres and Shatt el-Arab
3. adverse trade balance.

As in the lists of goals the ranking reflects the order of importance.

The intuitive concept of a fear is hard to capture. The following tentative definition remains somewhat unsatisfactory:

> A fear is the perception of a serious danger posed by the possibility of hostile actions by other actors or by the possibility of noxious accidental events.

Military attacks are hostile actions by other actors. An adverse trade balance may be caused by an accidental event like an unexpected substantial decrease of demand for oil.

It is not important whether a perceived danger is real or imaginary. A fear of military attack may persist without reason. This may be due to recent historical experience. The perceived danger must be serious in the sense that the government takes it seriously.

Whereas goals are concerned with desirable objectives, the content of a fear is a danger of undesirable future developments. Goals and fears can be regarded as opposite motivational forces. This does not exclude the possibility of close interrelationships. In fact a fear and a goal may be like mirror images of each other.

An illustrative example can be seen in the lists for China in the conference results. Here we find the goal of 'promotion and fostering of US-Soviet conflict' and the fear of 'US–Soviet collusion'. Both are obviously closely connected since both are concerned with the nature of the relations between the USA and the USSR. Nevertheless, the fact that both appear in the lists does not involve a duplication of information. The goal does not necessarily imply the fear and vice versa. Conflict may be perceived as desirable and collusion may be regarded as too improbable to be feared or, alternatively, collusion may be feared but conflict may not be seen as desirable.

4.3 Goal types

It is doubtful whether any classification scheme is flexible enough to do justice to the great variety of possible goals. Nevertheless, some distinctions may be useful.

The conference at Bad Salzuflen came to the conclusion that in the lists for the superpowers *global* goals must be separated from *regional* goals. In the conference results the rankings of both goal types are kept apart. Global goals are concerned with worldwide interests whereas regional goals are more specifically related to the area under consideration.

Sometimes it is useful to make the obvious distinction between *internal* and *external* goals. 'Internal stability' and 'economic development' are typical internal goals. In the following we look at several categories of external goals.

Territorial goals are an important source of international conflict. Country A may have a territorial claim on country B and country B may have the goal of 'territorial integrity'.

Hegemony over a specified region is often pursued as a goal. In terms of our characterization of goals as desired solutions of unsolved problems, hegemonial ambitions may be interpreted as arising from a real or imaginary need for more control over the international environment. Much weaker forms of control are intended by goals of *influence* on other countries, or even less than that *presence* in other countries. The rejection of outside control may find its expression in a goal of *independence*.

The desire to strengthen political, religious, ethnic or supranational ideologies leads to *ideological goals* like panaratism, panislamism or the spread of state-controlled economies.

Some governments want to achieve or to maintain a special power status like that of a superpower or a military middle power. Such goals may be called *power status* goals.

The distinction between internal and external goals is irrelevant for the *non-state goals* of statehood or autonomy.

Some goals are *intrinsic* in the sense that their attainment immediately satisfies basic aspirations and desires of the actor. The objective is something of intrinsic value. 'Internal stability' and 'territorial integrity' are typical examples. One hardly needs to ask why a government values the attainment of such objectives.

Other goals are not pursued as ends in themselves, but rather as means to enhance the overall capability to cope with the international environment. Some goals of the USA are of this kind. It is sufficient to mention the 'maintenance of overall and regional political and military balance', the 'avoidance of involvement in conflicts' and the 'creation of political, military and economic dependencies'. Generally, such goals presuppose specific views of the causal relationships governing the international system. In the examples given above it is not at all obvious that the overall capability to cope with the international environment is really enhanced in this way. It is possible to question the underlying convictions.

Causal assumptions about the nature of the international system may be described as systemic presumptions. Therefore, it is natural to use the term *systemic* in connection with goals as those discussed above. A systemic goal may be characterized as a non-intrinsic goal based on systemic presumptions.

The distinction between intrinsic and systemic goals is not a dichotomy. Some goals may be partly intrinsic and partly systemic. For the purposes of the scenario bundle method it is not necesssary to apply these categories to every special case. The significance of the distinction lies elsewhere. A failure to understand the nature of systemic goals may mislead the analyst's efforts. We might be tempted to try a reduction to more basic objectives. In principle, this may be possible, but one meets two difficulties. One is posed by the vague and non-operational character of objectives like the enhancement of the capability to cope with the international environment. The other is caused by the fact that different actors may have different systemic presumptions.

Sometimes an actor may be firmly convinced that a certain policy like the military support of friendly regimes has beneficial systemic consequences. The continuation of the policy may be a systemic goal. In such cases we speak of *policy goals*.

It is possible to make further classificatory distinctions but this shall not be done here.

4.4 Fear types

The classification of fears is as problematic as the classification of goals. The distinctions to be discussed in the following are similar to those in the previous subsection.

Fears may be classified as *internal* or *external*. 'Civil war' and 'ethnic unrest' are examples of internal fears.

An important class of external fears are the fears of military attack. Other examples are fears of new enemy bases, of new hostile alliances and of adverse political changes in other countries.

In the same way as goals, fears can be *intrinsic* or *systemic*. Fears of military attack are intrinsic, but it is at least questionable whether the same is true for the other external examples mentioned above. The basis of the belief underlying the fear of new enemy bases is the assumption that the resulting increase of the enemy's military strength is more important than the possibility of political reactions by third powers which are unfavourable to the enemy. This belief may be justified, but it has the character of a systemic presumption. Fortunately, for the purposes of the scenario bundle method, it is not necessary to decide whether a fear is intrinsic or systemic.

The conference results indicate that 'extinction' and 'schism' are typical *non-state fears*.

5 Military strength and protective relationships

The task of ranking the actors according to their military strength requires more than the comparison of figures on the number of tanks, airplanes, and so on. Other factors like past performance are important, too.

The ranking of military strength serves several purposes. It serves to assist the expert's judgements on expected outcomes of posssible military conflicts. Not only the ranking as such but also the exchange of information during the discussions is of importance in this respect.

Another purpose of the ranking is its preparatory function for the determination of the structure of protective relationships. The scenario bundle method makes use of a concept of a protective relationship which involves a military stronger power which extends some kind of protection to a military weaker power. This need not be a full fledged patron–client relationship. It is sufficient that the stronger power is willing to contribute significant resources in order to secure the survival of the weaker power, should this be necessary.

The structure of protective relationships is described by Figure A.1 in the appendix to this paper.

In the process of scenario bundle construction it is often important to judge who could consider to come to the aid of who in case of conflict. The structure of protective relationships has the purpose to facilitate such judgements.

The experiences with the conference at Bad Salzuflen indicate that the structure of protective relationships is a less important part of the scenario bundle method than the lists of goals and fears. Maybe the concept of a protective relationship is not yet the best instrument which can be used in order to achieve its purpose.

6 Plausible coalitions

Not only individual actors but also coalitions may have initial options which lead to scenario bundles. Generally, it is impossible to look at all combinatorially possible coalitions. For quite reasonable numbers of actors one receives a huge number of coalitions. The conference results list 22 actors; in this case the number of all combinatorially possible coalitions with at least two members is 4,194,281.

Fortunately, there is no need to look at all possible coalitions. Most of them are nonsensical combinations of actors, for example, the coalition of China, Israel and Qatar. We must concentrate our attention on those coalitions which may have a reason to unite for common action.

6.1 Coalitions

In the conceptual framework of the scenario bundle method the word coalition has a special meaning which must be explained before we can go on to discuss criteria of plausibility. *A coalition is a group of actors which cooperates in order to take a common action.* There need not be any cooperation beyond this action.

An action which requires the cooperation of several actors is called a *coalitional action*. The actors who participate in the coalitional action form a coalition.

The terminology introduced above is suggested by game theory and does not always conform to common language use. Ordinarily one would not think of a coalition in the case of two countries who agree on a territorial settlement about some disputed border region. Nevertheless, agreements of this kind are important examples of coalitional actions. Obviously, territorial settlements require the cooperation of the involved countries. Therefore these countries form a coalition if they reach an agreement.

Contrary to game theoretical terminology our use of the word coalition always refers to a group of more than one actor.

Coalitions should not be confused with alliances. Groups of states bound together by formal treaties do not necessarily take a common action. Only if an alliance takes a coalitional action, does it become a coalition.

6.2 Coalition types

It is useful to distinguish types of coalitions. An *agreement coalition* is formed in order to remove some source of conflict between its members in a peaceful way. Territorial settlements are typical examples.

An *offensive coalition* is formed in order to take military action against other actors.

It may happen that a group of strong powers agrees to enforce a peaceful solution of a regional conflict between weak powers who are unable to find a settlement among themselves. In such cases we speak of a *peace-imposing coalition*.

A group of actors who agree on a new defence treaty form a *treaty coalition*. A similar type is that of an *aid coalition* where some actors agree to render some sort of assistance to other actors. Even if this support is strictly unilateral it may involve a coalitional action, since it must be accepted by the receiving party. Only in cases where the recipients' willingness to accept aid is beyond doubt, the extension of support may be looked upon as an option of the donor alone.

It rarely happens, but it is possible, that a group of states decide to form a union where the individual countries lose their separate existence. In such cases we speak of a *merger coalition*. As the result of a merger coalition several actors cease to exist and a new actor enters the scene.

It is not the intention of our discussion of coalition types to provide an exhaustive classification. It might be necessary to consider possibilities which are not covered by any of the types.

6.3 The list of plausible coalitions

The list of plausible coalitions serves the purpose to prepare the judgements on initial coalitional options. Coalitions which do not have initial options are without interest for this list, even if they do occur in a scenario bundle at a later stage.

Experience seems to suggest that agreement coalitions and offensive coalitions are the most important ones. Peace imposing coalitions are of considerable interest, too. Treaty coalitions do not seem to lead to interesting scenario bundles. The same can be said about most aid coalitions. The conference results contain an exception to this rule where a very dramatic form of aid is considered, namely the transfer of nuclear capability from China to Egypt.

It is recommended that the following procedure is used in order to construct the list of plausible coalitions. For one actor after the other those plausible coalitions are determined which contain him as a member. It is best to begin with the local powers and to continue with the superpowers and to look at the peripheral powers last. Most of the important coalitional possibilities will involve at least some of the local powers.

For each actor one must examine one coalition type after the other. In order to find plausible agreement coalitions it is useful to ask the following questions: which sources of conflict involving the actor could be solved by agreement coalitions? If there are such coalitions, which of them are compatible with the goals and fears of the members?

In a similar way we look for plausible offensive coalitions. The goals and fears of the actor suggest possible targets of military action. Starting from this information one asks whether there are other actors who might serve their goals and reduce their fears by a successful participation in the campaign.

It is not yet necessary to describe the coalitional options in detail but it is unavoidable that some vague picture is formed already here.

Peace-imposing coalitions need not be considered unless the actor is strong enough to be a natural member of a coalition of this kind. Experience suggests that it is easy to find the very few plausible peace-imposing coalitions.

In order to find interesting aid coalitions we must try to think of strong forms of assistance. Weak forms of support are unlikely to cause a significant change in the situation. The structure of protective relationships may be helpful in suggesting who might assist whom.

Coalitional possibilities of less general importance like merger coalitions might be explored in a similar fashion. It may also happen that some plausible coalition not covered by any of the types suggests itself.

It is not clear whether we should separate the determination of a list of plausible coalitions from the task of constructing lists of initial options for coalitions. As has been mentioned before, we cannot avoid thinking at least vaguely about possible coalitional actions, if we look for plausible coalitions. If both tasks are lumped together we should approach them only after the determination of lists of initial options for individual actors.

7 Initial options

Applications of the scenario bundle method start from the situation in a specific geographical area at a specific point of time. We refer to this situation as the *initial situation*.

The word 'option' is used for the actions which may or may not be taken by an actor or a coalition in a scenario bundle. *Initial options* are options which are open in the initial situation before anything else has happened.

In the construction of scenario bundles we must concentrate our attention on possible actions with strong impact on the international system. The analysis should not be burdened by the inclusion of relatively unimportant political moves.

Peaceful actions of considerable impact tend to be coalitional actions. Without the cooperation of other actors an individual actor rarely has the opportunity to achieve significant advantages without the use of force. Therefore one should not be surprised if most of the initial options of an individual actor turn out to be military attacks.

As an example of a list of initial options we look at the list for Iraq in the conference results:

1. attack on Kuwait
2. cancellation of treaty with Iran
3. attack on Syria.

The initial options of an actor are suggested by his goals and fears. In the case of Iraq the conference came to the conclusion that the first two options are plausible in the light of the goal 'territorial claims on Iran and Kuwait'. The third one is connected to the goal 'domination of the Baath party' and to the fear of Syrian attack.

7.1 Favourable intervention assumptions

Later we discuss two criteria of plausibility for initial options. These criteria are based on the examination of a hypothetical favourable case concerning the intervention of third powers. We refer to this hypothetical case as the *favourable intervention assumption*. Many initial options involve actions whose purpose is best achieved without the intervention of third powers. The option 'attack on Kuwait' in the list for Iraq is an example of this kind. In such cases the absence of any third power intervention is the obvious favourable intervention assumption.

Some initial options do not make sense, unless they are taken in the intention to provoke the intervention of third powers. Thus a surprise attack may be made in the expectation that a peace-imposing intervention by stronger powers will lead to a consolidation of initial territorial gains which could not be defended in a longer lasting conflict. In such cases the intended intervention provides the favourable intervention assumption. Initial options of this kind are called *intervention-oriented*. In order to have a short name for those initial options which are not intervention-oriented we call them *autonomous*.

The determination of lists of initial options occur at an early stage in the construction of the scenario bundles. Therefore it cannot be the task of the criteria of plausibility for initial options to decide that a given favourable intervention assumption is reasonable. It is more important that the assumption is favourable. In the case of an autonomous initial option no further justification is needed for the assumption of no intervention. The situation is a little more difficult with respect to intervention-oriented initial options. Here the expected intervention must be such that it seems to be possible in the light of the goals and fears of the intervening powers.

7.2 The realism criterion

A plausible initial option should be realistic in the sense that at least under the adequate favourable intervention assumption it is reasonable to expect a success. This is the *realism criterion*. The word success must be understood in terms of those goals and fears which motivate the action.

In the case of a military action success does not necessarily require victory. Sometimes even an indecisive result may strengthen a territorial claim.

With respect to initial coalitional options success means success for every member of the coalition. Every member must be motivated to join the coalition.

7.3 The desirability criterion

The successful completion of an action may have undesirable side-effects on goals and fears other than those which motivate the action. The realism criterion does not yet look at these side-effects.

A victory in a military campaign motivated by territorial claims may increase the danger of a military coup and thereby decrease internal stability or it may involve a drain on economic resources which delays economic development.

Some side-effects are more easily visible than others. They can be detected without looking at the impact on third powers which are not directly involved. Such side-effects will be called *near*, and those which are not near, will be referred to as *remote*.

The side-effects of a victory on internal stability and economic development in the example given above are near side-effects. It may happen that the successful completion of an action has the consequence that the goals and fears of a third power change in an undesirable way. This is a remote side-effect.

Remote side-effects can be very important. They must be considered in the construction of scenario bundles, but the determination of plausible initial options is not the best place to do this.

A plausible initial option should have the property that under the adequate favourable intervention assumption the advantages of success outweigh the undesirable near side-effects. This is the desirability criterion. The word 'outweigh' must be understood as descriptive of the actor's preferences or in the case of a coalitional option, every member's preferences.

7.4 Types of initial options

The coalition types discussed in subsection 6.2 are actually types of initial coalitional options. There is no need to repeat this discussion here. We can restrict our attention to initial options of individual actors.

As pointed out before, an individual actor not in coalition with other actors in most cases has little scope for peaceful actions with considerable impact on the international system. Therefore military initial options are a very important class of individual initial options.

Military options are military attacks, the establishment or expansion of bases, substantial increases or decreases of armament or other options which involve the use or the deployment of military force.

Support options are options to give, increase or decrease, military or financial support to other actors whose willingness to accept such aid is beyond any doubt. (Otherwise the extension of aid would require an aid coalition.)

Economic warfare options involve such measures as oil embargoes, boycotts, and similar severe restrictions of international trade, with the purpose of inducing other actors to yield to political demands.

Political options concern serious political moves like the cancellation of an important treaty. Such actions may change the structure of protective relationships and may have influence on the goals and fears of other actors.

There may be individual initial options which do not fit in any of these categories. The types described above do not provide a complete classification.

7.5 Orchestration

The suggestive term *orchestration* was introduced by R. Kolkowicz during the conference at Bad Salzuflen as a characterization of propaganda campaigns and similar government activities. It seems to be desirable to look at such efforts as accompanying phenomena of more important political moves rather than options in their own right. This is vividly expressed by the term 'orchestration'. Governments can be expected to orchestrate any option they may take in an adequate way.

The exclusion of measures of orchestration does not seem to lead to serious omissions in the lists of initial options. It is important to avoid unnecessary complications in the construction of scenario bundles. Therefore, we should not consider options of orchestration unless there are compelling reasons to deviate from this principle.

8 Internal events

Revolutions, military coups, the death of a ruler or similar occurrences which change the initial situation in a significant way are called 'internal events'. Important future developments may arise from internal events. An actor's system of goals and fears may change radically and new plausible possibilities of action can lead to interesting scenario bundles.

Significant changes of an actor's goals and fears may result from slow social processes rather than dramatic developments. It is convenient to look at such cases as internal events, too. In this connection we speak of *slow events*.

It is conceivable that as a result of increasing military capabilities an actor gradually develops territorial goals. The possibility of a considerably increased importance of Iran's claims on Bahrain is an example taken from the conference results.

Generally, internal events are occurrences related to the internal affairs of a specific actor. Some important changes of the initial situations not caused by initial options do not fit this description. The following example is taken from the conference results:

> A high price of oil together with a serious deterioration of trade balances in industrialized countries and huge positive balances in oil countries.

In such cases we may speak of *transregional events*. As a matter of convenience transregional events are looked upon as a special category of internal events, even if this involves a slight abuse of terminology.

A tentative general characterization of internal events may serve to summarize what has been said above: *an internal event is a slow or dramatic development outside the control of the actors and connected to significant changes of goals and fears.*

The most important internal events are those which have their origin in the region under consideration. We may wish to concentrate our attention on these events. This has been done at the conference at Bad Salzuflen. Lists of internal events have been discussed for local powers and non-states, but not for other actors.

The task of compiling lists of internal events is best attacked actor by actor. At the end we may look at transregional events. The experts must be asked to provide qualitative probability judgements. It is convenient to use a verbal scale like that employed by the conference at Bad Salzuflen:

- very unlikely
- unlikely
- low probability
- at least moderate probability.

This scale is a very crude one. Finer distinctions between different degrees of at least moderate probabilities are intentionally avoided in order to facilitate consensus. We may, of course, want to force the experts to provide more information. In this case, one should use a different scale.

No attempt was made to connect verbal judgements with numerical estimates. In view of what is known about the psychology of probability judgements,[20] it is doubtful whether we should do this.

9 Scenario bundles

The graphical representation by a game tree is a natural way to describe a scenario bundle. The conventions used for extensive games with perfect information can be applied here, too. Unlike ordinary extensive games, scenario bundles permit coalitional choices. Apart from this, scenario bundles are very similar to extensive games with perfect information.

In the following section the construction of scenario bundles and the conventions of graphical representation will be explained with the help of examples taken from the conference results. Formal definitions will be given in a later section.

9.1 Construction and representation

Every game tree has a starting point, the *origin* of the tree. In the graphical re-presentation of a scenario bundle, the origin corresponds into the initial situation.

For scenario bundles generated by initial options, the origin is a decision point of an individual actor or a coalition. An example is provided by the beginning of scenario bundle 1 in the conference results (see Figure 1).

Here the origin is a decision point of Iraq. The decision has to be taken, whether to take the option 'attack Kuwait' or not. Both possibilities are represented by edges of the tree.

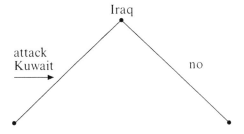

Figure 1   Beginning of scenario bundle 1

In order to continue the construction of the scenario bundle, we must ask the following question: suppose that the initial option has been taken; is there an actor or a group of actors under immediate pressure to make a decision how to react? If the answer is 'yes' we must ask a further question: what are the reactive options?

In the case of scenario bundle number 1, the conference came to the conclusion that an Iraqi attack on Kuwait immediately requires a decision by a coalition of Iran, Saudi Arabia and the USA whether to intervene or not.[20] Accordingly the tree was continued as in Figure 2:

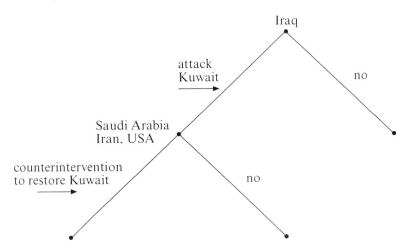

Figure 2   Part of the scenario bundle 1

The complete picture of scenario bundle 1 shows that a further decision point follows the counterintervention. There the USSR has two options, namely to support Iraq on a low level or a high level. Accordingly, the tree is continued by three edges corresponding to the two options and the alternative to take none of them.

A somewhat more complicated example is provided by scenario bundle 14. This bundle is generated by an internal event. The origin is not a decision point of an

308   Game Theory and Economic Behaviour I

individual actor or a coalition, but a point where an event may or may not take place. The event is symbolized by an edge of the tree (see Figure 3).

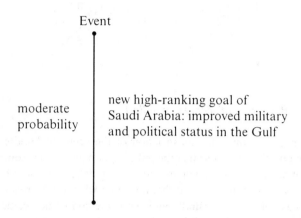

Figure 3   Beginning of scenario bundle 14

For the sake of simplicity the possibility that the event does not take place is left out of the picture. (Representation by an edge would not serve any useful purpose.)

A goal change like that at the beginning of scenario bundle 14 raises the following question: what can the actor do in order to reach his new goals? The conference decided that after the goal change 'take over Emirates' is a plausible option of Saudi Arabia. Scenario bundle 14 was continued accordingly (see Figure 4).

It is now necessary to look at the situation which arises if Saudi Arabia moves to take over the Emirates. Who is put under pressure to make a decision? The answer seemed to be obvious to the conference: Iran has to decide whether to counterintervene or not. Upon second thought a very plausible alternative was proposed: a coalition of Iran and Saudi Arabia where both agree to a defence pact and Iran tolerates the take-over of the Emirates. This treaty coalition would serve to strengthen the hegemonial position of Iran in the Gulf and it would also help Saudi Arabia to attain her new goals.

A difficulty arises at this point. The two options to be considered do not belong to the same actor or coalition. One option belongs to a coalition and the other belongs to an actor who is a member of the coalition.

The problem can be solved with the help of the following *sequential model*: first the coalition decides whether it takes its option or not; in the latter case the actor has to make a second decision in order to select one of his alternatives.

The sequential model must be understood as a description of the strategical structure of the situation; it is not meant to be a correct description of the temporal structure of the decision processes involved. One may think of the choices of the actor and the coalition as taking place at the same time. This is expressed graphically by drawing the no-branch of the coalitional decision as a horizontal line (see Figure 5). The decision points of the actor and the coalition are put on the same level in order to indicate that

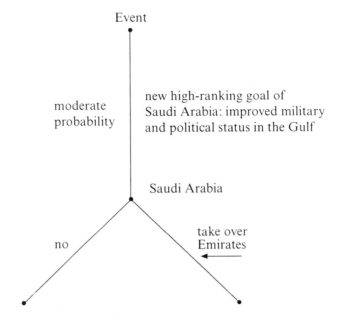

Figure 4   Part of scenario bundle 14

the sequential order is strategical rather than phenomenological. The game theoretical justification of the sequential model will be discussed later.

The complete picture of scenario bundle 14 (see Figure A.3 in the appendix) shows that the option 'counter-intervention-war' leads to a further decision point, where the USA may or may not mediate the conflict.

There is only one other scenario bundle in the conference results where the sequential model had to be applied. This is scenario bundle 8 (see Figure A.4 in the appendix). Here the USSR has several options. The sequential model covers such cases, too. More complicated situations with simultaneous reaction possibilities of several actors and coalitions did not arise at the conference.

There is never more than one option at a decision point of a coalition. Experience suggests that it is not necessary to consider more complicated cases. Therefore we restrict our attention to scenario bundles where a coalition has exactly one option at every decision point. A single actor may have several options.

The scenario bundle method permits extensions in several directions. In order to keep the analysis as simple as possible we should resist the temptation to introduce a conceptual framework which is more general than necessary. Therefore we require the following two properties as a part of the definition of a scenario bundle:

1. applicability of the sequential model
2. one option at every decision point of a coalition.

At the conference at Bad Salzuflen the necessity to go beyond properties (1) and (2) did not arise.

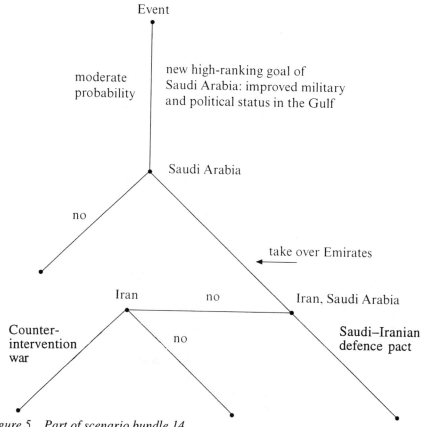

Figure 5   Part of scenario bundle 14

9.2   Stopping principles

The construction of a scenario bundle cannot be continued indefinitely. At first glance, this seems to raise a serious problem. Fortunately, natural stopping principles are not hard to find.

An *end-point* is a node beyond which the construction of a scenario bundle is not continued. Different stopping principles correspond to different types of end-points.

Occasionally, one may arrive at a node where the construction must stop, since no plausible options can be found. A node of this kind is called a *blind alley end-point*.

Sometimes it can be seen without any continuation of the construction of the scenario bundle that an alternative at a decision point will not be taken, no matter what reactions may be expected afterwards. This happens if another alternative is clearly better for the actor or the coalition who has to make the decision, no matter what reactions will follow there. In such cases we speak of an *inferior alternative dominated* by a *superior alternative*. It is a natural stopping principle *not to continue the construction of a scenario bundle after an inferior alternative*. An end-point reached in this way is called an *inferiority end-point* and the stopping principle is referred to as the *inferiority criterion*.

Up to now we have discussed two stopping principles. The first one leads to blind alley end-points and may be called the *blind alley criterion*. The second one is the inferiority criterion. These two criteria are much less important than the third principle which we will now discuss.

Consider a node which corresponds to a situation where an actor or a group of actors are under pressure to make a decision whether to react or not. A node of this kind is called a *node with reactive pressure*. Clearly, it is necessary to continue the construction at a node with reactive pressure unless the node is a blind alley end-point or an inferiority end-point. This is not yet a stopping principle, since there are some nodes without reactive pressure where we do not want to stop the construction of the scenario bundle.

A trivial example of a node of this kind is the origin. More interesting examples arise in the following way. Consider a slow internal event like that at the beginning of scenario bundle 14. The edge corresponding to the event leads to a node without reactive pressure where we do want to continue the construction of the scenario bundle.

The following *criterion of reactive pressure* excludes such examples: *the construction of a scenario bundle is not continued beyond a node without reactive pressure which comes after at least one decision point of an actor or a coalition.* An end-point reached by the application of this criterion is called a *normal end-point*.

The word 'normal' indicates that the reactive pressure criterion is the most important stopping principle. The other two criteria are less frequently used.

There is an obvious objection against the reactive pressure criterion. How can we ignore the possibility of important further developments after a normal end-point? Should we not try to continue the construction at least in some cases?

If we take this objection seriously, we must be aware of the fact that a normal end-point corresponds to a situation which is wide open in the sense that many new developments may start there. Generally, it will be necessary to think of a normal end-point as a new initial situation with a variety of new scenario bundles beginning there. The construction of these bundles may be desirable in some cases but the continuation of the original bundle requires more than that.

In this connection it is important to point out that the scenario bundle method does not try to combine the set of all bundles beginning at the same initial situation into one superstructure. This would involve arbitrary judgements on the time sequence of initial options and internal events. It seems to be better to avoid such judgements.

The difficulties of combining a set of bundles into one superstructure are a good reason not to continue the construction beyond a normal end-point. This limitation of the scenario bundle method is a very natural one. The farther one looks into the future, the more hazy the picture becomes. Therefore detailed modelling cannot be expanded beyond certain limits.

Normal end-points are characterized by the distinguishing feature that there the modelling process meets a specific barrier. The unavoidable uncertainties involved in any continuation of a scenario bundle are suddenly considerably increased. Moreover, the nature of the uncertainty undergoes a qualitative change caused by the bundle combination problem.

The discussions of this subsection may be summarized by the following *combined stopping principle*: *the construction of a scenario bundle is continued until a further continuation would have to go beyond a blind alley end-point, an inferiority end-point or a normal end-point.*

9.3 Consequences

The construction of a scenario bundle involves judgements about the expected consequences of hypothetical courses of action. Let us look at the example of scenario bundle 14 of the conference results. The first important judgement about consequences concerns the question whether Saudi Arabia will succeed in taking over the Emirates if there is no intervention. The answer was yes.

Another question of a similar nature arises in connection with Iran's option 'counterintervention-war'. Here the outcome in case of no further intervention was judged to be a success of Iran.

For the purposes of analysis, the situation at the end-points is of special significance. Therefore, the graphical representations in the conference results contains rectangular boxes near the end-points where some consequences of special importance are indicated by very short keyword descriptions.

Military success or failure is only one category of consequences which must be considered. Political results like 'Iran's hegemonial position weakened' are as important (see scenario bundle 14). Sometimes the most significant consequence is the change of an actor's goal system. Scenario bundle 12 is a case in point (see Figure A.5 in the appendix). Here an expected change of the USA's goals deters a Chinese–Egyptian option of nuclear proliferation.

9.4 Plausibility of reactive options

In principle it is possible to test the plausibility of reactive options in the same way as that of initial options. With some modifications the criteria of realism and desirability can be used here, too.

In most cases it seems to be relatively easy to decide which reactive options should be used. The selection of plausible initial options is more difficult. Therefore criteria of plausibility are less important with respect to reactive options.

In the same way as initial options, reactive options are suggested by the goals and fears of the actors. In this respect there is no difference between both types of option.

Sometimes it appears to be desirable to include options into the scenario bundle which do not really pass the tests of plausibility. The best way to make a somewhat hidden lack of plausibility explicit may be the formal inclusion in the bundle. It is easier to understand the reasoning behind the construction if seemingly obvious reaction possibilities are shown as options, even if they fail to be plausible in the strict sense of the criteria.

9.5 Preference judgements and analysis

Every actor is assumed to have an unambiguous preference ranking over the end-points of a scenario bundle. The theoretical possibility of indifference between two end-points is excluded by assumption. This is meant by the word *unambiguous*.[21]

It is very unlikely that a group of experts will reach the consensus that an actor is indifferent between two end-points. It is more likely that they disagree on the order of preference. Therefore the requirement of unambiguity hardly involves any practical restriction. Cases of unresolvable disagreement can be handled by analysing the bundle in several different ways.

It is advantageous to combine the collection of preference judgements with the analysis of the scenario bundle. In this way the number of judgements to be made is drastically reduced. Only those preference relationships need to be discussed which determine the equilibrium solution.

It is convenient to explain the combined process of analysis and preference judgement collection with the help of the example of scenario bundle 14 (see Figure A.3 in the appendix). The process begins at the end of the bundle and proceeds backwards in a dynamic programming fashion. In this way one determines a unique equilibrium solution, technically known as a perfect equilibrium point (see section 10).

The first decision points to be looked at are those followed by no other nodes than end-points. Scenario bundle 14 has only one such decision point, namely that of the USA. Here we must ask the question whether the USA prefers the result of taking the option 'mediation' to the situation at the end-point following the no-branch. According to the conference consensus 'mediation' is the preferred alternative. This is indicated graphically by 'crossing out' the no-branch at the decision point; the two little lines may be thought of as a barrier to the flow of the equilibrium solution.

At any stage of the combined process of analysis and preference judgement collection we must always look at those decision points which are followed by no other nodes than end-points or nodes which have been analysed already. The next decision point to be investigated in our example is that of Iran. Since the no-branch at the decision point of the USA has been crossed out, Iran has the choice between two end-points, namely that following the option 'mediation' and that following Iran's no-branch. The conference consensus was that Iran prefers the first alternative to the second one. Accordingly the no-branch at Iran's decision point has been crossed out.

The next decision point to be considered belongs to the coalition of Iran and Saudi Arabia. The coalition decides between two end-points, namely that following 'counterintervention-war' and 'mediation' and that which results from the option 'Saudi Iranian defence pact'.

As has been explained at the end of section 9.1 we exclude the possibility of more than one option at a coalitional decision point. *A coalitional option is taken, if and only if every member of the coalition prefers the result of taking it to the result of not taking it.* This is a very natural assumption on coalition formation. A coalition cannot form unless the coalitional option is advantageous for every member.

In the case at hand the conference consensus was that both members of the coalition prefer the Saudi Iranian defence pact to the other alternative. Accordingly, the no-branch at the coalitional decision point has been crossed out.

The last decision point to be investigated is that of Saudi Arabia. Here the conference came to the conclusion that the end-point following the coalitional option 'Saudi–Iranian defence pact' is more advantageous to Saudi Arabia than the situation which results from not taking the option 'take over Emirates'. Accordingly, the no-branch has been crossed out.

The equilibrium solution may be characterized as the collection of choices not crossed out. Note that the equilibrium solution describes more than the course of action resulting from equilibrium choices. This course of action is called the equilibrium play. In our example 'take over Emirates' followed by 'Saudi–Iranian defence pact' is the equilibrium play. The equilibrium solution also determines hypothetical equilibrium choices for the case that contrary to equilibrium expectations the coalitional option is not taken. These hypothetical choices are an important part of the equilibrium solution since they explain why the equilibrium play does not reach that part of the game.

Preference judgements should be justified in terms of the goals and fears of the actors. The goals and fears to be taken into account are not necessarily only those which are present at the decision point where the choice has to be made. Later developments must be considered, too. A rational decision maker will be aware of the fact that new problems arise from new developments and goal emphasis may shift accordingly. He must be able to compare situations which differ with respect to his system of goals and fears.

9.6 The sequential model of coalition formation

It is the purpose of this subsection to explain the game theoretical justification of the sequential model introduced in subsection 9.1.

Consider a situation where a coalition C has an option c and, simultaneously, a member A of C has options $a_1,\ldots,a_m$ (we do not exclude the special case $m = 1$, where A has just one option). Let a_o be the alternative that none of the options $a_1,\ldots,a_m$ and c is taken. Let $A, B_1,\ldots,B_r$ be the members of C.

In order to show that the sequential model adequately represents the strategic situation, we first introduce a description in terms of a more natural simultaneous model. Then we shall argue that both models are strategically equivalent.

In the simultaneous model the actors $A, B_1,\ldots,B_r$ have to make simultaneous and independent decisions. Each of the actors $A, B_1,\ldots,B_r$ must decide whether he agrees to take option c or not; in addition to this, A must decide which of the alternatives $aô,\ldots,a_m$, he wants to choose if option c is not taken. This selection is of no significance if every member of C agrees to option c (case 1 of Table 1).

Table 1 Simultaneous model

	Decisions	Result
Case 1	All members of C agree to c and A selects a_k	Option c
Case 2	At least one member of C does not agree to c and A selects a_k	Alternative a_k

On the other hand, if at least one of the members of C does not agree to c, the result is the alternative a_k chosen by A (case 2 of Table 1).

Suppose that every actor has an unambiguous preference ranking $a_o,\ldots,a_m$ and c. Obviously A must select that alternative a_j which is best among $a_o,\ldots,a_m$ with respect to his ranking. Otherwise he would exhibit a form of non-rational behaviour which is technically called a 'dominated strategy'. Taking this into account an actor in C will agree to c if and only if he prefers c to A's optimal choice a_j among $a_o,\ldots a_m$. The same result is obtained in the sequential model. This shows that the analysis of both models comes to the same conclusion. In this sense the two models are strategically equivalent.

The discussion can be generalized to a wider class of situations. Suppose that A's alternatives are $a_o,\ldots,a_m$ as before; in addition to this, several coalitional options $c_1,\ldots,c_s$ are available to coalitions $C_1,\ldots,C_s$, respectively. Assume that $A, C_1, C_2,\ldots, C_s$ is a nested sequence in the following sense: A is one of several members of C_1 and for $j = 1,...,s-1$, coalition c_j is a proper subset of C_{j+1}.

In the generalized simultaneous model every actor has to decide for every coalition C_i where he is a member whether he accepts c_i or not. In addition to this, A has to choose among $a_o,....,a_m$. Finally the largest coalition whose option is accepted by all its members takes this option; if there is no such coalition the result is A's choice among $a_o,...,a_m$.

In the generalized sequential model first C_s decides whether to take its option or not, then c_{s-1}, etc. In the same way as before it can be shown that both models are strategically equivalent. Since we do not need the more general model in the scenario bundles of the conference results this will not be done here.

10 Formal structure

In this section we introduce a formal game theoretical definition of the concept of a scenario bundle. For reasons of mathematical convenience we shall not adhere to terminology used up to now. Thus the actors will be numbered from 1 to n; they will be called players. A coalition will be any subset of $N = \{1,...,n\}$, including the empty set. The number of elements of a finite set S will be denoted by $|S|$. It is convenient to introduce some definitions and notations relating to game trees before we go on to define the concept of a scenario bundle.

10.1 Game trees

A game tree K is a tree in the sense of graph theory, with a distinguished node o, the origin of K. A node y follows a node x, if x is on the path from o to y and y is different from x. A node y immediately follows a node x, if y follows x and x and y are connected by an edge of K. A choice at x is an edge which connects x with a node immediately following x. (In games of perfect information it is convenient to think of a choice as an edge of the tree.)

The set of all nodes of K is denoted by X. The set of all nodes $x \varepsilon X$, such that there are exactly m choices at x, is denoted by X_m. The nodes in X_o are called end-points. The set of all choices at x is denoted by $A(x)$.

The subtree K_x of K at x is the game tree consisting of x and all nodes of K following x, together with the edges connecting such modes.

10.2 Scenario bundles

A scenario bundle $B = (N,K,c,a,h)$ is defined as follows:

(a) $N = \{1,...,n\}$ is the player set
(b) K is a game tree
(c) c is a function which assigns a coalition $C = c(x)$ to every $x \varepsilon X \backslash X_o$. The following conditions (1) and (2) must be satisfied:

$$\text{If } c(x) = \phi \text{ then } x = o \text{ and } o \varepsilon X_1 \tag{1}$$

$$\text{If } |c(x)| > 1 \text{ then } x \varepsilon X_2 \tag{2}$$

The function c is called decision point function.

(d) a is a function defined on the set of all x with $|c(x)| > 1$. This set is denoted by Y. The function a assigns an edge $a(x)$ at x to every $x \varepsilon Y$. This edge is called the option at x; the other edge at x is denoted by $b(x)$ and is called the no-branch at x. The function a is the option function.

(e) $h = (h_1,...,h_n)$ is a system of pay-off functions for the players $1,...,n$. The function h_i assigns a real number $h_i(z)$ to every end-point $z \varepsilon X_o$. The following condition (3) must be satisfied for $i = 1,...,n$:

$$h_i(y) \neq h_i(z) \text{ for } y \neq z \text{ and } y, z \varepsilon X_o \qquad (3)$$

Interpretation: the decision point function determines who has to make a decision at x. In the case of $c(o) = \phi$ nobody has to make a decision at o since the bundle begins with an event. Condition (1) excludes events at a later stage. Condition (2) secures that a coalition with more than one member does not have more than one option at a given node (see subsection 9.1). It is necessary to introduce an option function since in the case of coalitions with more than one member the edges corresponding to taking and not taking a coalitional option cannot be treated symmetrically. The pay-off functions h_i must be understood as ordinal utility indices representing preference rankings of the end-points. Condition (3) excludes ambiguous preference rankings (see subsection 9.5).

10.3 Strategies

A strategy combination is a function s which assigns a choice at x to every $x \varepsilon X \backslash X_o$. The choice $s(x)$ is called the local strategy at x. Let a be a choice at y. Then the strategy combination s' with

$$s'(x) = \begin{cases} s(x) & \text{for } x \varepsilon X \backslash X_o \text{ and } x \neq y \\ a & \text{for } x = y \end{cases} \qquad (4)$$

is denoted by s/a. Consider a node $x \varepsilon X \backslash X_o$ and a strategy combination s. Obviously there is a uniquely determined end-point z such that all edges on the path from x to z are selected by s. This end-point is denoted by $z(x,s)$. Define

$$h_i(x,s) = h_i(z(x,s)) \qquad (5)$$

The pay-off $h_i(x,s)$ is called the local pay-off of player i at x for s.

10.4 Optimality

Consider a node $x \varepsilon X \backslash X_o$ with $c(x) = \{i\}$. We say that $\tilde{a} \varepsilon A(x)$ is optimal with respect to s, if we have

$$h_i = (x,s/\tilde{a}) = \max_{a \varepsilon A(x)} h_i(x,s/a) \qquad (6)$$

The definition of optimality is a different one for $|c(x)| > 1$. Consider a node x with this property. There are exactly two choices at x, namely $a(x)$ and $b(x)$. We say that $a(x)$ is optimal with respect to s, if we have

$$h_i(x,s/a(x)) > h_i(x,s/b(x)) \quad (7)$$
for every $i \varepsilon c(x)$

and we say that $b(x)$ is optimal with respect to s, if we have

$$h_i(x,s/b(x)) > h_i(x,s/a(k)) \quad (8)$$
for at least one $i \varepsilon c(x)$.

For the sake of formal completeness the unique choice at o in the case of $c(o) = \phi$ is defined as optimal with respect to s.

Interpretation: obviously the definition of optimality must distinguish between $a(x)$ and $b(x)$. Every member of $c(x)$ has to agree to $a(x)$ whereas any one of them can enforce $b(x)$.

Lemma 1: let s be a strategy combination. At every $x \varepsilon X \backslash X_o$ there is exactly one optimal choice with respect to s.

Proof: the assertion is an immediate consequence of condition (3).

10.5 Perfect equilibrium point[22]

A strategy combination s^* is a perfect equilibrium point of G if for every $x \varepsilon X \backslash X_o$ the choice $s^*(x)$ is optimal with respect to s^*.

Let K_x be the subtree at x of K and let c_x, a_x, and h_x be the restrictions of c, a and h to K_x. Obviously $G_x = (N, K_x, c_x, a_x, h_x)$ is a scenario bundle. We call G_x the subgame of G at x. It follows by the definition of optimality that a perfect equilibrium point has the following property:

Lemma 2: let s^* be a perfect equilibrium point of G and let s_x^* be the restriction of s^* to K_x. Then s_x^* is a perfect equilibrium point of the subgame G_x at x.

Theorem: every scenario bundle $B = (N,K,c,a,h)$ has one and only one perfect equilibrium point.

Proof: The length L of a scenario bundle is defined as the number of edges in the longest path from o to an end-point. The theorem is proved by induction on L.

For $L = 1$ the assertion follows directly by lemma 1. Suppose that the assertion is true for $L = 1...,m$. Consider a scenario bundle G with $L = m + 1$. The subgames at the nodes immediately following the origin are of smaller length. Therefore these subgames have uniquely determined perfect equilibrium points. It follows by lemma 2 that the choices selected by these perfect equilibrium points are also selected by any perfect equilibrium point $\tilde{s}$ of G. The definition of optimality of a choice at x with respect to s does not depend on $s(x)$. Therefore it follows by lemma 1 that the optimal choice at

o with respect to any perfect equilibrium point $\tilde{s}$ is the same one. Denote this choice by $s^*(o)$. Obviously G has exactly one perfect equilibrium point s^* namely that strategy combination s^* which selects $s^*(o)$ at o and those choices at the other nodes of $X\backslash X_o$ which are chosen by the perfect equilibrium points of the subgames at the nodes immediately following the origin.

Remark: the notion of an equilibrium point has not been defined above. In the framework of the scenario bundle it is more natural to introduce the concept of a perfect equilibrium point without doing this. For an ordinary equilibrium point optimality would be required for nodes on the equilibrium play only. Imperfect equilibrium points are intuitively unsatisfactory. Rationality must be exhibited in all parts of the game, including those which are not reached by the equilibrium play.

11 Additional comments on the aim of the method

Scenario bundles are the final result of an elaborate judgemental process. The application of the method requires time and effort. We may ask whether it is worth while to go to the trouble. After all, as has been pointed out in subsection 2.8, predictive reliability cannot be promised. What are the benefits which can be expected from the use of the method?

The answer to this question becomes apparent if we look at the analogy of the chess player. Generally, a chess player who tries to plan ahead cannot really predict the future course of the game. Nevertheless, he will approach his decision problem in a predictive spirit. It will be his aim to explore the likely consequences of a selection of plausible moves .[23]

Obviously it is impossible to look at all possibilities. A selection has to be made. In order to do this, criteria of plausibility must be used. Predictive reliability cannot be achieved, since different players look at different possibilities.

The lack of predictive reliability does not mean that it is futile to analyse the situation. Analysis is necessary in order to play in an efficient way. Nobody can hope to be successful without it.

Human decision making in chess seems to be analogous to the construction and evaluation of scenario bundles. It may even be possible to pursue the analogy in detail. The stopping principles explained in subsection 9.2 are a case in point. A chess player needs similar rules in order to limit the depth of his explorations.

An outside observer who wants to gain insight into a game situation, must try to engage in the same kind of analysis as the decision makers themselves, otherwise he cannot hope to achieve understanding. The scenario bundle method tries to serve this purpose in the exploration of international conflict.

Notes
1. The conference was held under the auspices of SADAC (Verein zur Förderung der Systemanalyse des Arms Control) and was financed by a grant from the Volkswagen Foundation.
2. The introduction of the report contains a list of the participants and explains their roles in the cooperative effort to the conference.
3. See Guetzkow, H., C. Alger, R. Brody, R. Noel and R. Snyder, *Simulation in International Relations; Developments for Research and Teaching*, Englewood Cliffs: Prentice Hall, 1963.
4. Allison, G.T., *Essence of Decision: Explaining the Cuban Missile Crisis*, Boston: Little, Brown, 1971.

5. Holsti, K.J., *International Politics, a Framework for Analysis*, second edition, London: Prentice Hall, 1974.
6. von Neumann, John and Oskar Morgenstern, *Theory of Games and Economic Behavior*, second edition, Princeton: Princeton University Press, 1944.
7. Savage, L.H., *The Foundations of Statistics*, New York, New York: Wiley, 1954.
8. Simon, A.H., 'Theories of Decision Making in Economics and Behavioral Science', *American Economic Review*, 1959, pp. 253–83.
9. Sauermann, H. and R. Selten, 'Anspruchsanpassungstheorie der Unternehmung', *Zeitschrift für die gesamte Staatswissenschaft*, 1962, pp. 577–97.
10. For the technical concept of certainty equivalence see Simon, Herbert A., 'Dynamic Programming under Uncertainty with a Quadratic Criterion Function', *Econometrica*, 1956, pp. 74–81.
11. Shackle, George L., *Uncertainty and Expectations in Economics*, Oxford: Blackwell, 1972.
12. Holsti, K.J., *International Politics*, op. cit.
13. Deutsch, Karl, *Analyse internationaler Beziehungen*, Frankfurt: Europäische Verlagsanstalt, 1968.
14. Morgenthau, Hans, *Politics among Nations*, fourth edition, New York: Alfred A. Knopf, 1967.
15. Deutsch, Karl, *Analyse Internationaler Beziehungen*, op. cit., pp. 185–6.
16. Compiled by R. Avenhaus and R. Stoecker, assisted by B. von Essen, on the basis of judgements by R. Büren, P.H. Chang, M. Handel, R. Kolkowicz, A. Perlmutter, E. Ravenal and U. Steinbach, in the original report.
17. Holsti, K.J., *International Politics*, op. cit., Chapter 5.
18. Wolfers, Arnold, *Discord and Collaboration, Essays in International Politics*, Baltimore: Johns Hopkins Press, 1962, Chapter 5, The Goals of Foreign Policy.
19. Tversky, Amos and Daniel Kahneman, 'Judgement under Uncertainty: Heuristics and Biases', in D. Wendt and Ch. Vlek (eds), *Utility, Probability and Human Decision Making*, Dordrecht-Holland and Boston: Reidel, 1975.
20. The reader is reminded that in 1976 Iran was still ruled by the Shah.
21. Technical texts on the subject of preference orders use less easily intelligible terms like 'asymmetric', See Sen, A., *Collective Choice and Social Welfare*, San Francisco, Cambridge, London, Amsterdam: Holden-Day, 1970, Chapter 1, see p. 17.
22. The definition of perfectness given here is close to the original one: see Selten, R., 'Spieltheoretische Behandlung eines Oligopolmodells mit Nachfrageträgheit', *Zeitschrift für die gesamte Staatswissenschaft*, 121, pp. 301–24; 677–89; 1965. A more refined notion of perfectness has been introduced later; see Selten, R., 'Reexamination of the Perfectness Concept for Equilibrium Points in Extensive Games', *International Journal of Game Theory*, 4, 1975, pp. 25–55. The new definition requires cardinal pay-offs and therefore cannot be transferred to scenario bundles.
23. The views expressed here are based on the human problem-solving literature. See Newell, A. and H.A. Simon, *Human Problem Solving*, Englewood Cliffs: Prentice Hall, 1972. No attempt is made here to draw a detailed picture of decision making in chess. It is sufficient to point out some of the analogies to the scenario bundle method.

References

Allison, G.T., *Essence of Decision: Explaining the Cuban Missile Crisis*, Boston: Little, Brown, 1971.
Deutsch, Karl, *Analyse internationaler Beziehungen*, Frankfurt: Europäische Verlagsanstalt, 1968.
Guetzkow, H., C. Alger, R. Brody, R. Noel and R. Snyder, *Simulation in International Relations: Developments for Research and Teaching*, Englewood Cliffs: Prentice Hall, 1963.
Holsti, K.J., *International Politics, A Framework for Analysis*, 2nd edition, London: Prentice Hall, 1974.
Morgenthau, H., *Politics among Nations*, 4th edition, New York: Alfred A. Knopf, 1967.
Newell, A. and H.A. Simon, *Human Problem Solving*, Englewood Cliffs: Prentice Hall, 1972.
Sauermann, H. and R. Selten, 'Anspruchsanpassungstheorie der Unternehmung', *Zeitschrift für die gesamte Staatswissenschaft*, 1962, pp. 577–97.
Savage, L.H., *The Foundations of Statistics*, New York: Wiley, 1954.
Shackle, George, L., *Uncertainty and Expectations in Economics*, Oxford: Blackwell, 1972.
Selten, R., 'Spieltheoretische Behandlung eines Oligopolmodells mit Nachfrageträgheit', *Zeitschrift für die gesamte Staatswissenschaft*, 121, pp. 301–24; 667-89, 1965.
Selten, R., 'Reexamination of the Perfectness Concept for Equilibrium Points in Extensive Games', *International Journal of Game Theory*, 4, 1975, 25–55.
Sen, A., *Collective Choice and Social Welfare*, San Francisco, Cambridge, London, Amsterdam: Holden-Day, 1970, Chapter l*.
Simon, Herbert A., 'Dynamic Programming under Uncertainty with a Quadratic Criterion Function', *Econometrica*, 1956, pp. 74–81.

Simon, Herbert A., 'Theories of Decision Making in Economics and Behavioral Science', *American Economic Review*, 1959, pp. 253–83.

Tversky, Amos and Daniel Kahnemann, 'Judgement under Uncertainty: Heuristics and Biases', in D. Wendt and Ch. Vlek (eds), *Utility, Probability and Human Decision Making*, Dordrecht-Holland and Boston: Reidel, 1975.

von Neumann, John and Oskar Morgenstern, *Theory of Games and Economic Behavior*, second edition, Princeton: Princeton University Press, 1944.

Wolfers, Arnold, *Discord and Collaboration, Essays in International Politics*, Baltimore: Johns Hopkins Press, 1962, Chapter 5, The Goals of Foreign Policy.

Appendix: Selected results from the Research Conference on Strategic Decision Analysis focusing on the Persian Gulf

Ranking of military strength

1 USA
2 USSR
3 Europe
4 China
5 India
6 Israel
7 Egypt
8 Iran
9 Pakistan
10 Syria
11 Iraq
12 Jordan
13 Saudi-Arabia
14 Kuwait
15 PLO
16 South Yemen
17 Oman
18 Qatar
19 Bahrain
20 UAE
21 Kurds
22 Gulf Revolutionaries

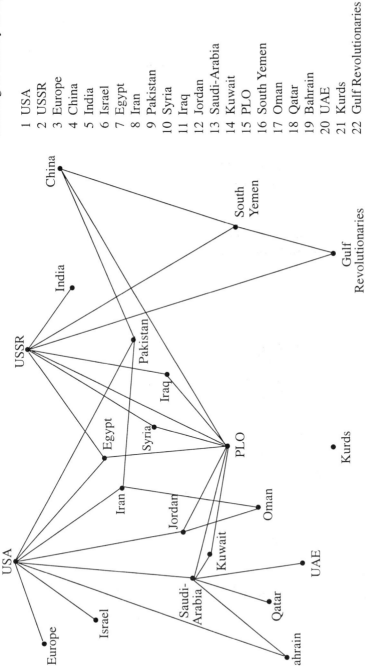

Figure A.1 Graph of protective relationships

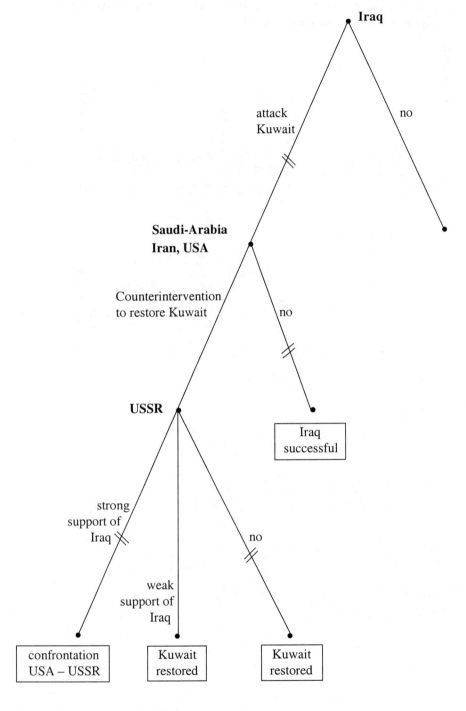

Figure A.2   Scenario bundle 1

Game Theory and Economic Behaviour I   323

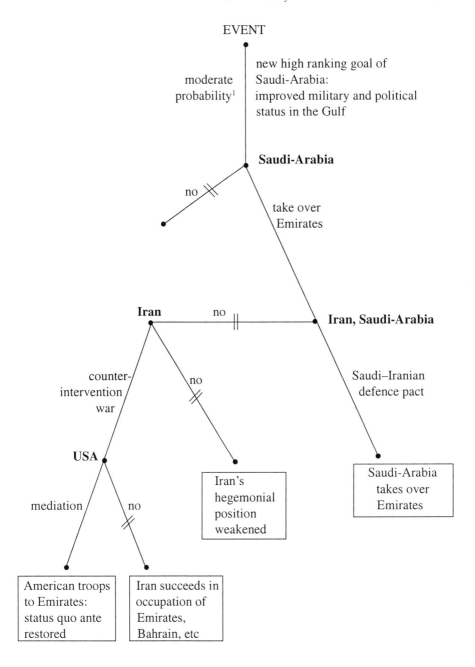

Note:
1. The probability judgement is based on the idea that with the increase of military power the army will become more ambitious. This can lead to the new goal of reaching an improved military and political status in the Gulf.

Figure A.3   Scenario bundle 14

324 *Game Theory and Economic Behaviour I*

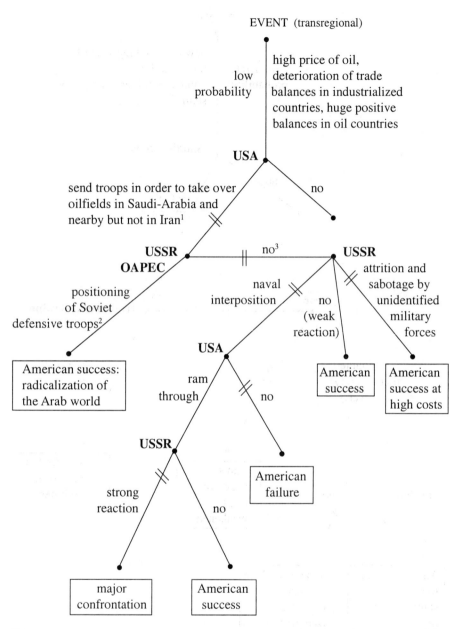

Notes:
1. The occupation of Iranian oil fields does not seem to be militarily feasible. These oil fields are too far from the American bases in the Indian Ocean.
2. There are Soviet naval facilities at Umm Qasr, Iraq, which permit the dispatch of military units to the conflict area.
3. The sequential model has been used here (see section 9 of this chapter). This is indicated by drawing the no-branch as a horizontal line.

Figure A.4   Scenario bundle 8

Game Theory and Economic Behaviour I   325

Figure A.5   Scenario bundle 12

[11]

BALANCE OF POWER IN A PARLOR GAME

R. Selten
University of Bonn, FRG[1]

1. Introduction

The idea of balance of power in a system of sovereign states has a long history. Already 300 B.C., Artha-Sastra Kautilya has written on the subject (Kautilya, 1957). A recent book by Bernholz (1985) looks at the historical facts in the light of the theory. A more detailed account of the literature can be found here.

The term *balance of power* is not always used in the same sense, but it seems to apply to a state of affairs, where potential aggressors are deterred from military actions whose immediate or remote probable consequences involve the elimination of one of the actors. This does not necessarily exclude wars which do not endanger the survival of any actor. Wars may happen, but the system of actors remains the same. It is also possible that a situation of long lasting stable peace is secured by a balance of power

Game theory is a general theory of rational behavior in interactive decision situations of conflict and cooperation. Balance of power theories (Kaplan 1957, Zinnes 1967. Bernholz 1985) try to explain the typical behavior of nation states as the rational pursuit of their interests. However, traditional verbal theories do not offer a logically stringent explanation. For this purpose, the use of game theory seems to be unavoidable.

An interesting game theoretic approach to the balance of power problem has been developed by Niou and Ordeshook (1986, 1987). Their model is a cooperative game, to which they apply a version of the bargaining set (Aumann and Maschler 1964). Strategies of individual players cannot be discussed in the framework of a cooperative game. Therefore, we think that a non-cooperative game model is preferable. The use of non-cooperative game theory does not mean that cooperation is excluded. On the contrary, even if this may seem to be paradoxical, the explanation of cooperation requires non-cooperative game theory.

---

[1] I am grateful to Peter Bernholz for stimulating discussions and valuable suggestions.

A non-cooperative model of a balance of power system has been proposed by Harrison Wagner (1986). His analysis is heuristic rather than precise, but nevertheless suggestive. In his model, the players have to decide on the deployment of continuous resources and on the military use in continuous time. The strategy space remains unclear. Regardless of how this gap is filled, a precise analysis is bound to be difficult.

In this paper, a non-cooperative game model will be presented which takes the form of a parlor game. Theoretical results need to be confronted with the behavior of subjects in experiments. Therefore, playability is an important modelling consideration.

The rules of the parlor game will be given in ordinary language. The playing board and other paraphernalia are described in detail and can easily be prepared by anybody who wants to play. The game is called *Changing Alliances*. Playing the game does not require any knowledge of game theory.

Practical experience shows that *Changing Alliances* is not without entertainment value, even if this aspect has not been a major concern in the construction of the game. The rules have been kept simple in order to enhance the usefulness of the game as a research tool.

The game is not meant to be a realistic description of a system of nation states, but rather a radically simplified picture which concentrates on the essence of the subject matter. Features of historical systems which are not essential for the reasoning underlying balance of power theories are intentionally neglected.

The players in the parlor game represent sovereign states in a world consisting of 30 *provinces* but without any geographical structure. Provinces are resources which can be won or lost in war. The game is played over 30 *rounds*. In each round an *alliance* and a *counteralliance* can be built up by formal moves on the board. Alliances may change from round to round.

After the description of the parlor game in Section 2, reasons for details of the rules will be given in Section 3. In particular, it will be discussed why a fixed number of rounds is specified. An infinite number of rounds is theoretically more convenient but practically infeasible.

In Section 4, a modified version of the game with only 6 provinces and an infinite number of rounds will be introduced; a long run average definition of payoffs is used. Unlike the much more complex parlor game with its 30 provinces, this simpler model permits a game theoretical analysis without serious difficulties. A "solution" determined by symmetry, stationarity and local non-cooperative equilibrium conditions is derived in Section 5. In order to clarify technical problems concerning global equilibrium conditions, discounted payoffs are considered in Section 6.

The solution of the 6-province model is discussed in Section 7. An interpretation of the game theoretical results will try to exhibit resons for stability and instability.

2. Changing Alliances - A Parlor Game

2.1 Inventory

  1 playing board (see Figure 1)
 12 player tokens, marked 1,...,12
 12 country boards (see Figure 2)
 12 player cards, marked 1,...,12 (see Figure 3)
 30 province cards, marked 1,...,30 (see Figure 2)
 30 province stones, marked 1,...,30 (see Figure 4)
  1 province stone box (see Figure 3)
360 resource chips (see Figure 2)

2.2 The Object

Each player represents a country. Countries can form changing alliances and counteralliances and fight wars. Provinces are won and lost as a result of war. A player should try to obtain as many provinces as possible, but it is even more important to secure continued existence. A small advantage is attached to staying out of wars which do not permanently increase one's territorial holdings.

Scores are computed for each player at the end of the game. It is not important to have a higher score than other players. Everybody should aim at a high score for himself without being concerned about the scores of other players.

2.3 Preparation for Play

The game is played by at least 3 and up to 12 players. The players sit around a table. The playing board is put on this table (see Figure 1). Each player receives a country board placed in front of his or her seat (see Figure 2). The number on the country board becomes the number of the player.

The provinces are distributed among the players. At first, each player receives one province card which is put on his or her country board. Then, each of the remaining province cards is allotted as follows. The player cards (see Figure 3) of those players who do not yet have 12 province cards are mixed and one is drawn at random. This player receives the card. The province card is put on his or her country board. For each province, the player cards of those players who do not yet have 12 province cards, are mixed again and the same procedure is repeated until all province cards have been distributed.

Finally, the number of province cards owned by every player must be written down on a slip of paper, since it will be important for the computation of scores at the end of the game.

Each player receives a token showing the player's number. The tokens are placed on the field "wait for decision", each on the square bearing its number.

The game begins with the selection of the first decision maker for the first round.

The game is played over 30 rounds. Every round is played in the same way. It will now be explained how a round is played.

WAIT FOR DECISION			
1	2	3	4
5	6	7	8
9	10	11	12

ALLIANCE			
1	2	3	4
5	6	7	8
9	10	11	12

NON-ALIGNED			
1	2	3	4
5	6	7	8
9	10	11	12

COUNTER-ALLIANCE			
1	2	3	4
5	6	7	8
9	10	11	12

Figure 1: The playing board

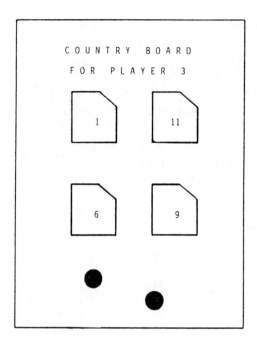

Figure 2: Player 3's country board with four province cards and 2 resource chips. (The upper right corner of province cards is cut off to distinguish 6 and 9.

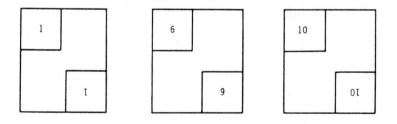

Figure 3: Player cards

2.4 Selection of a Decision Maker

As long as there are players with tokens on the field "wait for decision", the next decision maker is chosen among these players. For this purpose, their player cards are mixed and the upmost card in the deck is turned up. The player whose card has been turned up becomes the next decision maker.

2.5 Pre-Attack and Post-Attack Decision Maker

The choices which are open to a decision maker depend on previous decisions. As long as no player has attacked another one, the decision maker is in the position of a pre-attack decision maker. After an attack, the decision makers are post-attack decision makers, as long as there are tokens on the field "wait for decision".

Of course, the first decision maker is a pre-attack decision maker, randomly chosen among all players in the game.

2.6 Choices of a Pre-Attack Decision Maker

A pre-attack decision maker has the following options:

a) <u>Stay non-aligned</u>: In this case, the token of the decision maker is moved to its square on the field "non-aligned". Nothing else is changed.

b) <u>Attack another player</u>: The decision maker names the player he wants to attack. The attacked player can be any other player who owns at least one province, no matter whether the token of the attacked player is on the field "wait for decision" or on the field "non-aligned". The attack has the following consequences:
   - The token of the decision maker is moved to its square on the field "alliance".
   - The token of the attacked player is moved to its square on the field "counteralliance".
   - After the tokens of the decision maker and the attacked player have been moved, any tokens left on the field "non-aligned" are moved back to their squares on the field "wait for decision".

2.7 Choices of a Post-Attack Decision Maker

A post-attack decision maker has the following options:

a) <u>Stay non-aligned</u>: In this case, the token of the decision maker is moved to its square on the field "non-aligned". Nothing else is changed.

b) <u>Join the alliance</u>: The token of the decision maker is moved to its square on the field "alliance". - All tokens on the field "non-aligned" are moved back to their squares on the field "wait for decision".

c) <u>Join the counteralliance</u>: The token of the decision maker is moved to its square on the field "counteralliance". - All tokens on the field "non-aligned" are moved back to their squares on the field "wait for decision".

2.8 Stages of a Round

A round has two stages. The explanations given up to now concern the alignment stage, in which pre-attack decision makers and post-attack decision makers make their alignment decisions. Sooner or later, a situation is reached where no tokens are left on the field "wait for decisions". At this point, the alignment stage comes to an end and the round enters its allotment stage.

2.9 End in Peace

If at the end of the alignment stage all tokens on the board are on the field "non-aligned", the round ends in peace. In this case, the allotment stage is very short: Every player with a token on the field "non-aligned" receives one resource chip.

2.10 Allotment Procedure

If at the end of the alignment stage some tokens are on the fiels "alliance" and "counteralliance", the territorial consequences of the military confrontation have to be determined. One of both groups always emerges as the "winning party" and the other becomes the "losing party". A province is transferred from a member of the losing party to a member of the winning party. Which province is lost by whom and to whom is determined by an allotment procedure explained in the following.

- First, the strength of the alliance and the counteralliance is determined. The strength of each of both groups is measured by the number of provinces owned by its members. If one group is stronger, it becomes the winning party and the opposing group becomes the losing party.

- All province stones are put into the province stone box (see Figure 4). Then, one stone after the other is drawn at random without replacement until it becomes clear which province is lost by whom to whom.

- If one group is stronger, then the first stone with a number owned by a member of the winning party determines the "winner". The player who owns this province is the winner.

- If both groups are equally strong, then the first stone with a number of a province owned by a member of one of both groups determines the winner. The player who owns this province is the winner. His group becomes the winning party and the other group becomes the losing party.

- The first stone with a number of a province owned by a member of the losing party determines the province lost. The owner is the loser. The loser has to transfer the province card with this number to the winner.

Game Theory and Economic Behaviour I 333

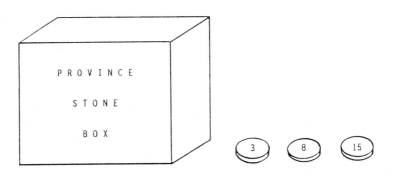

Figure 4: Province stone box and province stones.

2.11 Giving Away a Province Won

A player who has won a province can offer this province to another player if he wants to do this. The other player may be any other player, including the loser or a player who does not own a province. The player to whom the offer is made can accept or reject it. If the offer is accepted, the winner transfers the province card won to the concerning player. If the offer is rejected, the winner keeps the province. A second offer to give away the province is not permissible.

2.12 Resource Chips

At the end of the alignment stage of a round, each player with a token on the field "non-aligned" receives one resource chip.

A player who loses his last province does not lose his resource chips. Resource chips can be owned by players without provinces.

2.13 End of the Game

The game ends after 30 rounds. (Another fixed number of rounds can be agreed upon in advance.)

2.14 Beginning of a New Round

At the beginning of a new round, all tokens of players who own at least one province are put on their squares in the field "wait for decision". The tokens of players without a province are removed from the playing board. Then, the round is played according to the rules explained above.

2.15 Simplification of Random Procedures

In some cases it is not necessary to perform the random procedures exactly as prescribed by the rules given above. If only one potential decision maker is left, no card needs to be drawn. If the losing party has only one member, this player is the loser and it does not matter which of the loser's provinces is transferred to the winner.

2.16 Scores

At the end of the game, scores are computed for each player. The scores are based on the numbers of provinces owned at the beginning and the end of the game. The number of resource chips owned at the end also enters the determination of scores. The following symbols are used in the score formula:

a_i number of provinces owned by player i at the beginning of the game

m_i number of provinces owned by player i at the end of the game (after the last allotment)

r_i number of resource chips owned by player i at the end of the game

e_i this number is 1 if player i owns at least one province at the end of the game and zero otherwise

S_i score of player i.

The score formula is as follows:

$$S_i = 100 e_i + 30(m_i - a_i) + r_i. \qquad (1)$$

2.17 Game Director

The game runs more smoothly if one person does not actively participate, but, instead of this, functions as a *game director*, who performs the random choices which determine the order of moves in the alignment stage, moves the tokens on the board, runs the allotment stage, and pays out resource chips. The game director can concentrate on strict obeyance of the rules. Thus, he should not forget to move all tokens back from "non-aligned" to "wait for decisions" after a player has joined the alliance or the counteralliance. the greater the number of players, the more it is advisable to appoint a game director.

2.18 Playing for Money

Groups of players who want to play for money can agree to compute deviations from the average score and to determine wins and losses accordingly. However, if this is done, the payoff of a player does not only depend on his own score but also on the scores of the others. For theoretical reasons, it is preferable to employ a payoff scheme which avoids the dependence on other players' scores.

If one participant functions as a game director, one can proceed as follows: an agreed upon "base score" S_0 is subtracted from each players' score S_i in order to obtain a "net score" $S_i - S_0$. According to some fixed value of a score point, a player wins the corresponding amount if his net score is positive and loses the corresponding amount if his net score is negative. If there is a surplus of losses over gains, the game director receives it; otherwise he has to pay the surplus of gains over losses.

The net score S_0 must be determined before the beginning of the game. One way of doing this is an auction for the position of the game director. Every participant secretly writes his name and a proposed base score S_0 on a piece of paper. The position of the game director goes to the lowest bidder (if there are several lowest bidders, a random choice has to be made). The lowest bid becomes the base score S_0.

One could also agree to fix S_0 at 130 and to donate the surplus of losses over gains to a humanitarian cause. For $S_0 = 130$ there cannot be a surplus of gains over losses.

3. Comments on the Rules of *Changing Alliances*

3.1 The Number of Rounds

Practical experiences were gained with an earlier version of the game which involved a fixed stopping probability. After each round, two dice were thrown and the game stopped in the case of two sixes. In view of experiences with actual plays of the game, this rule was later abandoned in favor of a fixed length of 30 rounds.

A theoretical objection against a fixed stopping probability is based on the impossibility of playing such games in the laboratory. The subjects cannot be kept in the laboratory for an unlimited time. This imposes an upper bound on the number of rounds. It may not be clear how many rounds can be played at most, but a finite upper bound can be found below which the length of the game must remain. Eventually, the game must be stopped against the rules, if the random event required for stopping does not occur before the upper bound is reached.

In some games, e.g. in supergames of the prisoners' dilemma, it is of crucial theoretical importance whether the number of rounds has a finite upper bound or not. A finite upper bound supplies an anchor for the familiar induction argument, which shows that all equilibrium plays involve non-cooperative choices only (Luce and Raiffa, 1957). Contrary to this, the infinite supergame has subgame perfect equilibrium points whose equilibrium plays involve cooperative choices only.

It is quite common in parlor games that infinite plays are not excluded (e.g. in Parchesi). However, even if one ignores the theoretical objection raised above, one still has to face a completely different behavioral difficulty. In one of the actual plays with the stopping rule of a throw of two sixes, the game did not end in the first 36 rounds. After 36 rounds, several players expressed their belief that now the end must come soon. One of the players had a thorough training in statistics and knew that his confessed belief did not conform to probability theory. This kind of behavior is an example of the *gambler's fallacy*. The gambler's fallacy has a strong influence on decision making which should not be overlooked in the design of experiments.

The crucial theoretical difference between finite and infinite prisoner's dilemma supergames does not seem to be behaviorally relevant. This is shown by experimental studies (e.g. Selten and Stoecker, 1986). Apart from a few rounds near the end, observed behavior of experienced subjects achieves cooperation in sufficiently long finite prisoners' dilemma supergames. In view of these findings, it seems to be justifiable from the point of view of descriptive game theory to apply the analysis of infinite games to long finite games. We shall take this approach in the analysis of the model presented here.

3.2 Parameters

Numerical parameters like the number of provinces or the coefficients in the function for the determination of scores can be varied in the abstract model. It is of particular interest to look at cases with very few provinces, since there a complete theoretical analysis is easier to obtain.

3.3 Formation of Alliances and Counteralliances

It is important that all non-aligned players move back to "wait for decision" after a player has joined the alliance or the counteralliance. This rule has the purpose to make the random order of alignment decisions less important. A player can decide to enter an alignment only after some other player, even if he is the first one to make a decision.

3.4 Interpretation of a Round

Only one province can change hands during a round. This rule is based on the idea that alignments must be permitted to change after each change of the situation. Within a round, only a war episode takes place. The war can be continued in the next round with changed or unchanged alignments.

3.5 Allotment

If resources were infinitely divisible, the allotment procedure could be modelled deterministically as in Wagner's model (Wagner, 1986). In view of the indivisibility of the provinces, a random allotment is needed in order to make expected games and losses proportional to resources held. In his book, written 1832-34, Carl von Clausewitz emphasized the random element in military conflict (von Clausewitz, 1963).

The random element of the allotment procedure of Changing Alliances is rather weak, since the stronger bataillons always win.

3.6 Resource Chips

Resource chips provide an incentive not to try to gain a province whose ownership cannot be maintained in later rounds. The value of a resource chip is just small enough to exclude a disincentive to enter an alignment which is sure to win if in the case of winning a province, the concerning player can count on having one province more at the end. The minimum probability of winning a province for a member of the winning side is 1/29. This is the winning probability of a player with 1 province in an alliance with 29 provinces against a counteralliance of 1 province. The probability of 1/29 of having 1 province more at the end corresponds to an expected score gain of 30/29 which is slightly greater than 1, the value of 1 resource chip.

A player who loses all his provinces still keeps his resource chips earned in the past. One may think of resource chip as money accumulated on a Swiss bank account.

It can be argued that the first part of rule 2.12 should be changed as follows: At the end of a round, each player with a token on the field "non-aligned" receives one resource chip for every province earned. - Clearly, this rule would be more natural and it would have similar incentive effects. However, the simpler rule of 2.12 is easier to follow in the actual play of the parlor game.

3.7 Scores

As has been explained above, incentive considerations suggest a value of 30 resource chips for 1 province. It is reasonable to attach a positive value to existence, but there seem to be no theoretical reasons for a particular choice of this "existence value".

The score function is meant to be a reasonable representation of preferences of nation states in the simplified hypothetical environment of the game. Obviously, Changing Alliances is not a constant sum game. The score parts $30(m_i - a_i)$ due to the numbers m_i of provinces owned at the end sum up to zero, but each player who ends up without provinces decreases the score sum by 100 and each resource chip earned increases the payoff sum by 1. The greatest score sum is obtained if all 30 periods end in peace.

3.8 Playing for Money

A money payoff scheme based on deviations from average scores has the undesirable effect of transforming the game into a zero-sum game. The transformation creates an additional incentive for the elimination of players. Moreover, in the transformed game it is advantageous to prevent others from gaining resource chips. Therefore, it is preferable to use one of the alternative money payoff schemes proposed in 2.18. These proposals result in a money payoff which is a positive linear transformation of the score. Of course, one may prefer to determine the game director by another type of auction. The first price scaled bid auction proposed in 2.18 is only one of many possibilities.

4. A Theoretical Model

4.1 The Model

A modified version of *Changing Alliances* will now be introduced as a basis for a game theoretical analysis. In 3.1, it has been argued that there are good reasons for a fixed number of rounds in the parlor game. However, in a purely theoretical model, it is justifiable and maybe even preferable to take a different approach. The deviations of the modified model from the parlor game are as follows:

1) The number of rounds is infinite

2) There are only 6 provinces

3) There are no resource chips

4) There is no possibility to give away a province

5) Payoffs are based on the long run average of temporary payoffs for the number of provinces held in each round (see 4.3).

Apart from these modifications and simplifications, the rules for the alignment stage and the allotment stage are the same as in the parlor game.

4.2 Strategies

Technically, the model is an infinite extensive game with perfect information. A precise mathematical description of the game will be omitted here in order to avoid cumbersome formalism. The structure of the extensive game is sufficiently clear by the rules of the parlor game modified by 1) to 4). However, it is necessary to add some remarks to the formal definition of strategies and payoffs.

Our analysis will be focused on stationary behavior strategies. Stationarity means that decisions do not depend on anything else than the current province distribution and the board position. This means that a player's behavior is the same at all nodes of the game tree which agree with respect to the current province distribution and the board position.

Since there are only six provinces, it is not necessary to consider more than six players. Therefore, we assume that there are six *players* $1,\ldots,6$. A *province distribution*

$$m = (m_1,\ldots,m_6) \tag{2}$$

assigns a number of provinces m_i to every player $i = 1,...,6$. The m_i are non-negative integers which sum up to 6. The set of all province distributions is denoted by M. In order to describe the board position, we introduce the following notation:

W   the set of players with tokens on the field "wait for decision"

A   the set of players with tokens on the field "alliance"

C   the set of players with tokens on the field "counteralliance"

N   the set of players with tokens on the field "non-aligned".

These four sets form a *board position*:

$$B = (W,A,C,N). \tag{3}$$

A *decision situation* is a triple

$$D = (m,B,j) \tag{4}$$

composed of a province distribution, a board position, and a decision maker j. A player with at least one province is called *active*, the other players are called *passive*. Let I be the set of active players in m. A board position B partitions the set I into four sets W,A,C,N with the interpretation given above. W must be non-empty and j is one of the players in W. The set of all decision situations $D = (m,B,j)$ is denoted by Δ.

As in 2.5, we distinguish between *pre-attack* decision situations with $A = C = \phi$ and *post-attack* decision situations with $A \neq \phi$ and $C \neq \phi$. In a pre-attack decision situation $D = (m,B,j)$, the choice set has as many elements as there are active players. Player j can choose "non-aligned" or he can attack one of the other active players. In a post-attack decision situation, the decision maker has three choices, namely "non-aligned", "join the alliance", and "join the counteralliance". It is not necessary to add a formal description of the choice sets.

We shall always look at the case of an initial province distribution $(1,1,1,1,1,1)$, where each of the six players has one province. This does not involve any loss of generality since all other cases are subgames of this game.

A *local strategy* s_D at $D = (m,B,j)$ is a probability distribution over the choice set at D. We shall not be interested in stationary behavior strategies in general, but only in symmetric behavior strategies played by all players. We use the term *global strategies* for the sake of shortness. Formally, a *global strategy* s is a function which assigns a local strategy s_D at D to every $D \in \Delta$, such that two symmetry requirements (i) and (ii) are satisfied:

(i) <u>Symmetry with respect to players</u>: s is invariant with respect to a renumbering of players.

(ii) <u>Symmetry with respect to alignments</u>: s is invariant with respect to an exchange of the sets A and C, i.e., two board positions $B = (W,A,C,N)$ and $B' = (W,A',C',N)$ with $A' = C$ and $C' = A$ are treated symmetrically.

Of course, (i) and (ii) are not really mathematically precise statements, but since the meaning is sufficiently clear, we can avoid the formalism which would be necessary for a precise definition. The set of all global strategies is denoted by S.

A province distribution $m = (m_1,\ldots,m_6)$ is called *basic* if we have:

$$m_1 \geq m_2 \geq m_3 \geq m_4 \geq m_5 \geq m_6. \tag{5}$$

The set of all basic province distributions is denoted by M_0. A decision situation $D = (m,B,j)$ is called *basic* if $m \in M_0$ holds. The set of all basic decision situations is denoted by Δ_0. In view of (i), a global strategy s is completely determined by its restriction s_0 to Δ_0. In fact, s_0 is still redundant as a description of s, since (ii) permits a further reduction of the set of decision situations which need to be considered. A *basic strategy* is restricted to the set Δ_0 but otherwise defined in the same way as a global strategy. We say that a global strategy s and a basic strategy s_0 *correspond* to each other if s_0 is the restriction of s to Δ_0.

4.3 Payoffs

We first assign a *temporary payoff vector*

$$h(m) = (h_1(m),\ldots,h_6(m)) \tag{6}$$

to every province distribution $m = (m_1,\ldots,m_6)$. The *temporary payoffs* are given by

$$h_i(m) = \begin{cases} m_i & \text{for } m_i > 0 \\ -F & \text{for } m_i = 0 \end{cases}. \tag{7}$$

Here, F is a positive constant called the *existence value*. Payoffs will be defined as long-run averages of temporary payoffs. It will not be sufficient to look at payoffs in the whole game. It is also necessary to define local payoffs for every node which can be reached by a play. Consider first the situation at the beginning of a round t. Assume that the current province distribution is m and that in rounds t, t+1,... all players always play the same global strategy s. This stochastic process yields a sequence m(t), m(t+1),... of province distributions. The sequence can be looked upon as the realization of a finite Markoff chain (Feller, 1968). Depending on m(t) = m, we obtain uniquely determined stationary distributions over M. Expected temporary payoffs $H_i(m,s)$ can be computed with the help of these distributions. The vector

$$H(m,s) = (H_1(m,s),...,H_6(m,s)) \qquad (8)$$

is the *payoff vector* for m if s is played. We now define local payoffs $L(D,r_D,s)$ for every triple D,r_D,s, where $D = (m,B,j) \in \Delta$ is a decision situation, r_D is a local strategy at D and s is a global strategy:

$$L(D,r_D,s) = E(H_j(m(t+1),s)) \qquad (9)$$

with m(t) = m. Here, E is the expectation operator with respect to r_D. It is not necessary to express the identity of the player by an index at L, since the local payoff is always a payoff for the decision maker in D.

4.4. Equilibrium Strategies

The set of all local strategies at a decision situation $D = (m,B,j)$ is denoted by S_D. Local strategies which cannot be used by global strategies in view of the symmetry conditions (i) and (ii) in the definition of a global strategy are not excluded from S_D. A local strategy q_D at $D \in \Delta$ is a *local best reply* to a global strategy s if the following is true:

$$L(D,q_D,s) = \max_{r_D \in S_D} L(D,r_D,s). \qquad (10)$$

A global strategy s in an *equilibrium strategy* if for every $D \in \Delta$ the local strategy s_D assigned to D by s is a local best reply to s:

$$L(D,s_D,s) = \max_{r_D \in S_D} L(D,r_D,s). \qquad (11)$$

The equilibrium condition (11) is purely local. It is necessary but not sufficient for subgame perfect equilibrium in the usual sense. This is due to the definition of payoffs as a long run average of temporary payoffs. If the same decision situation is reached infinitely often, it is possible that nothing is gained by deviating just once, but payoffs can be improved by infinitely many deviations.

In finite games with perfect information, local equilibrium conditions like those expressed by (11) are sufficient for subgame perfect equilibrium. The same is true for the infinite game if payoffs are defined as discounted sums of temporary payoffs. This payoff definition seems to be the more adequate one, if one looks at the infinite game as an approximation of long finite games. In Section 6, we shall investigate payoffs defined as discounted sums of temporary payoffs in order to justify the neglect of global equilibrium conditions in the definition of an equilibrium strategy.

4.5 Classification of Province Distributions

Let m be a basic province distribution. We say that a province distribution m' is of *type* m if m can be obtained from m' by a renumbering of the players. In order to determine the payoff vector $H(m',s)$ for m', it is sufficient to determine $H(m,s)$ for the basic province distribution m of the same type.

A province distribution $m \in M$ is called *stable with respect to the global strategy* s if for all pre-attack decision situations $D = (m,B,j)$ the local strategies at D choose "non-aligned" with probability 1. Obviously, in this case we have:

$$H_i(m,s) = m_i \quad \text{for} \quad m_i > 0. \tag{12}$$

If m is stable, then every round ends in peace after m is reached as long as s is played.

We say that a province distribution m is *viable with respect to the global strategy* s if with probability 1 the set of active players remains the same for all remaining rounds after m is reached. In view of $F > 0$, this is the case if and only if we have

$$\sum_{i \in I} H_i(m,s) = 6 \tag{13}$$

where I is the set of active players.

The total number of provinces is always 6. If m is not viable, then the sum of the left hand side is diminished by the expected number of players which will lose their existence in the future multiplied by F.

A province distribution which is not viable with respect to s is called *transitory* with respect to s. After a transitory province distribution m has been reached, the set of active players must be eventually diminished with probability 1 if s is played. This is a consequence of the stationarity of s.

Obviously, stability implies viability, but not vice versa. A province distribution m which is viable but not stable with respect to s is called *fluid* with respect to s. The classification into stable, fluid and transitory province distributions with respect to s partitions the set M into three classes. Every province distribution belongs to exactly one of these classes. We say that s assigns the *stability status* "stable", "fluid" or "transitory" to a province distribution m, if m is stable, fluid or transitory resp. with respect to s. In view of the symmetry properties of global strategies, the stability status of m with respect to s depends only on s and the type of m.

4.6 Conservative Equilibrium Strategies

The game does not have a uniquely determined equilibrium strategy. However, some equilibrium strategies seem to be unreasonable and therefore should be eliminated by some equilibrium selection criterion. Recently, a general equilibrium selection theory has been developed (Harsanyi and Selten, 1988). However, this theory is restricted to finite games and therefore cannot be applied here. Instead of this, a criterion will be applied which is suggested by the context of our analysis. We shall make use of the following notation:

M_k the set of province distributions with k active players (players with at least one province);

$M_{<k}$ the set of province distributions with less than k active players;

$P_k(s)$ the set of all province distributions with k active players which are viable with respect to s.

The subgames beginning with province distributions in M_k cannot move to province distributions with more than k active players. Therefore, it is natural to search for an equilibrium selection by first looking at M_1, then at M_2, then at M_3, etc., and finally at M_6. The analysis of equilibrium conditions for province distributions in M_k can rely on results obtained for $M_{<k}$. However, usually other province distributions in M_k can be reached from a province distribution in M_k. Therefore, the analysis of the game is not completely recursive as in the search for subgame perfect equilibrium points in finite games with perfect information. All province distributions in M_k have to be considered simultaneously.

We say that an equilibrium strategy s is *conservative* if for k = 2,...,6 there is no other global strategy r_k which satisfies the following three conditions (a), (b), and (c):

(a) The global strategy r_k agrees with s for decision situations $D = (m,B,j)$ with $m \in M_{<k}$.

(b) The global strategy r_k assigns local best replies to r_k to all decision situations $D = (m,B,j)$ with $m \in M_k$.

(c) $P_k(s)$ is a proper subset of $P_k(r_k)$.

The concept of a conservative equilibrium strategy is based on the idea that the players form expectations recursively in the order of increasing numbers of active players and that at each inductive step as much priority as possible is given to the assumption of viability.

Viability means that no active player is in danger to be eliminated. In this sense, viability is a weak form of balance of power. Even if the province distribution keeps changing, the actors in the system always remain the same. Conservativeness postulates a process of expectation formation which favors balance of power in the weak sense of viability. To some extent, balance of power needs the expectation of balance of power. This can be shown by a simple example which arises in the analysis of our game.

Consider the province distribution (3,3,0,0,0,0). If one of the active players attacks the other, each of them has half a chance to conquer the world and to obtain a payoff of 6 and half a chance to be eliminated and to receive -F. The attacker has a local payoff of 3-F/2 for the decision to attack. Obviously, it is better for both active players not to attack each other. Nevertheless, the local equilibrium conditions applied to decision situations with province distributions with two active players do not exclude a global strategy where both of them attack the other one whenever there is an opportunity. If the other one is going to attack anyhow, one can just as well attack oneself.

As we shall see, equilibrium strategies exist, with respect to which (3,3,0,0,0,0) is stable. Conservative equilibrium strategies must have this property. Conservativeness excludes the possibility of unreasonable distrust in the balance of power between the two active players in (3,3,0,0,0,0).

5. The Solution of the Theoretical Model

5.1 Character of the Solution

In view of the symmetry properties of global strategies, the equation $H_i(m,s) = H_j(m,s)$ holds for every province distribution m with $m_i = m_j$. The payoff $H_i(m,s)$ of player i does not depend explicitly on i, but only on m_i, the number of provinces held by i. It will be useful to look at $H_i(m,s)$ as a function of m_i. Therefore, the following notation is introduced:

$$V_k(m,s) = H_i(m,s) \quad \text{with} \quad m_i = k > 0. \tag{14}$$

Of course, $V_k(m,s)$ remains undefined for $k = 0$ and in the case that $m_i \neq k$ holds for $i = 1,\ldots,6$. Let S^* be the set of all conservative equilibrium strategies. We write $V_k(m)$ instead of $V_k(m,s)$ with $s \in S^*$, if $V_k(m,s)$ does not depend on s within S^*. In this case, we call $V_k(m)$ the *value for k provinces in* m. We say that values for conservative equilibrium strategies are *uniquely determined*, if for every $m \in M$ and every k, for which $V_k(m,s)$ is defined, $V_k(m,s)$ does not depend on s within S^*. Similarly, we say that conservative equilibrium strategies *uniquely determine* the stability status, if every $s \in S^*$ assigns the same stability status to each $m \in M$.

We say that the game is *solvable* for a given existence value F if for this F conservative equilibrium strategies exist and uniquely determine values and stability status. If this is the case, the values $V_k(m)$ and the classification of province distributions according to their stability status form *the solution* of the game. In order to avoid unnecessary notation, a more formal definition of the solution is not given here.

5.2 The Range of Existence Values

In this paper, attention will be concentrated to a plausible middle range of existence values:

$$1/3 < F < 3. \tag{15}$$

An extension to existence values outside this interval is possible but will not be attempted here, since additional lengthy arguments would become necessary.

It is our intention to show that the game is solvable for every F in the interval (15). Within this range of F, the stability status does not depend on F and the values are linear functions of F. We shall refer to the interval (15) as the *middle range*.

5.3 The Solution in the Middle Range

Table 1 gives an overview over the solution in the middle range. Only basic province distributions are listed in Table 1, since values and stability status depend only on the type of the province distribution.

In view of the symmetry of global strategies with respect to players, it is convenient to introduce a manner of speaking which permits us to focus on numbers of provinces held instead of identification numbers. Therefore, the number of provinces held will be called the *size* of a player and a player of size k will be referred to as a *size-k player*. The *size compositions* of alignments are described by the juxtaposition of the sizes of their numbers; thus, 221 stands for an alignment of two players who own 2 provinces and one player who owns 1 province.

We have to show that conservative equilibrium strategies exist and that values and stability status are uniquely determined. Existence is proved by the construction of a particular conservative equilibrium strategy. This strategy is described by Tables 2, 3, and 4. We shall look at province distributions essentially in the order of increasing numbers of active players. At each step, we shall show for the province distributions under consideration that the decisions listed in Tables 2, 3, and 4 satisfy conditions required by the definition of a conservative equilibrium strategy and that values and stability status are uniquely determined.

The decisions specified by the particular conservative equilibrium strategy of Tables 2, 3, and 4 do not depend on N and W. Therefore, in these tables, board positions are described by the size compositions of both alignments. It does not matter which alignment is A or C. Therefore, only one of both possibilities is listed. There are 11 types of province distributions (see Table 1) and altogether 56 essentially different decision situations in the game (Tables 2, 3, and 4).

The particular conservative equilibrium strategy described by Tables 2, 3, and 4 is not the only one. Many other conservative equilibrium strategies exist, which deviate from this one with respect to unimportant details. However, in Section 6, we shall argue that under the definition of payoffs as discounted sums of temporary payoffs, r remains as equilibrium strategy if the discount rate is sufficiently small. In this respect, some of these details are not unimportant.

It will be shown that the game is solvable in the sense of 5.1 for $1/3 < F < 3$ and that the strategy described by Tables 2, 3, and 4 is a conservative equilibrium strategy, not only for the game with long run average payoffs, but also for the discounted payoff game if the discount rate is sufficiently small. The result for the game with long run average payoffs will be proved step by step in a sequence of 12 assertions. It will always be assumed that F is in the middle range $1/3 < F < 3$. This will not be explicitly mentioned in Assertions 1 to 12.

5.4 Province Distributions with a Player Owning at least 4 Provinces

If one player owns at least 4 provinces, he can conquer the world and he must do this eventually if he plays an equilibrium strategy. It is clear as far as these province distributions are concerned that Table 2 describes a conservative equilibrium strategy and that values and stability status are uniquely determined as shown in Table 1.

PROVINCE DISTRIBUTION						VALUES				STABILITY STATUS
PLAYERS						4, 5 or 6 provinces	3 provinces	2 provinces	1 province	
1	2	3	4	5	6					
6	-	-	-	-	-	6	-	-	-	stable
5	1	-	-	-	-	6	-	-	-F	transitory
4	2	-	-	-	-	6	-	-F	-	transitory
4	1	1	-	-	-	6	-	-	-F	transitory
3	3	-	-	-	-	-	3	-	-	stable
3	2	1	-	-	-	-	$3\frac{7}{8} - \frac{3F}{16}$	$1\frac{5}{8} - \frac{9F}{16}$	$\frac{1}{2} - \frac{3F}{4}$	transitory
2	2	2	-	-	-	-	-	2	-	stable
3	1	1	1	-	-	-	4	-	$\frac{2}{3} - \frac{F}{2}$	transitory
2	2	1	1	-	-	-	-	2	1	stable
2	1	1	1	1	-	-	-	$1\frac{1}{5}$	$1\frac{1}{5}$	fluid
1	1	1	1	1	1	-	-	-	1	stable

Table 1: The solution for $1/3 < F < 3$.

PROVINCE DISTRIBUTION TYPE	BOARD POSITION according to number of provinces held			DECISION
	A	C	decision maker	
(5,1,0,0,0,0)	-	-	5	attack other player
	-	-	1	non-aligned
(4,2,0,0,0,0)	-	-	4	attack other player
	-	-	2	non-aligned
(4,1,1,0,0,0)	-	-	4	attack one of both other players *)
	-	-	1	non-aligned
	4	1	1	join A
	1	1	4	join A or C *)
(3,3,0,0,0,0)	-	-	3	non-aligned
(3,2,1,0,0,0)	-	-	3	attack size 2 player
	-	-	2	non-aligned
	-	-	1	non-aligned
	3	2	1	join C
	3	1	2	join C
	2	1	3	join C
(2,2,2,0,0,0)	-	-	2	non-aligned
	2	2	2	join A or C *)

Table 2: Equilibrium decisions for province distributions with up to 3 active players and for $1/3 < F < 3$. The local strategies do not depend on N and W.

*) With equal probabilities

PROVINCE DISTRIBUTION TYPE	BOARD POSITION according to number of provinces held			DECISION
	A	C	decision maker	
(3,1,1,1,0,0)	-	-	3	attack one of the other players *)
	-	-	1	non-aligned
	3	1	1	join C
	1	1	3	join A or C *)
	1	1	1	non-aligned
	31	1	1	join A
	11	1	3	join C
	11	3	1	join A
(2,2,1,1,0,0)	-	-	2	non-aligned
	-	-	1	non-aligned
	2	2	1	join A or C *)
	2	1	2	join C
	2	1	1	join C
	1	1	2	join A or C *)
	22	1	1	join A
	21	2	1	join A
	21	1	2	join A
	11	2	2	join A

<u>Table 3</u>: Equilibrium decisions for province distributions with 4 active players and for $1/3 < F < 3$.

*) With equal probabilities

PROVINCE DISTRIBUTION TYPE	BOARD POSITION according to number of provinces held			DECISION
	A	C	decision maker	
(2,1,1,1,1,0)	-	-	2	non-aligned
	-	-	1	attack size-2 player
	2	1	1	join C
	1	1	2	join A or C *)
	1	1	1	non-aligned
	21	1	1	join A
	11	2	1	join A
	11	1	2	join C
	11	1	1	non-aligned
	21	11	1	join A
	11	11	2	join A or C *)
	211	1	1	join A
	111	2	1	join A
	111	1	2	join A
(1,1,1,1,1,1)	-	-	1	non-aligned
	1	1	1	join A or C *)
	11	1	1	join A
	11	11	1	join A or C *)
	111	1	1	join A
	111	11	1	join A
	1111	1	1	join A

Table 4: Equilibrium decisions for (2,1,1,1,1,0) and (1,1,1,1,1,1) and for $1/3 < F < 3$.

*) With equal probabilities.

5.5 The Province Distribution (3,3,0,0,0,0)

Values for all other province distributions with at most two active players have been determined already. Conservativeness requires stability of (3,3,0,0,0,0) if this is possible. The choice of "non-aligned" in all pre-attack decision situations connected to (3,3,0,0,0,0) is in equilibrium. The requirements imposed by the definition of a conservative equilibrium strategy are satisfied: As has been pointed out in 4.6, the equilibrium conditions for province distributions with two active players do not exclude the possibility that (3,3,0,0,0,0) is unstable, since each of both active players attacks in the expectation that the other one wants to attack anyhow.

5.6 Province Distributions With Three Active Players

There are two types of province distributions with 3 active players, namely (3,2,1,0,0,0) and (2,2,2,0,0,0). Both of them have to be considered simultaneously, since one can be reached from the other. We first determine values and stability status on the basis of the assumption that the game is solvable. Then, we shall see that the decisions specified by Table 2 for these province distributions meet the requirements imposed by the definition of a conservative equilibrium strategy. In order to shorten formulas, we introduce the following notation:

$$p = V_2((2,2,2,0,0,0),s) \qquad (16)$$

$$q_k = V_k((3,2,1,0,0,0),s) \qquad \text{for} \quad k = 1,2,3 \qquad (17)$$

where s is an equilibrium strategy. Since at most two active players can be eliminated after (2,2,2,0,0,0) has been reached, the sum $3p$ of the values of the three active players in (2,2,2,0,0,0) is at least 6-2F. In view of $F < 3$, this yields:

$$p > 0. \qquad (18)$$

We now prove the following assertion:

Assertion 1: Let s be an equilibrium strategy. Then, (3,2,1,0,0,0) is not viable with respect to s.

Proof: Assume that (3,2,1,0,0,0) is stable. If the size-3 player behaves as specified by Table 2, he can conquer the world with a probability of at least 1/2. In the remaining cases, he always receives something positive, namely 2 if he becomes a size-2 player in a different province distribution of the same type or p if he becomes a size-2 player in (2,2,2,0,0,0). This shows that the size-3 player in (3,2,1,0,0,0) can enforce an expected payoff greater than 3. It follows that $m = (3,2,1,0,0,0)$ is not stable and that $V_3(m,s)$ is greater than 3 for this m. This has the consequence that with positive probability the size-3 player obtains first 4 and eventually 6 provinces. Consequently, (3,2,1,0,0,0) is not viable with respect to s.

<u>Assertion 2</u>: Let s be a conservative equilibrium strategy. If (2,2,2,0,0,0) is stable with respect to s, then for $k = 1,2,3$ and $m = (3,2,1,0,0,0)$ the values $V_k(m,s)$ agree with those listed in Table 1.

<u>Alignment graph</u>: The proof of Assertion 2 will make use of an "alignment graph" for $m = (3,2,1,0,0,0)$. This graph is shown in Figure 5. The nodes of the alignment graph stand for incompletely described board positions. Only the size compositions of both alignments are indicated, separated by a dash. The lines represent possible changes in the further development of the round. Higher nodes represent earlier situations. The nodes on the bottom describe terminal board positions, without any tokens on "wait for decision". Thus, 3-2 above means, that the size-1 player still can choose to stay non-aligned or to join the size-3 player or the size-2 player. 3-2 below means that the size-1 player has already chosen to stay non-aligned. Below the bottom nodes, the corresponding payoffs are indicated for each number of provinces held in the current round. These payoffs are the expectations of the values obtained for the province distribution of the next round, based on the assumptions of Assertion 2 that a conservative equilibrium strategy is played and that (2,2,2,0,0,0) is stable. This implies $p = 2$. The assumption that s is a conservative equilibrium strategy permits us to make use of Table 2 as far as province distributions with less than 3 active players or with a size-4 player are concerned.

As an example for the payoff computation, we look at the payoff of the size-3 player for 31-2. With probability 3/4 he obtains 4 provinces worth 6 and with probability 1/4 he becomes the size-3 player in a different province distribution of the type (3,2,1,0,0,0). This yields an expected payoff of $(18+q_3)/4$. Payoffs depend on the unknown variables q_1, q_2, and q_3 which in turn depend on the strategy s being played.

The three arrows next to three lines of the alignment graph do not really belong to the graph, but indicate results of its analysis. As we shall see, changes of the alignments must occur in these directions if an equilibrium strategy is played.

<u>Proof of Assertion 2</u>: If the size-3 player in (3,2,1,0,0,0) behaves as specified by Table 2, he conquers the world with a probability of at least 1/2. The worst which can happen to him in the remaining cases is loss of existence connected to a payoff of -F. In view of $F < 3$, this yields:

$$q_3 \geq 3 - \frac{F}{2} > \frac{3}{2} . \tag{19}$$

Game Theory and Economic Behaviour I 353

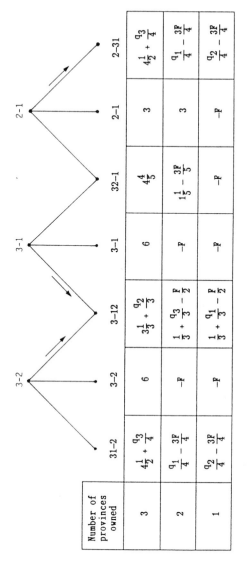

Figure 5: The alignment graph for (3,2,1,0,0).

Consequently, we have

$$4\tfrac{1}{2} + \frac{q_3}{4} > 4\tfrac{7}{8} . \qquad (20)$$

It follows by this inequality that in a situation described by the upper node 2-1 of the alignment graph of Figure 5, the size-3 player must join the alignment of the size-1 player. The arrow from 2-1 to 2-31 indicates a necessary change of the alignment situation after a board position corresponding to the upper node 2-1.

Since loss of existence is the worst which can happen, we have:

$$q_2 \geq -F \qquad (21)$$

$$q_1 \geq -F . \qquad (22)$$

In view of F < 3, this yields

$$3\tfrac{1}{3} + \frac{q_2}{3} \geq 3\tfrac{1}{3} - \frac{F}{3} > 2\tfrac{1}{3} . \qquad (23)$$

Since (3,2,1,0,0,0) is unstable with respect to s, eventually a terminal board position corresponding to one of the bottom nodes of the alignment graph must be reached. We already have shown that the bottom node 2-1 cannot be reached. It follows by (20) and (23) that the payoff of the size-3 player at all the other bottom nodes is greater than 7/3. Consequently, we have:

$$q_3 \geq 3\tfrac{1}{3} - \frac{F}{3} > 2\tfrac{1}{3} . \qquad (24)$$

We now show that after a board position corresponding to the upper node 3-1 the alignment situation must move to 3-12 as indicated by an arrow in Figure 5. For this purpose, it is sufficient to show that we have:

$$\tfrac{1}{3} + \frac{q_3}{3} - \frac{F}{2} > 1\tfrac{1}{5} - \frac{3F}{5} . \qquad (25)$$

Obviously, the right-hand side of this inequality is greater than -F. In a situation corresponding to the upper node 3-1, it does not pay for the size-2 player to stay non-aligned. With the help of (24) it can be seen that (25) holds for F > 8/9. Assume 1/3 < F ≤ 8/9. Then we have:

$$q_3 \geq 3\tfrac{1}{3} + \frac{q_2}{3} \geq 3\tfrac{1}{27}. \qquad (26)$$

This has the consequence that (25) holds for $1/3 < F \leq 8/9$, too. The arrow from 3-1 to 3-12 indicates that the size-2 player must join the alignment of the size-1 player after hostilities between the size-3 player and the size-1 player have been opened.

The next step in the proof of Assertion 2 justifies the third arrow in Figure 5. In order to show that the size-1 player must join the alignment of the size-1 player, after hostilities between both other players have been opened, it is sufficient to establish the validity of the following inequality:

$$\frac{1}{3} + \frac{q_1}{3} - \frac{F}{2} > \frac{q_2}{4} - \frac{3F}{4} . \tag{27}$$

Assume that (27) does not hold. Then, it follows by (22) that the right-hand side of (27) is greater than $-F$. From what we already know, it follows that only the terminal nodes 31-2, 3-12, and 2-31 can be reached in the alignment graph. the left-hand side of (27) is the minimum of the payoffs the size-1 player receives at these nodes. Therefore, we must have:

$$q_1 \geq \frac{1}{3} + \frac{q_1}{3} - \frac{F}{2} . \tag{28}$$

This yields:

$$q_1 \geq \frac{1}{2} - \frac{3F}{4} \tag{29}$$

and consequently:

$$\frac{1}{3} + \frac{q_1}{3} - \frac{F}{2} \geq \frac{1}{2} - \frac{3F}{4} . \tag{30}$$

Since it is assumed that (27) does not hold, this yields:

$$\frac{q_2}{4} - \frac{3F}{4} \geq \frac{1}{2} - \frac{3F}{4} \tag{31}$$

or, equivalently:

$$q_2 \geq 2 . \tag{32}$$

With the help of (32) we obtain:

$$3\frac{1}{3} + \frac{q_2}{3} \geq 4 .$$

This shows that at the terminal nodes not yet excluded the size-3 player receives at least 4. Therefore, we have

$$q_3 \geq 4. \tag{33}$$

All terminal nodes not yet excluded involve a probability of at least 1/2 for the elimination of the size-2 player. In view of (33), his expected number of provinces can be at most 2. This yields:

$$q_2 \leq 2 - \frac{F}{2}. \tag{34}$$

Inequality (34) contradicts (32). We can conclude that (27) holds and that the arrow from 3-2 to 3-12 in Figure 5 is justified.

Inequality (27) together with the conclusions expressed by the three arrows in Figure 5 has the consequence that in a pre-attack decision situation it is better for the size-1 player to attack the size-3 player than to attack the size-2 player. As we shall see, it is also better for the size-2 player to attack the size-3 player rather than the size-1 player. Figure 5 shows that this is the case if we have:

$$\frac{1}{3} + \frac{q_3}{3} - \frac{F}{2} > \frac{q_1}{4} - \frac{3F}{4}. \tag{35}$$

In view of (24), we must have:

$$\frac{q_1}{4} - \frac{3F}{4} \leq \frac{6 - 2\frac{1}{3}}{4} - \frac{3F}{4} = \frac{11}{12} - \frac{3F}{4}. \tag{36}$$

On the other hand, (24) shows that the following is true

$$\frac{1}{3} + \frac{q_3}{3} - \frac{F}{2} > \frac{10}{9} - \frac{F}{2}. \tag{37}$$

Inequalities (36) and (37) yield (35). We can conclude that both the size-1 player and the size-2 player do not attack each other in pre-attack decision situations. This has the consequence that only the terminal node 3-12 can be reached. It follows that q_1, q_2 and q_3 must be the payoffs at this node:

$$q_3 = 3\frac{1}{3} + \frac{q_2}{3} \tag{38}$$

$$q_2 = \frac{1}{3} + \frac{q_3}{3} - \frac{F}{2} \tag{39}$$

$$q_1 = \frac{1}{3} + \frac{q_1}{3} - \frac{F}{2}. \tag{40}$$

The equation system (38), (39), and (40) yields the values for $m = (3,2,1,,0,0,0)$ listed in Table 1:

$$q_3 = 3\frac{7}{8} - \frac{3F}{16} \tag{41}$$

$$q_2 = 1\frac{5}{8} - \frac{9F}{16} \tag{((OU(42}$$

$$q_1 = \frac{1}{2} - \frac{3F}{4}. \tag{43}$$

We have proved Assertion 2.

<u>Assertion 3</u>: Let s be a conservative equilibrium strategy. Then, for all province distributions with at most three active players values and stability status are as described by Table 1. Moreover, a global strategy r which assigns the local strategies specified by Table 2 to decision situation $D = (m,B,j)$ with $m \in M_{<4}$ satisfies the requirements imposed by the definition of a conservative equilibrium strategy as far as these decision situations are concerned.

<u>Proof</u>: Let r be a global strategy as described above. Consider a round with the province distribution $m = (2,2,2,0,0,0)$ and assume that one of the three active player attacks another one in a pre-attack decision situation and that the other players use r. Then, the local payoff of the attacker is as follows:

$$\frac{1}{2} q_1 + \frac{1}{4} q_2 + \frac{1}{4} q_3 = 1\frac{5}{8} - \frac{9F}{16}. \tag{44}$$

Since the right-hand side of (44) is smaller than 2, it is better tho choose "non-aligned". After an attack has happened, the third active player receives the local payoff q_2 for choosing "non-aligned" and the local payoff $(q_2+q_3)/2$ for joining one of both alignments. For him, too, the choice specified in Table 2 is the only local best reply to r.

With the help of Figure 5, it can be seen without difficulty that the local strategies specified by r for decision situations $D = (m,B,j)$ with $m = (3,2,1,0,0,0)$ are local best replies to r. The province distribution $(2,2,2,0,0,0)$ is stable with respect to r. In view of Assertion 1, the province distribution $(3,2,1,0,0,0)$ cannot be viable for any equilibrium strategy. Therefore, the set $P_3(r)$ is maximal in M_3 as required by the definition of conservativeness. This shows that the second part of Assertion 3 holds.

Consider a conservative equilibrium strategy s. If (2,2,2,0,0,0) is not stable with respect to s, then a province distribution of the type (3,2,1,0,0,0) must be reached eventually starting from (2,2,2,0,0,0) if s is played. This has the consequence that the set of active players is reduced with positive probability, which means that (2,2,2,0,0,0) is not viable. If both (2,2,2,0,0,0) and (3,2,1,0,0,0) are not viable, then $P_3(s)$ is a proper subset of $P_3(r)$ and s fails to be conservative. Therefore, (2,2,2,0,0,0) must be stable with respect to s. In view of Assertion 2 together with our earlier results on values and stability status for province distributions with up to 3 active players, it is now clear that Assertion 3 holds.

Comment: In a round with m = (3,2,1,0,0,0), the size-3 player takes the risk to lose one province for an even chance to conquer the world. The two other players form an alignment against him. Other types of military confrontations cannot happen on the equilibrium path, if (3,3,0,0,0,0) and (2,2,2,0,0,0) are stable with respect to the equilibrium strategy played. This result is hardly surprising, but in spite of its plausibility lengthy and tedious arguments were necessary in order to reach the conclusion. This may be due to our failure to find a short proof, but it is also possible that the seemingly obvious is not obvious at all. Difficulties arise from the fact that the values q_3, q_2, and q_1 appear in the payoffs of the alignment graph.

5.7 Province Distributions with Four Active Players

There are two types of province distributions with four active players, namely (3,1,1,1,0,0) and (2,2,1,1,0,0). As we shall see, (3,1,1,1,0,0) cannot be viable and (2,2,1,1,0,0) must be stable with respect to a conservative equilibrium strategy. In this respect, the case of four active players is similar to the case of three active players.

In order to shorten our formulas, we introduce the following notation:

$$u_k = V_k((3,1,1,1,0,0),s) \quad \text{for} \quad k = 1 \quad \text{and} \quad k = 3 \tag{45}$$

$$v_k = V_k((2,2,1,1,0,0),s) \quad \text{for} \quad k = 1,2 \tag{46}$$

where s is an equilibrium strategy.

Transition graph: It will be useful to support our reasoning by a *transition graph* which shows the possible transitions from province distributions with four active players to other province distributions with three or four active players. This graph is shown by Figure 6. A node corresponds to the province distribution indicated next to it and a connecting line represents a possible transition in the direction showing the arrow at one of its ends. If an equilibrium strategy s is played, these transitions occur with the probabilities x, y, z, µ, and η. Near to each connecting line the graph shows the symbol for the corresponding transition probability. (2,2,1,1,0,0) can be reached from (2,2,1,1,0,0) by a round which ends in peace (probability µ) or by a round where a size-1 player wins a province lost by a size-2 player (probability η). The transition graph of Figure 6 is based on the assumption that (3,1,1,1,0,0) is stable. Therefore, no transitions from (3,1,1,1,0,0) to other province distributions are shown.

Assertion 4: Let s be a conservative equilibrium strategy. If (3,1,1,1,0,0) is stable with respect to s, then we have

$$v_2 > 0. \qquad (47)$$

Proof: Assume (3,1,1,1,0,0) is stable with respect to s. We first show:

$$v_2 = 2x + \frac{q_2+q_3}{2} y + 2z + \frac{v_1+v_2}{2} \eta + v_2\mu \qquad (48)$$

$$v_1 = x + \frac{q_1-F}{2} y + \frac{2-F}{2} z + \frac{v_1+v_2}{2} \eta + v_1\mu . \qquad (49)$$

In a transition from (2,2,1,1,0,0) to (3,1,1,1,0,0), one size-2 player wins a province lost by the other size-2 player. Since (3,1,1,1,0,0) is stable and s is symmetric with respect to players, the payoff expectation connected to this transition for a size-2 player is 2 (see Figure 6). This yields the first term 2x in (48). Similarly, the second term in (48) is due to the fact that in the transition from (2,2,1,1,0,0) to (3,2,1,0,0,0) a size-2 player becomes a size-3 player or a size-2 player in (3,2,1,0,0,0) with equal probabilities. In this way, all terms in (48) and (49) can be derived easily by looking at the possibilities for the corresponding transitions in Figure 6.

With the help of (48) and (49) we shall show that v_2 must be positive. Consider first the case that we have:

$$x = y = z = 0. \qquad (50)$$

If in addition to this η = 0 holds, (2,2,1,1,0,0) is stable and we have v_2 = 2. If η > 0 holds, then (2,2,1,1,0,0) is fluid and the sum of the values of the active players is 6. This, together with (48) and (49), yields $v_2 = v_1$ = 3/2. Consequently, v_2 is positive in the case (50). From now on, assume that (50) does not hold. Obviously, we must have:

$$x + y + z + \eta + \mu = 1. \qquad (51)$$

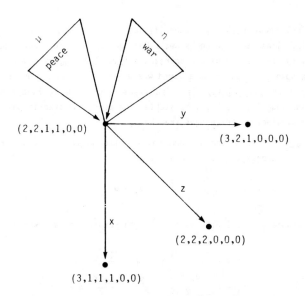

Figure 6: Transition graph for province distributions with four active players if $(3,1,1,1,0,0)$ is stable

Moreover, since (50) does not hold, the following is true:

$$x + y + z > 0. \tag{52}$$

With the help of (41), (42), (43), and (51), addition of (48) and (49) and division by $x+y+z$ yields:

$$v_2 + v_1 = 3 - \frac{5/4y + 1/2z}{x+y+z} F. \tag{53}$$

Similarly, subtraction of (49) from (48) leads to the following conclusion:

$$v_2 - v_1 = \frac{x + \frac{5}{2}y + z + (y+z)\frac{F}{2}}{x + y + z + \eta}. \tag{54}$$

Equation (54) shows that we have $v_2 > v_1$. Therefore $v_2 > 0$ holds for $v_1 + v_2 \geq 0$. In the following, we shall assume $v_1 + v_2 \leq 0$. In view of $F < 3$ and (53) the inequality $v_1 + v_2 \leq 0$ implies

$$\tfrac{5}{4}y + \tfrac{z}{2} \geq x+y+z. \tag{55}$$

A transition from (2,2,1,1,0,0) to a different province distribution of the same type cannot happen unless one of the following military confrontations takes place: 2-11 or 21-2 or 21-21 or 211-2. In all these cases the probability of reaching a province distribution of the type (3,1,1,1,0) is at least as great as the probability of reaching a province distribution of the type (2,2,1,1,0,0). Therefore, we have $x \geq \eta$. This together with (54) yields:

$$v_2 - v_1 \geq 1 + \frac{x + y + z + (y+z)\frac{F}{2}}{2x + y + z} .\qquad(56)$$

Equation (53) together with $F < 3$ implies $v_1 + v_2 \geq -3/4$. This inequality added to (56) yields the conclusion that v_2 is at least 1/8. Therefore the assertion holds.

<u>Assertion 5</u>: Let s be a conservative equilibrium strategy. Then, (3,1,1,1,0,0) is not viable with respect to s.

<u>Proof</u>: Assume that in a round with the province distribution (3,1,1,1,0,0) the size-3 player behaves as specified by Table 3. Then, he obtains 6 with a probability of at least 1/2. In the remaining cases, he becomes a size-2 player in (2,2,1,1,0,0) or (3,2,1,0,0,0); this means that in the remaining cases he obtains either 2 or v_2. His expected payoff is greater than 3 if v_2 is positive. Therefore, (3,1,1,1,0,0) cannot be stable for $v_2 > 0$. It follows by Assertion 4 that (3,1,1,1,0,0) must be unstable with respect to s.

The instability of (3,1,1,1,0,0) has the consequence that in a round with this province distribution an attack happens with a positive probability if s is played. After an attack, the size-3 player cannot have a higher probability than 1/2 of being at the losing side. Therefore, after an attack one of the size-1 players is eliminated with a probability of at least 1/2. This shows that (3,1,1,1,0,0) cannot be viable with respect to s.

<u>Alignment graphs for (3,1,1,1,0,0)</u>: Assume that s is a conservative equilibrium strategy and that (2,2,1,1,0,0) is stable with respect to s. Later, it will be shown that (2,2,1,1,0,0) is actually stable in our solution of the game. It is convenient to split the alignment graph for (3,1,1,1,0,0) into two parts shown in Figures 7 and 8. The graphical conventions are essentially the same as in Figure 5. However, in Figure 8 the situation 1-1 at the start may be changed by a decision of the size-3 player to join one of both alliances or by a decision of the remaining size-1 player to join one of both alliances. These two players may also stay non-aligned. Accordingly, five edges connect 1-1 to lower nodes. In the computation of payoffs, it is now necessary to distinguish between size-1 players in the alignment on the left of the dash (1 left), in the alignment on the right of the dash (1 right) and those which are non-aligned at the end of the round (1 in N).

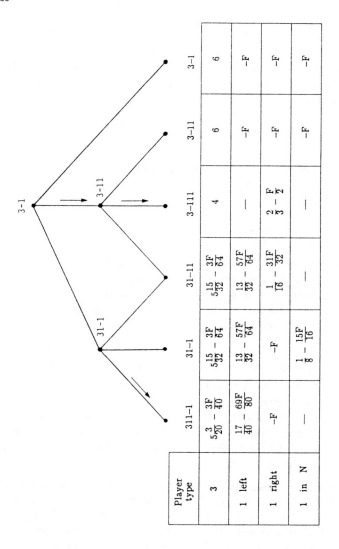

Figure 7: Alignment graph for (3,1,1,1,0,0) after 3-1. It is assumed that (2,2,1,1,0,0) is stable.

Game Theory and Economic Behaviour I

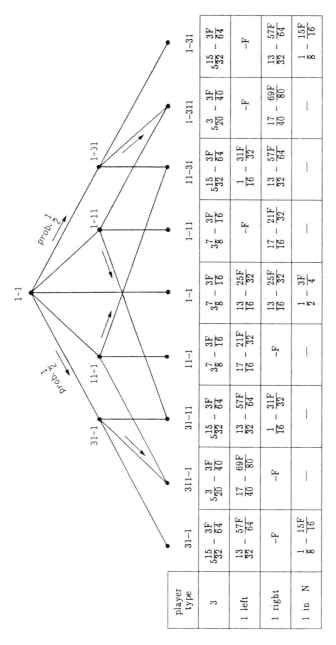

Figure 8: Alignment graph for (3,1,1,1,0,0) after 1-1.
It is assumed that (2,2,1,1,0,0) is stable.

The computation of payoffs is straightforward. It is easy to see that the arrows in Figure 7 represent the only possible equilibrium changes after 3-1. The same can be said on the lower arrows in Figure 8. With the help of these lower arrows, it can be seen immediately that it would be unwise for the remaining size-1 player to join one of both alignments; if he does this, he ends up on the losing side and receives (2-31F)/32. It is better to stay non-aligned until the size-3 player enters one of both alignments in order to join the stronger batallions afterwards.

Obviously, the size-3 player has no incentive to stay non-aligned after 1-1. This would result in a terminal board position described by the bottom node 1-1. In view of the symmetry properties of s, he must enter each of both alignments with probability 1/2 as indicated in Figure 8.

<u>Attacker's payoff</u>: With the help of Figures 7 and 8, local payoffs $L(r_D, D, s)$ for an attack r_D in a pre-attack situation $D = (m, B, j)$ with $m = (3,1,1,1,0,0)$ can be determined. We continue to assume that s is a conservative equilibrium strategy and that (2,2,1,1,0,0) is stable with respect to s. The local payoffs for the 3 possible types of attacks are shown by Table 5.

<u>Assertion 6</u>: Let s be a conservative equilibrium strategy. If (2,2,1,1,0,0) is stable with respect to s, then we have:

$$u_3 = 4 \qquad (57)$$

$$u_1 = \frac{2}{3} - \frac{F}{2} . \qquad (58)$$

(For the definition of u_1 and u_3 see (45).)

<u>Proof</u>: Table 5 shows that a size-1 player prefers the decision to attack the size-3 player to the decision to attack another size-1 player. Consequently, the situation 1-1 at the start of Figure 8 cannot occur on the equilibrium path. In view of this fact, it can be seen that a confrontation of the type 3-1 must be reached with positive probability in a round with the province distribution (3,1,1,1,0,0), since otherwise (3,1,1,1,0,0) would be stable with respect to s. This is excluded by Assertion 5. It follows that eventually a terminal board position corresponding to the bottom node 3-1II will be reached after (3,1,1,1,0,0) has been reached for the first time. This leads to the conclusion that (58) and (59) hold.

<u>Assertion 7</u>: Let s be a conservative equilibrium strategy. If (2,2,1,1,0,0) is viable with respect to s, then (2,2,1,1,0,0) is stable with respect to s.

<u>Proof</u>: The transition graph of Figure 6 shows which province distributions can be reached from (2,2,1,1,0,0). In the case of a transition to (3,2,1,0,0,0) or (2,2,2,0,0,0), one active player is eliminated. Assertion 5 shows that (3,1,1,1,0,0) is not viable. Therefore, active players are eliminated with positive probability

Number of Provinces		Attacker's local payoff for the attack
of attacker	of attacked player	
3	1	4
1	3	$\frac{2}{3} - \frac{1}{2} F$
1	1	$\frac{17}{64} - \frac{117}{128} F$

Table 5: Attacker's local payoff for an attack in a pre-attack decision situation $D = (m,B,j)$ with $m = (3,1,1,1,0,0)$. It is assumed that $(2,2,1,1,0,0)$ is stable.

after a transition to $(3,1,1,1,0,0,0)$. If $(2,2,1,1,0,0)$ is viable with respect to s, then a province distribution of the same type must be reached with probability 1 at the end of a round with $(2,2,1,1,0,0)$. Assume that $(2,2,1,1,0,0)$ is fluid with respect to s. Then, in a round with $(2,2,1,1,0,0)$ there must be a positive probability for a terminal board position with non-empty alignments A and C. It can be seen without difficulty that no such board position results in a new province distribution of the type $(2,2,1,1,0,0)$ with probability 1. A different province distribution of this type can be reached only if a size-1 player wins a province lost by a size-2 player. In order to exclude other types to be reached, only size-1 players should be on the winning side and only size-2 players should be on the losing side; moreover, the alignment with the size-1 players should be stronger than the other one. This is impossible. Consequently, Assertion 7 holds.

Alignment graphs for $(2,2,1,1,0,0)$: Figures 9, 10 and 11 show alignment graphs for $(2,2,1,1,0,0)$ starting with the three types of possible attacks 2-1, 2-2, and 1-1. The payoffs are computed on the basis of the assumption that $(2,2,1,1,0,0)$ is stable with respect to a conservative equilibrium strategy s. This implies that we assume:

$$v_2 = 2 \tag{59}$$

$$v_1 = 1. \tag{60}$$

Moreover, we can make use of the Eqs. (58) and (59) for u_3 and u_1 and, of course, of the Eqs. (41), (42), and (43) for q_3, q_2, and q_1. The computation of payoffs in Figs. 9, 10, and 11 is tedious, but straightforward. It can be seen without difficulty that for $1/3 < F < 3$ the arrows indicate the possible transitions on the equilibrium path.

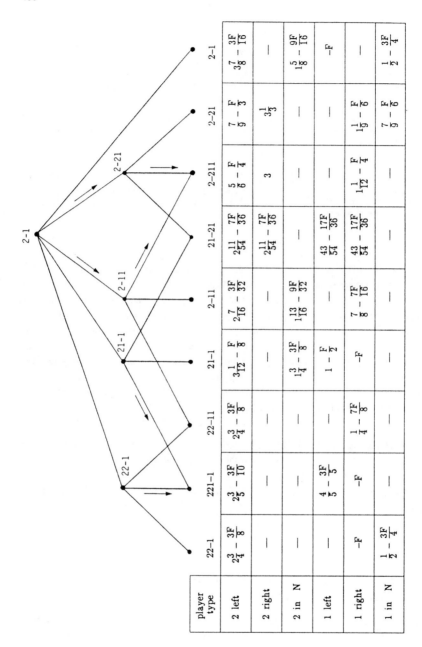

Figure 9: Alignment graph for (2,2,1,0,0) after 2-1.

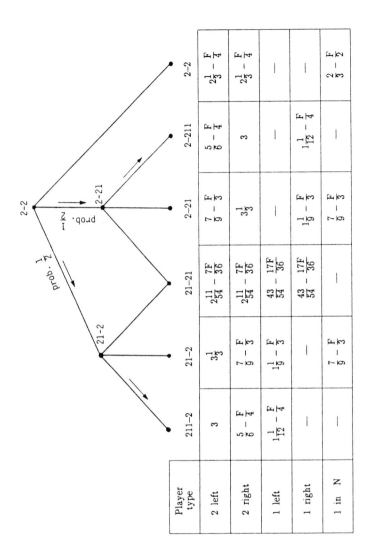

Figure 10: Alignment graph for (2,2,1,1,0,0) after 2-2.

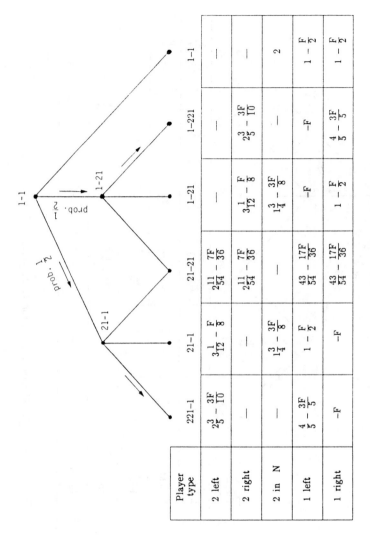

Figure 11: Alignment graph for (2,2,1,1,0,0) after 1-1.

Number of Provinces		Attacker's local payoff for the attack
of attacker	of attacked player	
2	2	$1\frac{11}{12} - \frac{F}{8}$
2	1	$\frac{5}{6} - \frac{F}{4}$
1	2	$1\frac{1}{12} - \frac{F}{4}$
1	1	$\frac{2}{5} - \frac{4F}{5}$

<u>Table 6</u>: Attacker's local payoff in a pre-attack decision situation $D = (m,B,j)$ with $m = (2,2,1,1,0,0)$.

<u>Attacker's payoff for (2,2,1,1,0,0)</u>: With the help of Figs. 9, 10, and 11, the entries in Table 6 can be determined easily. It can be seen immediately that for $F > 1/3$ an attack does not pay if nobody else attacks in a pre-attack situation. For $F < 1/3$, a size-1 player would have an incentive to attack a size-2 player, since then the expected payoff for this attack would be greater than 1. For $F < 1/3$, the province distribution (2,2,1,1,0,0) could not be stable with respect to a conservative equilibrium strategy s.

<u>Assertion 8</u>: Let s be a conservative equilibrium strategy. Then, for all province distributions with at most four active players values, and stability status are as described by Table 1. Moreover, a global strategy r which assigns the local strategies specified by Tables 2 and 3 to decision situations $D = (m,B,j)$ with $m \in M_{<5}$ satisfies the requirements imposed by the definition of a conservative equilibrium strategy, as far as these decision situations are concerned.

<u>Proof</u>: Let r be a global strategy as described above. With the help of Figs. 7 to 11, it can be seen that r assigns local best replies to r to all decision situations $D = (m,B,j)$ with $m = (3,1,1,1,0,0)$ or $m = (2,2,1,1,0,0)$. Moreover, (2,2,1,1,0,0) is stable with respect to r. Since (3,1,1,1,0,0) cannot be viable in view of Assertion 5, every conservative equilibrium strategy s must have the property that the set $P_4(s)$ of viable province distributions with respect to s agrees with the set $P_4(r)$ of viable province distributions with respect to r. It follows by Assertion 7 that (2,2,1,1,0,0) is stable with respect to every conservative equilibrium strategy s. This together with Assertions 5 and 6 shows

that as far as (3,1,1,1,0,0) and (2,2,1,1,0,0) are concerned values and stability status are uniquely determined and are correctly presented by Table 1.

5.8 Province Distributions With Five Active Players

There is only one type of province distribution with five active players, namely (2,1,1,1,1,0). As we shall see, these province distributions are fluid (viable but not stable) with respect to conservative equilibrium strategies. We shall use the following notation:

$$w_k = V_k((2,1,1,1,1,0),s) \quad \text{for} \quad k = 1,2. \tag{61}$$

Assertion 9: Let s be a conservative equilibrium strategy. Then, (2,1,1,1,1,0) is not stable with respect to s.

Proof: Assume that s is a conservative equilibrium strategy and that (2,1,1,1,1,0) is stable with respect to s. Then, we have $w_2 = 2$ and $w_1 = 1$. The payoffs in the alignment graphs of Figure 12 are computed on the basis of this assumption. The arrows in Figure 12 show that a size-1 player has an incentive to attack the size-2 player. He can count on all other size-1 players to join him and his local payoff for the attack is 5/4. Consequently, (2,1,1,1,1,0) cannot be stable.

Assertion 10: Let s be a conservative equilibrium strategy. If (2,1,1,1,1,0) is viable with respect to s, then we have:

$$w_2 = w_1 = 1\tfrac{1}{5}. \tag{62}$$

Proof: The transition graph of Figure 13 shows which province distributions can be reached from (2,1,1,1,1,0). The graphical conventions are the same as in Figure 6. Assume that s is a conservative equilibrium strategy and that (2,1,1,1,1,0) is viable with respect to s. Let x, y, α, and β be the transition probabilities generated by s belonging to the transitions indicated by the connecting lines next to which the x, y, α, and β resp. are shown. We must have

$$x = y = 0 \tag{63}$$

since otherwise one of the active players would be eliminated with positive probability after (2,1,1,1,1,0) has been reached. In view of Assertion 9, this province distribution is not stable with respect to s. Therefore, α must be positive. Whenever non-empty alignments are formed in a round with (2,1,1,1,1,0), one of the size-1 players wins a province lost by the size-2 player.

Game Theory and Economic Behaviour I 371

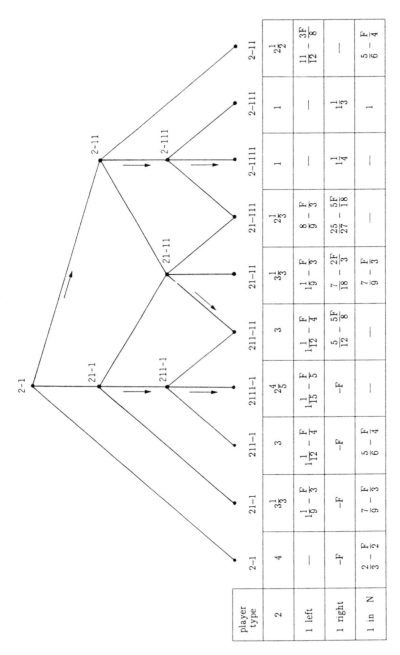

Figure 12: Alignment graph for (2,1,1,1,1,0) after 2-1 if $w_2 = 2$ and $w_1 = 1$ holds.

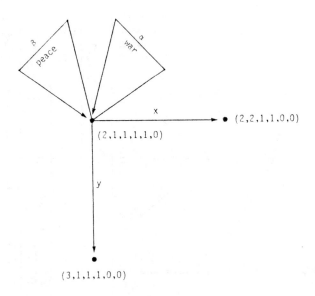

Figure 13: Transition graph for $(2,1,1,1,1,0)$

Consequently, we have:

$$w_2 = \alpha w_1 + \beta w_2 \tag{64}$$

In view of $\alpha + \beta = 1$, this yields

$$w_1 = w_2. \tag{65}$$

Since $(2,1,1,1,1,0)$ is viable, the sum w_2+4w_1 of the values of all five players is equal to 6, the total number of provinces. This, together with $\alpha > 0$ and Eq. (65), yields (62).

Alignment graphs for $(2,1,1,1,1,0)$: Figures 14 and 15 show alignment graphs for $(2,1,1,1,1,0)$ starting with the two possible types of attacks, namely 2-1 and 1-1. Payoffs in these alignment graphs are computed on the basis of (62), since nothing else is possible if $(2,1,1,1,1,0)$ is viable with respect to a conservative equilibrium strategy s. Of course, the values u_3 and u_1 of $(3,1,1,1,0,0)$ and v_2 and v_1 of $(2,2,1,1,0,0)$ which have been determined earlier are used, too. In order to simplify Figure 15, possible changes also shown by Figure 14 are omitted if in Figure 14 these changes are excluded by the arrows shown there.

Game Theory and Economic Behaviour I 373

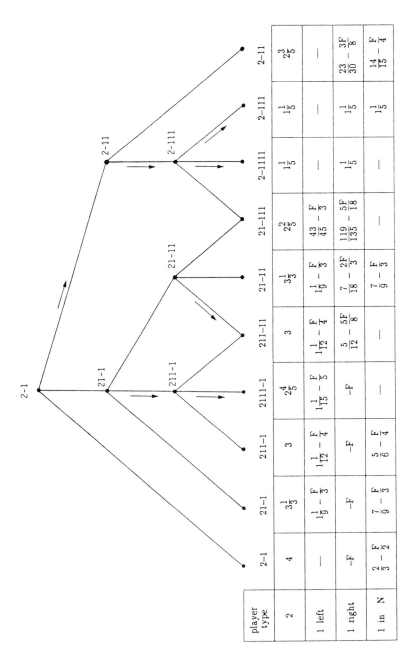

Figure 14: Alignment graph for (2,1,1,1,1,0) after 2-1 if $w_2 = w_1 = 1\frac{1}{5}$.

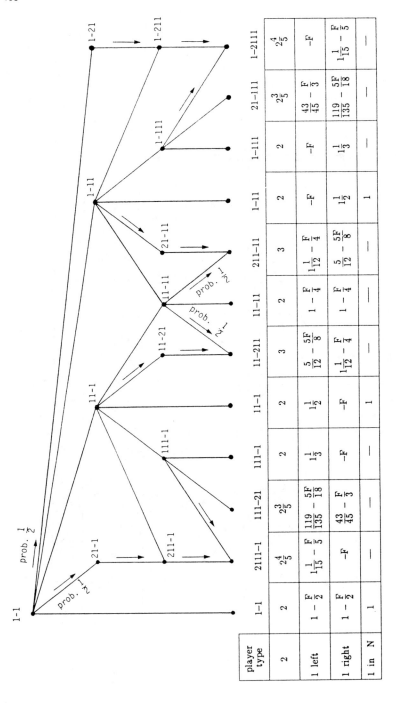

Figure 15: Alignment graph for (2,1,1,1,0) after 1-1 if $w_2 = w_1 = 1\frac{1}{5}$. Changes which cannot happen according to Figure 14 are omitted.

In Figure 15, it is unwise for a size-1 player to join one of both alignments in the situation 1-1. If he does this, he ends up on the losing side. He should wait until the size-2 player has entered one of both alignments and then join the side of the size-2 player.

Assertion 11: Let s be a conservative equilibrium strategy. Then, for (2,1,1,1,1,0) values and stability status are as described by Table 1. Moreover, a global strategy which assigns the local strategies specified by Tables 2, 3, and 4 to decision situations $D = (m,B,j)$ with $m \in M_{<6}$ satisfies the requirements imposed by the definition of a conservative equilibrium strategy, as far as these decision situations are concerned.

Proof: Let r be a global strategy as described above. With the help of Figures 14 and 15 and Table 7, it can be seen that r assigns local best replies to r to all decision situations $D = (m,B,j)$ with $m = (2,1,1,1,1,0)$. Moreover, (2,1,1,1,1,0) is viable with respect to r. Therefore, the definition of conservativeness requires that (2,1,1,1,1,0) is viable with respect to every conservative equilibrium strategy s. In view of Assertions 9 and 10, the province distributions of the type (2,1,1,1,1,0) are fluid with respect to every conservative equilibrium strategy s and the values for (2,1,1,1,1,0) are uniquely determined by (62).

Remark: Since the definition of an equilibrium strategy has been based on local conditions only, a conservative equilibrium strategy may involve a positive probability of a round with (2,1,1,1,1,0) ending in peace. If this is the case, a player can increase his payoff by a permanent change of his strategy: If in the role of a size-1 player he always attacks the size-2 player in a pre-attack decision situation $D = (m,B,j)$ with $m = (2,1,1,1,1,0)$, then his stationary probability of owning 2 provinces will be greater than 1/5. This results in an expected payoff of more than 6/5.

Number of Provinces		Attacker's local payoff for the attack
of attacker	of attacked player	
2	1	$1\frac{1}{5}$
1	2	$1\frac{1}{5}$
1	1	$\frac{8}{15} - \frac{3F}{5}$

Table 7: Attacker's local payoff in a pre-attack decision situation $D = (m,B,j)$ with $m = (2,1,1,1,1,0)$ for $w_2 = w_1 = 1\frac{1}{5}$.

5.9 The Province Distribution (1,1,1,1,1,1)

There is only one province distribution with 6 active players, namely (1,1,1,1,1,1). Fortunately, it is not necessary to draw an alignment graph in this case.

Assertion 12: Let s be a conservative equilibrium strategy. Then, (1,1,1,1,1,1) is stable with respect to s. Moreover, the strategy r described by Tables 2, 3, and 4 is a conservative equilibrium strategy.

Proof: If in a round with the province distribution (1,1,1,1,1,1) an attack takes place, the next province distribution is of the type (2,1,1,1,1,0). Therefore, all players who survive obtain the same payoff $w_2 = w_1 = 6/5$. This has the consequence that the local strategies specified for decisions situations $D = (m,B,j)$ with $m = (1,1,1,1,1,1)$ are local best replies to r. If follows by Assertion 11 that r is an equilibrium strategy. Obviously, (1,1,1,1,1,1) is stable with respect to r. It follows by the definition of conservativeness together with Assertion 11 that r is a conservative equilibrium strategy and that (1,1,1,1,1,1) is stable with respect to every other conservative equilibrium strategy.

5.10 The Result of the Analysis

The analysis of the game is now complete. We have obtained the following result:

Result 1: The game is solvable for every existence value F in the interval $1/3 < F < 3$. Values and stability status for each basic province distribution are as shown by Table 1. The strategy r described by Tables 2, 3, and 4 is a conservative equilibrium strategy.

Remark: For $F < 1/3$, the province distribution (2,2,1,1,0,0) cannot be stable with respect to a conservative equilibrium strategy. This can be seen with the help of Table 6. Therefore, our result cannot be extended to $F < 1/3$. The assumption $F < 3$ has been used in the proof of Assertion 4. With the help of Assertion 4, we have proved that (3,1,1,1,0,0) cannot be viable with respect to a conservative equilibrium strategy. The assumption $F < 3$ has also been used at other places in our proofs. However, it is possible that nevertheless our results can be extended to $F > 3$ by different proofs.

6. Discounted Payoffs

6.1 The Problem

The definition of an equilibrium strategy in Section 4 has been based on local optimality conditions only. Therefore, a conservative equilibrium strategy does not necessarily have the global equilibrium properties of a subgame perfect equilibrium point. This problem does not arise for a different payoff definition based on discounted sums of temporary payoffs. In this section, we shall show that the strategy described by Tables 2, 3, and 4 remains an equilibrium strategy under the changed payoff definition if the discount rate is sufficiently small. This shows that conservative equilibrium strategies exist which under the changed payoff definition do not only have local but also global equilibrium properties. As the discount rate approaches zero, the local payoffs approach those of the game with long-run average payoffs. The solution derived in Section 5 can be interpreted as a description of the limit of conservative equilibrium strategies under the changed payoff definition obtained for a discount rate approaching zero.

We shall not ask the question whether for sufficiently small discount rates under the changed payoff definition all conservative equilibrium strategies are approximately described by the solution derived in Section 5. It is the purpose of this section to show that this is the case for at least one conservative equilibrium strategy.

6.2 The Discounted Payoff Game

In the following, we shall change the theoretical model introduced in Section 4 by the introduction of a different payoff definition. Everything else will remain the same.

The payoff definition depends on a parameter ρ, called the *discount factor*. The discount factor satisfies the following inequality

$$0 < \rho < 1. \tag{66}$$

We are mainly interested in the case that ρ is very near to 1. The following notation will be used

$$\eta = 1-\rho. \tag{67}$$

We refer to η as the *discount rate*.

As in 4.3, assume that we have $m(t) = m$ and that in rounds $t, t+1, \ldots$ the same global strategy s is always played by all players. A player's expected payoff $H_{\rho i}(m,s)$ is now defined as follows

$$H_{\rho i}(m,s) = \eta \, E[\sum_{r=t}^{\infty} \rho^{r-t} h_i(m_i(r))] \tag{68}$$

where $m(t), m(t+1), \ldots$ is a realization of the stochastic process over M described in 4.3 and where E is the expectation operator; of course, $m_i(t)$ is player i's component in $m(t)$. The expected discounted payoff sum is multiplied by the discount rate η in order to make the payoff comparable to a one-round payoff.

For example, if $m = (m_1,\ldots,m_6)$ is stable with respect to s, we obtain $H_{\rho i}(m,s) = m_i$. The *payoff vector* $H_\rho(m,s)$ is defined analogously to (8) and *local payoffs* are defined as in (9) by

$$L_\rho(D,r_D,s) = E(H_{\rho j}(m(t+1),s)) \tag{69}$$

with $m(t) = m$ and $D = (m,B,j)$. We refer to the changed model as the *discounted payoff game* with the discount factor ρ. Local best replies and equilibrium strategies are defined as before with L_ρ instead of L. Conservativeness is also defined as before.

6.3 Local and Global Optimality

Without going into too much detail, we want to indicate why in the case of the discounted game local equilibrium conditions are sufficient for subgame perfect equilibrium in the usual sense. Consider an equilibrium strategy s for the discounted payoff game and assume that player i plays a behavior strategy b_i (not necessarily a stationary one) whereas all other players always play s. The definition of payoffs for this situation is straightforward and will not be given here. Assume that player i obtains a better payoff for b_i than for s if the other players use s.

For every $T > 1$, we construct a *truncated game* which ends after T rounds, but otherwise has the same structure as the discounted payoff game. The payoffs in this truncated game are those which would be obtained in the discounted payoff game, if in all rounds $T+1, T+2, \ldots$ the players always play s. It can be seen without difficulty that for sufficiently great T player i's payoff for using b_i in rounds $1,\ldots,T$ must be greater than player i's payoff for using s in rounds $1,\ldots,T$ if all other players play s in rounds $1,\ldots,T$. However, the truncated game is finite and the restriction of s to the truncated game satisfies the local equilibrium conditions for this finite game; this implies that the restriction of s to the truncated game being played by all players is a subgame perfect equilibrium point of the truncated game, contrary to our assumption. Therefore, in the discounted game local equilibrium conditions are sufficient for subgame perfect equilibrium in the usual sense.

6.4 The Strategy Described by Tables 2, 3, and 4

Let r be the strategy described by Tables 2, 3, and 4. In the following, we shall show that for sufficiently small discount rate η the strategy r is an equilibrium strategy of the discounted payoff game; in other words, a $\rho_0 < 1$ exists such that for $1/3 < F < 3$ and $\rho_0 < \rho < 1$ the strategy r is an equilibrium strategy for the discounted payoff game with the discount factor ρ.

Values: We shall use the notation

$$V_{\rho k}(m) = H_{\rho i}(m,r) \quad \text{with} \quad m_i = k > 0 \tag{70}$$

where m_i is player i's component of m. If m is stable with respect to r, then $V_{\rho k}(m)$ agrees with the corresponding value listed in Table 1. For those province distributions which are not stable with respect to r, it can be seen without difficulty that we have:

$$\lim_{\rho \to 1} V_{\rho k}(m) = V_k(m,r) \tag{71}$$

wherever values are defined.

Province distributions where one player has at least 4 provinces: Obviously, a player with at least 4 provinces optimizes his payoff in the discounted payoff game if he behaves according to r. The same is true for weaker players who face a player with at least 4 provinces in a pre-attack decision situation. If the player with at least 4 provinces is in N, "non-aligned" is the only local best reply; if he is in W, it does not hurt to choose "non-aligned". Only one post-attack decision situation arises for a weaker player faced by a player with 4 or more provinces: after an attack of the size-4 player in (4,1,1,0,0,0), the remaining size-1 player has to decide whether he wants to join A or C or whether he wants to stay non-aligned. The expected temporary payoffs for the decision maker are shown by Table 8. It is

DECISION	EXPECTED TEMPORARY PAYOFFS			
	ROUNDS			
	t	t+1	t+2	>t+2
join A	1	$1\frac{1}{5}$	$\frac{1-4F}{5}$	-F
join C	1	$\frac{1-F}{2}$	-F	-F
non-aligned	1	1	-F	-F

Table 8: Expected temporary payoffs for the remaining size-1 player after an attack of the size-4 player in a round t with the province distribution (4,1,1,0,0,0).

clear that for every discount factor ρ with $0 < \rho < 1$ the decision "join A" yields the highest expected payoffs.

Local payoff comparisons: In view of (71), local payoff comparisons in the long-run average payoff case based on strong inequalities remain valid in the discounted payoff case, if ρ is sufficiently near to 1. Similarly, equalities between local payoffs for different choices in the long-run average case are preserved in the discounted payoff case for ρ sufficiently near to 1, if they are due to symmetry or to the same outcome being reached at the end of the round.

Additional strong inequalities which are not present in the long-run average payoff case must be considered in the discounted payoff case. A player whose value for the current province distribution is lower than his number of provinces should stay non-aligned in a pre-attack decision situation where all other players are in N. A player whose value for the current province distribution is higher than his number of provinces should attack with probability 1 rather than with a smaller positive probability. These requirements of payoff maximization in the discounted payoff game are satisfied by the strategy r of Tables 2, 3, and 4.

Province distributions with up to 4 active players and with at most 3 provinces for each player: In view of what has been said on local payoff comparisons, it can be seen with the help of Figures 5, 7 to 11 and Tables 5 and 6 that for ρ sufficiently near to 1 the local strategies specified by r are local best replies to r in the discounted payoff game at decision situations with such province distributions.

The province distribution (2,1,1,1,1,0): For $k = 1,2$, let w_k be the value of a size-k player for (2,1,1,1,1,0) and r in the discounted payoff game. Since the outcome of a round with (2,1,1,1,1,0) is 2-1111 if r is played (see Figure 12), the following equations hold:

$$w_2 = 2\eta + \rho w_1 \tag{72}$$

$$w_1 = \eta + \rho(\tfrac{3}{4}w_1 + \tfrac{1}{4}w_2). \tag{73}$$

This yields:

$$w_2 = \frac{6}{5} + \frac{4\eta}{5 + \tfrac{5}{4}\rho} \tag{74}$$

$$w_1 = \frac{6}{5} - \frac{\eta}{5 + \tfrac{5}{4}\rho}. \tag{75}$$

For every discount factor ρ the value w_2 is greater than w_1. The arrows in Figures 14 and 15 are based on strong inequalities between local payoffs. In view of what has been said on local payoff comparisons, it can be seen with the help of Figures 14 and 15 and Table 7 that the local strategies specified by r for decision situations $D = (m,B,j)$ with $m = (2,1,1,1,1,0)$ are local best replies to r in the discounted payoff game for every discount factor ρ with $0 < \rho < 1$.

<u>The province distribution (1,1,1,1,1,1)</u>: In the long-run average payoff case, all survivors after an attack in a round with (1,1,1,1,1,1) receive the same expected payoff 6/5. In view of $v_2 > v_1$, the expected payoff of a survivor in the discounted payoff game depends on his probability of obtaining 2 provinces at the end of the round; the greater this probability is, the greater is his expected payoff. After an attack in a round with (1,1,1,1,1,1), the local strategies specified by r maximize the probability of obtaining 2 provinces at the end of the round. For every ρ with $0 < \rho < 1$, the local strategies specified by r for decision situations $D = (m,B,j)$ with $m = (1,1,1,1,1,1)$ are local best replies to r. We have obtained the following result:

<u>Result 2</u>: A discount factor ρ_0 with $0 < \rho_0 < 1$ exists such that the following is true for $\rho_0 < \rho < 1$ and $1/3 < F < 3$: The strategy described by Tables 2, 3, and 4 is an equilibrium strategy for the discounted payoff game with the discount factor ρ.

7. The Interpretation of the Solution

In Section 5, we have derived a unique solution for our theoretical model with 6 provinces and the assumption that the existence value is in the middle range $1/3 < F < 3$. In the following, we shall look at the reasons for stability and instability in this solution.

Figure 16 shows the transitions among types of province distributions permitted by the solution. Types of province distributions are represented by nodes, and transitions are represented by connecting lines. Arrowheads indicate the direction of the transition. The numbers written next to the connecting lines are the probabilities of the corresponding transitions if the equilibrium strategy described by Tables 2, 3, and 4 is played. Circles surrounding nodes indicate stable province distribution types and the fluid province distribution type is graphically distinguished by a square surrounding its node. Figure 16 provides an overview over important qualitative features of the solution.

Five of the eleven types of province distributions are stable. We shall now look at each of these types separately in order to discuss the reason for its stability.

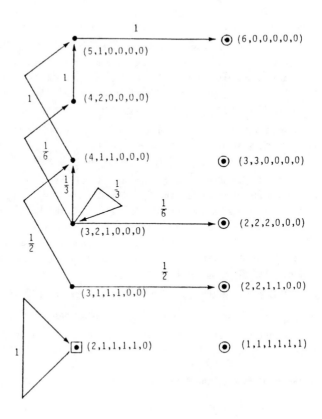

- • transitory
- ⊙ stable
- ⊡ fluid

Figure 16: Transitions among types of province distributions for the equilibrium strategy described by Tables 2, 3, and 4.

The province distribution type (6,0,0,0,0,0): This type is stable for the trivial reason that wars cannot arise if there is only one power left.

The province distribution type (3,3,0,0,0,0): Here, a potential attacker is deterred by half a chance to lose his existence which in view of $F > 0$ has a greater weight than half a chance to conquer the world. As has been pointed out, the expectation of stability is necessary for the stabilization of (3,3,0,0,0,0) in the case of long-run average payoffs. This is different in the discounted payoff game. There, "non-aligned" is the only possible equilibrium choice of one of the action players if the other one has already chosen "non-aligned". This makes it optimal to choose "non-aligned" in all pre-attack decision situations in a round with (3,3,0,0,0,0).

The province distribution type (2,2,2,0,0,0): Here, all three active players hold the same number of provinces. This has the consequence that after an attack of one active player on another the remaining active player must join both sides with equal probability. Therefore, an attacker has half a chance to be on the losing side. If there were a riskless way to build up a coalition of two active players against the third one, such a coalition would be formed. However, the symmetry properties of global strategies exclude this possibility. After an attack, both sides must be treated symmetrically by the remaining player. One may say that symmetry is a source of stabilizing uncertainty on the formation of alignments.

The province distribution type (2,2,1,1,0,0): It is of crucial importance for the stability of this type that the existence value is assumed to be greater than $1/3$. For $F < 1/3$, a size-1 player would have an incentive to attack a size-2 player (see Table 6). A size-1 player who attacks a size-2 player can be sure to be on the winning side (see Figure 9). However, with probability $1/2$ he becomes a size-1 player in an unstable province distribution of the type (3,2,1,0,0,0). This creates a risk of being eliminated. We may say that the danger of instability in the next round is a stabilizing force.

The province distribution type (1,1,1,1,1,1): Here, an attacker has half a chance to be on the losing side, since after an attack both sides are treated equally by the next decision maker. In this case, too, symmetry is a source of stabilizing uncertainty on the formation of alignments.

Reasons for stability and instability: The discussion of the five types of stable province distributions has revealed several possible reasons for stability. In the world ownership case (6,0,0,0,0,0), the reason for stability is trivial. the *bipolar case* (3,3,0,0,0,0) is stabilized by mutual deterrence. the discussion of the three stable province distributions with at least three active players has revealed two stabilizing forces:

   (a) Symmetry as a source of uncertainty on the formation of alignments.
   (b) The danger of instability in the next round.

In the solution of our model, transitory instability is always connected to an attempt of a player with at least 3 provinces to conquer the world. However, other reasons of instability are possible. For an existence value smaller than 1/3, the stability of (2,2,1,1,0,0) breaks down, because a weak player, who can be sure to be on the winning side, has an incentive to attack one of the stronger players. The danger of instability in the next round and the associated risk of losing existence destroy this incentive if the existence value is greater than 1/3.

Coalitions against actors who threaten to become too strong: Informal balance of power theories convey the impression that the most important stabilizing force is a tendency to form coalitions against actors who threaten to become too strong. Our solution does not seem to support this view. Coalitions against actors who threaten to become too strong are formed in the cases of the two unstable types of province distributions (3,2,1,0,0,0) and (3,1,1,1,0,0). These coalitions are desperate attempts with at best half a chance of success to prevent the strong player from reaching world ownership and to establish a balance of power among the existing active players. It is true that the coalitions against the strong player in rounds with (3,2,1,0,0,0) and (3,1,1,1,0,0) decrease the incentive for the stronger players to attack in a round with (2,2,1,1,0,0), but they also increase the incentive for an attack by a weaker player, and as we have seen, an attack of a weaker player is the real danger to the stability of (2,2,1,1,0,0).

The fluid province distributions of the type (2,1,1,1,1,0): Here, we can speak of a balance of power even if these province distributions are not stable. Peace is never achieved, but none of the actors must fear to become eliminated. Whenever a war occurs, a coalition against the strongest actor forms, but not in order to prevent him to become too strong, but rather for the purpose of taking one province away from him. However, the impossibility of building up a successful alliance including the current owner of two provinces may be attributed to the danger that this player threatens to become too strong. This suggests that in some situations exploitative coalitions against the strongest player are facilitated by the fact that other coalitions are too risky for the weaker players who must fear that the strongest player becomes too strong.

Limitations of the analysis: It is not difficult to analyze analogous games with a smaller number of provinces. 6 is the minimum number which yields an interesting case. It must be expected that additional phenomena emerge for greater numbers of provinces.

Our analysis is based on assumptions of stationarity and symmetry. Experiences with the parlor game suggest that human players do not play stationary strategies. Informal coalitions are built up which may last for many rounds. Expectations on the behavior of other players may crucially depend on past history. Informal agreements can destroy the inherent symmetry of a situation like (1,1,1,1,1,1). This province distribution must be expected to be unstable in a game played by human players.

Our observations of plays of the parlor game are not yet sufficiently extensive to permit the construction of a descriptive theory. It is clear that a descriptive theory must take account of the limited rationality of human players. Unfortunately, a precise general theory of limited rationality did not yet emerge from empirical evidence. Therefore, a game theoretical approach based on strong rationality assumptions must serve as a substitute for a more realistic approach. Moreover, the contrast between game theoretical analysis and observed behavior provides insights into the problem of limited rationality.

References

Aumann, R.J., and M. Maschler (1964). The Bargaining Set for Cooperative Games. In: Dresher et al. (eds.), Advances in Game Theory. Princeton, N.J.: Princeton University Press.

Bernholz, P. (1985). The International Game of Power. Berlin: Mouton.

Clausewitz, C. von (1963). Vom Kriege. Leck (Schleswig): Clausen und Bosse.

Feller, W. (1968). An Introduction to Probability Theory and its Applications, Vol. I, 3rd ed. New York: Wiley.

Harsanyi, J.C., and R. Selten (1988). A General Theory of Equilibrium Selection in Games. Cambridge, Mass.: MIT Press.

Kaplan, M.A. (1957). System and Process in International Politics. New York: Wiley.

Kautilya, A.-S. (1957). Excerpts. In: S. Radakrishan and Ch.A. Moore: A Source Book of Indian Philosophy. Princeton: Princeton University Press.

Luce, D., and H. Raiffa (1957). Games and Decisions. New York: Wiley.

Niou, E.M. and P.C. Ordeshook (1986). A Theory of the Balance of Power in International Systems. J. Conflict Resol. **30(4)**: 685-715.

Niou, E.M., and P.C. Ordeshook (1987). Preventive War and the Balance of Power. J. Conflict Resol. **31(3)**.

Selten, R., and R. Stoecker (1986). End Behavior in Sequences of Finite Prisoner's Dilemma Supergames. J. Econ. Beh. Organ. **7**: 47-70.

Wagner, H. (1986). The Theory of Games and the Balance of Power. World Politics **37(4)**: 546-576.

Zinnes, D.A. (1967). The Analytical Study of the Balance of Power Theories. J. Peace Res. **4**: 270-328.

[12]

The Distribution of Foreign Language Skills as a Game Equilibrium

Reinhard Selten

Jonathan Pool

1. Introduction

The birth, death, growth, and shrinkage of languages over millennia has given us a world containing about three thousand living languages, whose speakers number from 1 up to several hundred million. Our current knowledge of what causes a language to gain more speakers than it loses or lose more speakers than it gains is limited to a few generalizations about bivariate, more–less effects (see Dressler, 1982; Laponce, 1984; Lieberson, 1982). One important generalization is that the children of two native speakers of the same language tend to acquire that native language unless outside the home the language is rarely used or is despised. In addition, persons who spend a few years or more in a milieu (e.g., neighborhood, school, or workplace) where a language other than their native language is the main language tend to add the other language to their repertoire. Persons tend to learn a language through deliberate study (in contrast with immersion in its milieu) when the language is spoken by many persons, has widely distributed speakers, has wealthy and powerful speakers, and has a prestigious literature, art, and history. Languages tend to lose speakers through death, of course, but also through forgetting by their native and nonnative speakers. Forgetting tends to take place among persons who are not in contact with other speakers of the language or whose rewards for using the language are small or negative. There is little evidence as to whether the difficulty of a language or its effectiveness as an instrument of thought and communication influences its acquisition of new speakers or its loss of former speakers.

As Vaillancourt (1985, p. 18) points out, almost all attempts to model aspects of the distribution of language skills in a population have started with the assumption that this distribution is fixed. Given a population distribution of language skills and some mechanism whereby costs of production or benefits of consumption depend on this distribution, one can derive predictions about the production and consumption of linguistically specialized products (Hočevar, 1975; Vaillancourt, 1985a), the earnings of persons with different repertoires of language skills (Sabourin, 1985; Vaillancourt & Lacroix, 1985), and the distributions of language skills within labor markets and firms (Breton & Mieskowski, 1975; Sabourin, 1985).

In the short term, this assumption of a fixed language-skill distribution in a population is realistic, but over a period of years the distribution of language skills changes, partly as a consequence of choices that persons make about which languages to impart to their children, which languages to learn, and how thoroughly to learn languages. Grenier (1985) and Lang (1986) have attempted to model such choices.

Grenier's (1985) model of a two-language economy assumes that each person's earnings level is a function of (1) the person's native language, (2) whether the person is bilingual, and (3) other attributes of the person. A person is assumed to be bilingual if and only if the present value of the increase in earnings for that person resulting from being bilingual exceeds the present value of the cost to that person of becoming bilingual. Using some additional assumptions about the form of the earnings function, the model estimates what the earnings of any member of a population would be if that member's monolingual/bilingual status were reversed and all the other attributes of the person were held constant. The Grenier model uses population data to estimate the parameters of the prediction equation, and for any set of population data it derives parameters. Thus, the model does not make falsifiable predictions of the distribution of language skills in a population. It does not predict that any distribution of language skills is by itself impossible, and it does not predict that any distribution of language skills is incompatible with some distribution of other attributes.

Lang's (1986) first model of a two-language labor market in equilibrium assumes that all employers have the same native language. Employers choose whether to hire workers with the same native language as theirs or the other native language. In the latter case, either these workers must be bilingual or the employer must become bilingual. The wage of bilingual workers is assumed to compensate them exactly for the cost of becoming bilingual. In a more elaborate model, Lang assumes that most but not all employers have the same native language, and that employers hire not only workers but also supervisors. The workers are assumed to be monolingual, and the supervisor in any firm must be bilingual if the employer, the supervisor, and the workers do not all have the same native language. Supervisors' wages compensate them exactly for their supervisorial training, and bilingual supervisors are further compensated exactly for the cost of learning the second language. Lang finds in both models that only native speakers of the language that has few or no employers become bilingual. Their workers are paid less, while their supervisors are paid more, than native speakers of the language that has most or all of the employers. One peculiarity of Lang's models is their treatment of the linguistic distribution of employers as fixed. The overrepresentation of employers with one of the two native languages, combined with the assumed need for employers and their employees to understand one another, creates linguistically differentiated wage rates. The lower wages paid to workers of one native language makes it possible for the employers having that native language to make higher profits. Lang's models do not assume that such an advantage leads to the elimination of the disproportional representation of language groups among employers and

| 1   1+a_1 | 2   2+a_2 | 3   3+a_3 | 4   4+a_4 |

Figure 1. Partition of hypothetical population into native-language communities

a consequent elimination of the wage disparity. In addition, Lang's assumption of uniform learning costs precludes any prediction of an equilibrium in which at least some persons with each native language learn the other language. Yet we know that reciprocal language learning takes place, and it would be interesting to speculate about the assumptions that would predict an equilibrium that exhibits it.

In this paper we do not restrict ourselves to earnings as a mechanism or to firms as a milieu of the incentive to learn languages. We also abandon the two-language restriction imposed by both Grenier and Lang. Our aim is to predict some general features of distributions of nonnative language skills. We shall impose no limit on the number of languages in existence. We shall permit languages with native speakers and languages without native speakers to exist. The most important restriction from others' models that we keep intact is the treatment of the distribution of **native** languages as fixed.

Our model assumes that there is a positive number of languages in the world. At least one of these languages is natural, i.e. a language that has native speakers. There may also be any number of auxiliary languages, i.e. languages with no native speakers. Since everyone is assumed to have one and only one native language, the world population can be partitioned into communities, each community consisting of the native speakers of a particular language. We represent the members of the world population as a continuum of size 1. We represent the ith community as a continuum of size a_i. We choose to identify each member of the ith community with a number s such that $i \leq s \leq i + a_i$. Geometrically, the world population can be represented as a set of n line segments, where n is the number of communities. The ith community is represented as the line segment from i to $i + a_i$. Figure 1 gives an example of such a geometric representation for a world of four communities. The sum of the lengths of the four line segments is 1.

We model this continuum as a continuum of players of a noncooperative normal-form game. Each player chooses what set of additional (non-native) languages to learn. We consider only pure strategies. Thus, if there are m languages in the world, each player (represented as a point on one of the line segments) has a set of 2^{m-1} pure strategies among which to choose. A player in figure 1, for example, has eight possible pure strategies if there are no auxiliary languages, sixteen if there is one auxiliary language, and so forth.

The players' payoffs are the net benefits they derive from learning the languages that they learn. The positive term in a payoff is the player's communicative benefit. The negative term is the player's learning cost. A person's payoff is the difference between

	2+t							
	None		1		3		1 & 3	
None	0	0	$b_{12}(t)$	$b_{21}(s)$	0	0	$b_{12}(t)$	$b_{21}(s)$
1+s 2	$b_{12}(t)$	$b_{21}(s)$	$b_{12}(t)$	$b_{21}(s)$	$b_{12}(t)$	$b_{21}(s)$	$b_{12}(t)$	$b_{21}(s)$
3	0	0	$b_{12}(t)$	$b_{21}(s)$	$b_{12}(t)$	$b_{21}(s)$	$b_{12}(t)$	$b_{21}(s)$
2 & 3	$b_{12}(t)$	$b_{21}(s)$	$b_{12}(t)$	$b_{21}(s)$	$b_{12}(t)$	$b_{21}(s)$	$b_{12}(t)$	$b_{21}(s)$

Figure 2. Weight of two players in one another's communicative benefits

these: the communicative benefit minus the learning cost.

A player's communicative benefit is proportional to the importance-weighted fraction of the world population that shares at least one language with the player. Any language community has an importance function that assigns some importance to each member of the world population. To compute a player's communicative benefit, we use the importance function of the player's community to give a weight to everyone in the world with whom the person shares knowledge of at least one language. We give a weight of 0 to everyone with whom the person does not have a common language. The integral of the world population, thus weighted, is the player's communicative benefit.

To see how the communicative benefit is defined, consider a world with 2 native languages and one auxiliary language. Let $1 + s$ be one of the players in community 1, and $2 + t$ be one of the players in community 2. Suppose that community 1's importance function assigns an importance of $b_{12}(t)$ to $2 + t$, while community 2's importance function assigns an importance of $b_{21}(s)$ to $1 + s$. Then figure 2 shows the weight that each of these two players has in the other's communicative benefit, depending on the pair of strategies that they adopt. The rows are labeled with the four possible strategies for player $1 + s$ (learn no additional language, learn language 2, etc.), and the columns are labeled with the four possible strategies for player $2 + t$. The left entry in each cell is the weight of player $2 + t$ in player $1 + s$'s communicative benefit, and the right entry in each cell is the weight of player $1 + s$ in player $2 + t$'s communicative benefit.

The importance function is intended to reflect the fact that economic, geographical, and other attributes of a person affect the value of being able to communicate with the person. When a person has unequal importance to several other persons, a typical reason is differences in geographical proximity. Since language communities tend to be geographically concentrated, we allow importance functions to differ among language communities but not within them. With this definition of communicative benefit, any two players who belong to the same community and adopt the same strategy must get the same communicative benefit.

A player's learning cost is the product of two factors: a linguistic factor and a personal factor. The linguistic factor depends only on a person's languages: native and additional. Any two players who have the same native language and learn the same set of additional languages share the same linguistic factor. By assumption, the linguistic factor is always nonnegative, and it is 0 for any player who learns no additional language. These assumptions about the linguistic factor reflect three facts: (1) some languages are more difficult to learn than others, (2) the effort required in learning a combination of languages is not necessarily the sum of the efforts of learning the languages in the combination, and (3) the difficulty of learning a combination of languages depends on the native language of the learner.

The other component of a player's learning cost is a personal factor, which reflects the differences among language learners in aptitudes and motivations. The personal factor is a nonnegative number assigned to a player. For simplicity of analysis, it is assumed that within any language community no player has the same personal factor as any other player. Thus, any two players with the same native language who learn no additional language have the same learning cost (0), but if they learn the same additional language or languages they must have different learning costs. We have already assumed that their communicative benefit is identical. Therefore, with unequal learning costs they must have different payoffs.

Our first substantively interesting result describes a feature of the players' payoffs. The payoff to any player is a function of the strategy of that player and the strategies of all players who are members of other communities. So, a player's payoff is independent of the strategies of the other members of that player's own community. The reason is that a player's communicative benefit depends on the subset of the world population that shares at least one language with the player. But every member of the player's community shares a language with the player (their native language). This fact cannot be changed by any strategy that another member of this community adopts. Thus, the only way in which a player can affect the payoff of another player in the same community is indirectly: by affecting the choices of players in other communities. For example, in our model a native speaker of Japanese does not directly affect the payoff of another native speaker of Japanese. But, if one native speaker of Japanese learns English, this choice may reduce the incentive of a native speaker of English or some other language to learn Japanese, and the reduced learning of Japanese that results from this incentive reduction may affect the payoffs of other native speakers of Japanese.

Our analysis proceeds by examining the features of best replies and group best replies. A best reply is a strategy that maximizes the adopting player's payoff, given the combination of strategies adopted by all other players. The combination of the strategies adopted by all the members of a community is called a group strategy. If every player in a community adopts a best reply to the combination of all other players' strategies, then we say that the community's group strategy is a group best reply. Referring to group strategies and group best replies is only an analytical device, since the players in our game are individuals, not communities.

Our next result describes a regularity in best replies. Consider the members of one language community. For any combination of the other communities' members' strategies, each member of this community has at least one strategy that is a best reply. A strategy that is a best reply for one member may not be a best reply for some other member. We show, however, that whenever a strategy is a best reply for two members it is also a best reply for all members whose personal factors of their learning costs lie between those of the two members. In terms of a geometric representation, we can arrange the members of a community in order of increasing personal factors on the line segment that represents their community. If we do this, then any strategy that is a best reply at more than one point in a community is a best reply everywhere in some interval within the line segment, and nowhere else in the community. This result leads to an empirical prediction that, among the native speakers of any language, those who learn the same set of foreign languages will tend to be more similar in the extent of their (dis)inclination and (in)ability to learn languages than is true for any more loosely defined set of language learners. For example, the variance in scores on a language aptitude test should be greater among all those native speakers of German who have learned English and any one other foreign language than among the subsets of native German speakers who have learned English and French, who have learned English and Russian, etc.

We build on this result to show that any group best reply exhibits a negative association between the two factors of the learning cost. When a community adopts a group best reply, for any two members in the community whose learning costs have unequal linguistic factors, the one with the higher linguistic factor (the member who learns the more expensive set of languages) has the lower personal factor (finds it easier to learn languages). Suppose we arrange the members of a community from left to right in order of increasing personal factors, and suppose the community adopts a group best reply which involves more than one strategy. As we move from left to right, we find personal factors everywhere rising, while at each point where the linguistic factor changes it declines. An example of this pattern for a world with two natural languages and one auxiliary language is illustrated in figure 3.

As we move from left to right in a community that adopts a group best reply, we must also find communicative benefit decreasing whenever there is an decrease in the linguistic factor. If this relationship did not exist, then the community would not be adopting a group best reply, because some member of the community could get a higher payoff by changing to a strategy with a lower linguistic factor.

As the linguistic factor decreases moving from left to right, not only do communicative benefits decrease, but so do payoffs. It might be imagined that this is not always true.

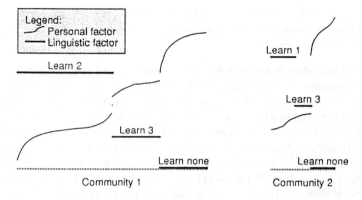

Figure 3. Illustration of group best replies in a two-community, three-language world

Under some conditions, the drop in the linguistic factor of the learning cost might be steep enough to counteract both the drop in the communicative benefit and the rise in the personal factor. But we show that there are no conditions that allow payoffs to rise or remain constant as the linguistic factor decreases. This result can be interpreted as saying that, in a community adopting a group best reply, those who learn more difficult sets of languages get higher profits, despite the difficulty of the languages they learn.

Having determined some features of best replies, we move on to show that it is possible for every player to adopt a best reply at the same time. In other words, at least one equilibrium point exists. There are several steps in proving this result.

First, we arbitrarily resolve any ties that may exist in the linguistic costs of the possible strategies for the players in any community. In other words, we label the possible strategies in a community with the numbers 1 to 2^{m-1}, in order of decreasing linguistic cost. Whenever we encounter two or more strategies having the same linguistic cost, we number them in any order.

The next step is to construct a new game, whose players are the communities. In this new, or aggregated, game, we require the members of any community to adopt individual strategies that are numbered in the same order as the members' personal costs are ordered. Thus, if any two members of a community adopt different individual strategies, the member with the higher personal factor must adopt the higher-numbered strategy. This requirement limits the set of possible group strategies for the communities playing the aggregated game. A community must be partitioned into 2^{m-1} intervals, such that the kth interval is the locus of members adopting the individual strategy numbered k. We limit the communities' strategy sets even further by defining a group strategy only as the set of upper bounds of these 2^{m-1} intervals. To see that this definition restricts the set of strategies, consider a point x in a community and suppose that the members to the immediate left of x adopt strategy k and the members to the immediate right of x adopt strategy $k + w$. Then the member precisely at point x can adopt any of the strategies $k, k + 1, ..., k + w$. But it makes no difference to the group strategy what the individual strategy of this border member is, since the upper bounds of the intervals are unaffected. A community i's strategy in the aggregated game is thus a nondecreasing set of 2^{m-1} border points, each located between i and $i + a_i$ and the last of them being equal to $i + a_i$.

The payoffs to the community players in the aggregated game are the respective integrals of the payoffs to their individual members in the original game. Each combination of strategies in the aggregated game induces a unique payoff to each community, despite the fact that the group strategies may leave the individual strategies of their border members incompletely specified. The reason is that the border members are only an infinitesimal fraction of all the members. The payoffs of the 2^{m-1} border members of a community (or of any finite number of its members) do not affect the integral of the payoffs of all the members of the community.

The last step in proving the existence of an equilibrium in the original game involves establishing relationships between the original and the aggregated games. We first show that whenever there is an equilibrium point in an aggregated game there is also an equilibrium point in the original game to which the aggregated game corresponds. We then show that for each aggregated game there is always at least one equilibrium point. But for each original game at least one corresponding aggregated game exists. These findings imply that there is always at least one equilibrium point in any original game.

In this paper we say nothing about the number and nature of the equilibria whose existence we prove. As an aid to intuitions, however, we picture one equilibrium in figure 4. The example is a world with three language communities, whose shares of the population are 0.5, 0.35, and 0.15, respectively. One auxiliary language exists, implying that there are four languages. For each player, three of the languages are open to choice, so each player has eight available strategies. The importance weight function of each community assigns identical weights to all players; thus, everyone is equally important for everyone else as a potential communication partner. The linguistic factors of the learning costs are additive functions of the languages contained in the learned sets of languages, with each natural language contributing 1.8 and the auxiliary language contributing 0.6 to the linguistic factor, regardless of the learner's native language. Finally, the personal factors of the learning cost are distributed uniformly in each community from 0 to 1, shown as diagonal dashed lines in the figure.

In the equilibrium shown for this world in figure 4, only three of the eight possible strategies in each language community are chosen. These are (1) to learn both nonnative natural languages, (2) to learn the nonnative natural language with the larger number of

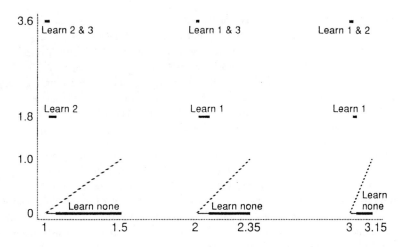

Figure 4. Illustration of equilibrium in a three-community, four-language world

native speakers, or (3) to learn nothing. The players are arranged in order of increasing personal factors of learning cost, with this personal factor plotted by dashed lines. When the players are so arranged, each community is partitioned into three convex sets of players, all players in the ith set choosing the ith strategy. The boundaries between the adjacent sets of players are, in distance from the community's origin, 0.0292 and 0.0729 in community 1, 0.0205 and 0.0875 in community 2, and 0.0219 and 0.0448 in community 3. The sets of players choosing these strategies are plotted at heights equal to the linguistic factors of their learning costs.

In this example, the proportion of each community that learns two nonnative languages is largest in the smallest community and smallest in the largest community. The same is true for the proportion that learns at least one nonnative language. We also notice that no player learns the auxiliary language. These results match well-known tendencies: those who learn nonnative languages tend to choose the most widely spoken ones; members of small language communities are more likely to learn foreign languages than members of large communities; and few learn auxiliary languages, despite their relatively low learning costs.

Another insight we may derive from our results concerns the phenomenon of apparently deliberate foreign-language incompetence. It has been claimed that the native speakers of certain languages (such as English and French) succeed in getting the native speakers of other languages to learn their languages. The strategy allegedly used to attain this effect is the intentional neglect of foreign-language learning and the ostentatious exhibition of low language-learning aptitudes. In the terms of our model, this claim can be rephrased into a suggestion that the native speakers of a language benefit from having high personal factors and hence high learning costs. The predictions from our model do not support this idea. Our predictions are compatible with the possibility that a player's payoff can be an increasing function of the learning costs of the player's fellow native speakers. But payoffs to a member of a community cannot, in any equilibrium, be an increasing function of the member's learning cost. Thus, even if the language community with the highest payoffs is the one whose members have the greatest average difficulty learning foreign languages, it is the easy learners within this group who have the highest payoffs in their group.

2. The model

The model describes a world with m languages, 1, ..., m. Every individual has one and only one of these languages as his *native language*. He is a *native speaker* of that language. Every language other than an individual's native language is a *foreign language* for that individual. Some languages may be foreign to all individuals. Languages with native speakers will be called *natural* languages, and the others will be referred to as *auxiliary* languages. Let languages 1,..., n, with $n \leq m$, be the natural languages. Let M be the set $\{1, ..., m\}$ of all languages, and let N be the set $\{1, ..., n\}$ of all natural languages.

Players and strategies. The set of all individuals with native language i is the *language community i* or, for short, *community i*. Each language community i is described as a continuum of individuals represented as points in an interval $[i, i + a_i]$, with $a_i > 0$. The measure of all individuals is normalized to 1:

(1) $$\sum_{i=1}^{n} a_i = 1.$$

We call a_i the *size* of language community i. Our model may be looked upon as a game with a continuum of players. The players are the members of the language communities 1, ..., n. The pure strategies are possible sets C of languages to be learned. Of course, a member of language community i always learns language i and has no choice in this respect. The set of all pure strategies C of a member $i + s$ of community i is as follows:

(2) $\qquad M_i = \{C \mid i \in C \subseteq M\}.$

A *group strategy* of community i is a Borel measurable function which assigns a pure strategy $\varphi_i(s)$ to every member $i + s$ of community i:

(3) $\qquad \varphi_i \colon [0, a_i] \to M_i.$

It is notationally convenient to define φ_i on $[0, a_i]$ rather than on $[i, i + a_i]$. The set of all group strategies of community i is denoted by Φ_i. A *strategy combination*

(4) $$\varphi = (\varphi_1, \ldots, \varphi_n)$$

is an n-tuple of group strategies $\varphi_i \in \Phi_i$. The set of all strategy combinations is denoted by Φ.

An *i-incomplete* strategy combination φ_{-i} is an $(n-1)$-tuple of group strategies $\varphi_j \in \Phi_j$ with $j \neq i$:

(5) $$\varphi_{-i} = (\varphi_1, \ldots, \varphi_{i-1}, \varphi_{i+1}, \ldots, \varphi_n).$$

The set of all i-incomplete strategy combinations is denoted by Φ_{-i}. The i-incomplete strategy combination φ_{-i} whose components agree with the corresponding components of a strategy combination φ is called the i-incomplete combination *in* φ.

Learning costs. There are two influences on a player's payoff: "communicative benefit" and "learning cost". We first turn our attention to learning cost. Consider a member $i+s$ of community i who chooses C. His *learning cost* $K_{i+s}(C)$ is modeled as follows:

(6) $$K_{i+s}(C) = f_i(s)g_i(C),$$

where f_i is a bounded monotonically increasing nonnegative function defined on $[0, a_i]$ and g_i is a function defined on M_i with

(7) $$g_i(C) \geq 0 \text{ for all } C \in M_i$$

and

(8) $$g_i(\{i\}) = 0.$$

Learning costs are the product of a *personal factor* $f_i(s)$ and a *linguistic factor* $g_i(C)$. The personal factor summarizes influences like aptitude. We think of the members of community i as arranged according to their personal factors. Therefore the monotonicity assumption on f_i is not very restrictive.

Communicative benefit. We now turn our attention to the communicative benefit of language learning. Every other individual $j+t$ with whom a player $i+s$ has a common language provides a communication opportunity, which contributes to the communicative benefit of player $i+s$. The importance of the opportunity to communicate with $j+t$ may depend on various characteristics of $j+t$ such as income or geographical distance. Therefore, in the computation of communicative benefit, communicative opportunities are weighted by an importance factor $b_{ij}(t)$. In order to express the communicative benefit we make use of the following notation. For any two subsets C and D of M define:

(9) $$\delta(C, D) = \begin{cases} 1 \text{ for } C \cap D \neq \emptyset \\ 0 \text{ for } C \cap D = \emptyset \end{cases}.$$

Let $\varphi = (\varphi_1, ..., \varphi_n)$ be a strategy combination. The *communicative benefit* $U_i(C, \varphi)$ obtained by a member of community i from C if φ is played by all other individuals is defined as follows:

$$(10) \quad U_i(C, \varphi) = \sum_{j=1}^{n} \int_0^{a_j} b_{ij}(t)\delta(C, \varphi_j(t))dt.$$

It is assumed that for $i = 1, ..., n$ and $j = 1, ..., n$ the *importance weight functions*

$$(11) \quad b_{ij}: [0, a_j] \to \mathbb{R}_+$$

are bounded, nonnegative and piecewise continuous with finitely many discontinuities. This assumption together with the measurability of the φ_j guarantees the existence of the integrals on the right-hand side of (10).

Since all $C \in M_i$ contain i, we have:

$$(12) \quad U_i(C, \varphi) = \int_0^{a_i} b_{ii}(t)dt + \sum_{\substack{j=1 \\ j \neq i}}^{n} \int_0^{a_j} b_{ij}(t)\delta(C, \varphi_j(t))dt.$$

This shows that $U\backslash s\backslash do6(i)(C, \varphi)$ does not depend on $\varphi\backslash s\backslash do6(i)$, but only on the i-incomplete combination φ_{-i} in φ. Accordingly, we shall also use the notation $U_i(C, \varphi_{-i})$:

$$(13) \quad U_i(C, \varphi_{-i}) = U_i(C, \varphi) \text{ with } \varphi_{-i} \text{ in } \varphi.$$

Payoffs. We now define the *payoff* $H_{i+s}(C, \varphi_{-i})$ to a member $i + s$ of community i obtained against an i-incomplete strategy combination $\varphi_{-i} \in \Phi_{-i}$ for $C \in M_i$:

$$(14) \quad H_{i+s}(C, \varphi_{-i}) = U_i(C, \varphi_{-i}) - f_i(s)g_i(C)$$

for $i = 1, ..., n$ and $s \in [0, a_i]$. The *payoff function* H assigns $H_{i+s}(C, \varphi_{-i})$ to every triple $(i + s, C, \varphi_{-i})$, where $i + s$ is a member of community i, where C is one of his pure strategies, and where φ_{-i} is an i-incomplete strategy combination. The pure strategy sets M_i of the members of the language communities together with H constitute a game in normal form with a continuum of players. The measurability requirement imposes a restriction on the region where payoffs are defined, but apart from this a game in normal form with a continuum of players is a straightforward modification of the usual concept of a game in normal form.

Summary of assumptions. Let $a = (a_1, ..., a_n)$ be the vector of the a_i and let $f = (f_1, ..., f_n)$ and $g = (g_1, ..., g_n)$ be the vectors of the functions f_i and g_i, respectively. Moreover, let b be the $n \times n$ matrix of the functions b_{ij} with $i = 1, ..., n$ and $j = 1, ..., n$. The sextuple (m, n, a, b, f, g) summarizes the elements which have to be fixed in a specification of the model. The notation $\Gamma(m, n, a, b, f, g)$ indicates the normal form game resulting from the specification (m, n, a, b, f, g). In the following we shall give a summary of the assumptions made about the elements to be specified.

1. m, the number of languages, and n, the number of natural languages, are positive integers with $m \geq n$.

2. The components a_i of a, the sizes of language communities, are positive and sum to 1.

3. For $i = 1, ..., n$ and $j = 1, ..., n$, the importance weight functions b_{ij} are bounded, nonnegative, and piecewise continuous with finitely many discontinuities.

4. For $i = 1, ..., n$, the personal factor $f_i(s)$ of the learning cost is a bounded monotonically increasing nonnegative function.

5. For $i = 1, ..., n$, the linguistic factor g_i of the learning cost has the properties $g_i(\{i\}) = 0$ and $g_i(C) \geq 0$ for all $C \in M_i$.

3. Equilibrium points and the structure of group best replies

The model will be analyzed as a noncooperative game. In this section we shall prepare for the investigation of equilibrium points in pure strategies. Definitions and statements will refer to an arbitrarily fixed game $\Gamma(m, n, a, b, f, g)$.

Best replies. Loosely speaking, a best reply is a strategy which maximizes the payoff of a player given the behavior of all others. The payoff that $i + s$ obtains by a pure strategy $C \in M_i$ depends only on $\varphi_i(s)$ and φ_{-i}. Therefore, player $i + s$ does not have to know $\varphi_i(t)$ for $t \neq s$ in order to determine his best reply. Let φ_{-i} be an i-incomplete strategy combination. Formally, $D \in M_i$ is a *best reply* of $i + s$ to φ_{-i} if we have:

$$(15) \qquad H_{i+s}(D, \varphi_{-i}) = \max_{C \in M_i} H_{i+s}(C, \varphi_{-i}).$$

We say that D is a *best reply* to $\varphi = (\varphi_1, ..., \varphi_n)$ if D is a best reply to the i-incomplete strategy combination φ_{-i} in φ.

A group strategy $\psi_i \in \Phi_i$ is a *group best reply* or simply a *best reply* to an i-incomplete combination φ_{-i} if for every member $i + s$ of community i his strategy $\psi_i(s)$ is a best reply to φ_{-i}. We say that ψ_i is a *group best reply* or simply a *best reply* to $\varphi = (\varphi_1, ..., \varphi_n)$ if ψ_i is a best reply to the i-incomplete strategy combination φ_{-i} in φ. A strategy combination $\psi = (\psi_1, ..., \psi_n)$ is a *best reply combination* for $\varphi = (\varphi_1, ..., \varphi_n)$ or simply a *best reply* to φ if for $i = 1, ..., n$ the group strategy ψ_i is a best reply to φ.

A group strategy ψ_i is an *almost best reply* to an i-incomplete strategy combination φ_{-i}

or a strategy combination φ if for $0 \le t \le a_i$ with the exception of finitely many values of t the strategy $\psi_i(t)$ is a best reply of player $i + t$ to φ_{-i} or φ, respectively.

Equilibrium points. Since mixed strategies are not considered in this paper, the term "equilibrium point" will always refer to an equilibrium point in pure strategies. A strategy combination $\varphi = (\varphi_1, ..., \varphi_n)$ is an equilibrium point if φ is a best reply to itself or, in other words, if for $i = 1, ..., n$ the group strategy φ_i is a best reply to φ_{-i} in φ. Analogously, an almost-equilibrium point is a strategy combination which is an almost best reply to itself.

Lemma 1. Let $\varphi = (\varphi_1, ..., \varphi_n)$ be an almost-equilibrium point and let $\psi = (\psi_1, ..., \psi_n)$ be a strategy combination such that, for $i = 1, ..., n$ and $0 \le t \le a_i$, wherever $\varphi_i(t)$ is a best reply of $i + t$ to φ the strategy $\psi_i(t)$ agrees with $\varphi_i(t)$, and wherever $\varphi_i(t)$ is **not** a best reply of $i + t$ to φ the strategy $\psi_i(t)$ **is** a best reply of $i + t$ to φ. Then ψ is an equilibrium point.

Proof. The transition from φ to ψ involves only a finite number of changes. Therefore, we have

(16) $\qquad H_{i+t}(C, \psi_{-i}) = H_{i+t}(C, \varphi_{-i})$

for $i = 1, ..., n$, for $0 \le t \le a_i$, and for every $C \in M_i$, with φ_{-i} in φ and ψ_{-i} in ψ. Consequently, the assertion holds.

Comment. Lemma 1 shows how an almost-equilibrium point can be changed into an equilibrium point by the adjustment of the strategies of finitely many players. In order to prove the existence of equilibrium points and to explore their structural properties, it is sufficient to investigate almost-equilibrium points.

We shall now derive a useful result on the set of the members of a language community i for whom a given strategy C is a best reply to an i-incomplete strategy combination φ_{-i}. Then we shall prove a theorem which exhibits an interesting structural feature of group best replies: those who learn more difficult sets of languages have lower personal learning costs.

Lemma 2. Let φ_{-i} be an i-incomplete strategy combination and let $C \in M_i$ be a strategy for a member of language community i. Moreover, let I be the set of all t with $0 \le t \le a_i$ such that C is a best reply of $i + t$ to φ_{-i}. If the set I contains at least two points, then I is a closed, open, or half-open interval.

Proof. Assume that s and t with $s \ne t$ belong to I. For every $D \in M_i$ with $D \ne C$ we have:

(17) $\qquad U_i(C, \varphi_{-i}) - f_i(s)g_i(C) \ge U_i(D, \varphi_{-i}) - f_i(s)g_i(D);$

(18) $\qquad U_i(C, \varphi_{-i}) - f_i(t)g_i(C) \ge U_i(D, \varphi_{-i}) - f_i(t)g_i(D).$

For every t' with $s < t' < t$, the personal cost factor $f_i(t')$ is a convex linear combination of $f_i(s)$ and $f_i(t)$; in other words, for some α with $0 < \alpha < 1$ the following is true:

(19) $\quad f_i(t') = (1-\alpha)f_i(s) + \alpha f_i(t)$.

The sum of (17) multiplied by $(1-\alpha)$ and (18) multiplied by α yields

(20) $\quad U_i(C, \varphi_{-i}) - f_i(t')g_i(C) \geq U_i(D, \varphi_{-i}) - f_i(t')g_i(D)$.

Consequently, C is a best reply of $i + t'$ to φ_{-i} for every t' with $s < t' < t$. It follows that the assertion of the lemma holds.

Theorem 1 (on group best replies). Let φ_{-i} be an i-incomplete strategy combination and let φ_i be a group best reply to φ_{-i}. Moreover, let $i + s$ and $i + t$ be two members of community i with

(21) $\quad g_i(\varphi_i(s)) > g_i(\varphi_i(t))$.

Then we have:

(22) $\quad s < t$.

Proof. The following inequalities are consequences of the fact that $\varphi_i(s)$ and $\varphi_i(t)$ are best replies to φ_{-i}:

(23) $\quad U_i(\varphi_i(s), \varphi_{-i}) - f_i(s)g_i(\varphi_i(s)) \geq U_i(\varphi_i(t), \varphi_{-i}) - f_i(s)g_i(\varphi_i(t))$;

(24) $\quad U_i(\varphi_i(s), \varphi_{-i}) - f_i(t)g_i(\varphi_i(s)) \leq U_i(\varphi_i(t), \varphi_{-i}) - f_i(t)g_i(\varphi_i(t))$.

Subtraction of (24) from (23) yields

(25) $\quad [f_i(t) - f_i(s)]g_i(\varphi_i(s)) \geq [f_i(t) - f_i(s)]g_i(\varphi_i(t))$

or, equivalently,

(26) $\quad [f_i(t) - f_i(s)][g_i(\varphi_i(s)) - g_i(\varphi_i(t))] \geq 0$.

In view of (21), the second factor on the left hand side of (26) is positive. Therefore, the first factor must be nonnegative, giving us:

(27) $\quad f_i(s) \leq f_i(t)$.

Since $\varphi_i(s)$ and $\varphi_i(t)$ must be different, we have $s \neq t$. Moreover, f_i is monotonically increasing. Consequently, (22) holds.

Remark. It can be seen immediately that the theorem has the following consequence: $s < t$ implies $g_i(\varphi_i(s)) \geq g_i(\varphi_i(t))$. Moreover, it follows with the help of (23) that $g_i(\varphi_i(s)) > g_i(\varphi_i(t))$ implies

(28) $$U_i(\varphi_i(s), \varphi_{-i}) > U_i(\varphi_i(t), \varphi_{-i}).$$

It is also worth pointing out that we have

(29) $$H_{i+s}(\varphi_i(s), \varphi_{-i}) > H_{i+t}(\varphi_i(t), \varphi_{-i})$$

for $s < t$ and $g_i(\varphi_i(t)) > 0$, since under these conditions $i + s$ can obtain a higher payoff than $i + t$ by choosing the same strategy, $\varphi_i(t)$.

4. Monotone group strategies and aggregated games

Theorem 1 shows that group best replies have the property that the two factors of learning costs are inversely connected. Higher personal factors are connected with lower or equal linguistic factors. Loosely speaking, we may say that a group best reply must respect the order of decreasing linguistic factors. The order of decreasing linguistic factors may be looked upon as a ranking of the individual strategies for the members of a language community. In the presence of equal linguistic factors for different individual strategies, this ranking has ties. For every possible strengthened order that one can create by arbitrarily breaking all ties, we shall define a narrower "monotonicity class" of group strategies which respect this strengthened order of decreasing linguistic factors. We shall prove that every monotonicity class contains a group best reply to every i-incomplete strategy combination. This result will be the basis of our construction of "aggregated games".

Within a monotonicity class a group strategy is almost completely characterized by the border points of the intervals where the same individual strategy is used. These border points form a "border point vector". Within the monotonicity class the group strategies with the same border point vector form a "border equivalence class". Two members of the same border equivalence class differ at most at finitely many points, namely the border points.

In an *aggregated game*, the players are the language communities. Payoffs are "group payoffs" defined as integrals over individual payoffs. Strategies correspond to border equivalence classes within a particular monotonicity class. Formally, strategies are border point vectors.

In the presence of equal linguistic factors for different strategies for the individuals in a language community, more than one aggregated game exists. For every way of selecting a monotonicity class for each player (i.e. for each language community), we obtain a different aggregated game.

An equilibrium point of the original game is associated with every equilibrium point of every aggregated game. Therefore, the task of proving the existence of equilibrium points can be shifted from the original game to the aggregated games.

We continue to use the definitions and notations introduced in the previous sections. All statements refer to an arbitrarily selected fixed game $\Gamma(m, n, a, b, f, g)$ whose specification satisfies assumptions 1 to 5.

Monotonicity classes. A member of community i can learn up to $m - 1$ foreign languages. Therefore, the number of pure strategies in the set M_i is 2^{m-1}. Let r_i be a function which assigns one of the numbers $1, ..., 2^{m-1}$ to every $C \in M_i$, a different one for each $C \in M_i$. We call r_i a ranking over M_i. We say that r_i is *cost compatible* if, for every pair of strategies C and D in M_i, we have:

(30) $\qquad r_i(C) < r_i(D)$ for $g_i(C) > g_i(D)$.

Let r_i be a cost compatible ranking over M_i. We say that a group strategy $\varphi_i \in \Phi_i$ is *monotone* with respect to r_i if for s and t in $[0, a_i]$ the following is always true:

(31) $\qquad r_i(\varphi_i(s)) \le r_i(\varphi_i(t))$ for $s < t$.

The set of all group strategies $\varphi_i \in \Phi_i$ which are monotone with respect to r_i is denoted by $\Psi_i(r_i)$. We call $\Psi_i(r_i)$ the *monotonicity class* of r_i.

Lemma 3. Let $\Psi_i(r_i)$ be the monotonicity class of a cost compatible ranking r_i over M_i. Then $\Psi_i(r_i)$ contains a group best reply φ_i to every i-incomplete strategy combination φ_{-i}.

Proof. For every t with $0 \le t \le a_i$, let $\varphi_i(t)$ be that best reply C of $i + t$ to φ_{-i} which has the lowest rank $r_i(C)$ among all best replies of $i + t$ to φ_{-i}. Obviously, φ_i is a group best reply to φ_{-i}. We have to show that φ_i is monotone with respect to r_i. In view of theorem 1, inequality (31) holds if $g_i(\varphi_i(s))$ is different from $g_i(\varphi_i(t))$. Suppose that $g_i(\varphi_i(s)) = g_i(\varphi_i(t))$, that $s < t$, and that (31) does not hold. Then $r_i(\varphi_i(s))$ is greater than $r_i(\varphi_i(t))$. Since $\varphi_i(s)$ is the lowest ranking best reply of $i + s$ to φ_{-i}, it follows that $\varphi_i(t)$ is not a best reply of $i + s$ to φ_{-i}. Therefore, we must have:

(32) $\qquad U_i(\varphi_i(s), \varphi_{-i}) - f_i(s)g_i(\varphi_i(s)) > U_i(\varphi_i(t), \varphi_{-i}) - f_i(s)g_i(\varphi_i(t))$.

In view of the fact that $g_i(\varphi_i(s))$ and $g_i(\varphi_i(t))$ are equal, we can conclude:

(33) $\qquad U_i(\varphi_i(s), \varphi_{-i}) > U_i(\varphi_i(t), \varphi_{-i})$.

It follows by (29) and the equality of $g_i(\varphi_i(s))$ and $g_i(\varphi_i(t))$ that $\varphi_i(t)$ cannot be a best reply of $i + t$ to φ_{-i}. This is a contradiction. Therefore, the assertion of the lemma is true.

Border point vectors. In the following we shall look at a fixed cost compatible ranking r_i. For $k = 1, ..., 2^{m-1}$, the strategy $C \in M_i$ with $r_i(C) = k$ is denoted by C_{ik}. For every group strategy φ_i in the monotonicity class $\Phi_i(r_i)$ of r_i, let I_{ik} be the inverse image $\varphi_i^{-1}(C_{ik})$ of C_{ik} with respect to φ_i or in other words the set of all $t \in [0, a_i]$ with $r_i(\varphi_i(t)) = k$. We use the notation z_{ik} for the upper border point of I_{ik}; moreover, we define $z_{i0} = 0$. Of course, several border points z_{ik} may coincide. Obviously, I_{ik} is a closed, open, or half open interval between the border points $z_{i(k-1)}$ and z_{ik}. For $k = 2^{m-1}$, we always have $z_{ik} = a_i$. The vector

(34) $\quad z_i = (z_{i0}, ..., z_{i2^m-1})$

is called the *border point vector* of φ_i. A border point vector always satisfies the following condition:

(35) $\quad 0 = z_{i0} \leq z_{i1} \leq ... \leq z_{i2^m-1} = a_i$.

The set of all vectors z_i of the form (34) with (35) is denoted by Z_i. We call Z_i the *border point vector set* of language community i. The set of all group strategies $\varphi_i \in \Psi_i(r_i)$ which have the same border point vector z_i is denoted by $\Lambda_i(z_i)$. We call $\Lambda_i(z_i)$ the *border equivalence class* of z_i. We say that two group strategies in $\Psi_i(r_i)$ are *border equivalent* if they are in the same border equivalence class. If $z_i = (z_{i0}, ..., z_{i2^m-1})$ is the border point vector of a group strategy $\varphi_i \in \Psi_i(r_i)$, then we have:

(36) $\quad \varphi_i(t) = C_{ik}$ for $z_{i(k-1)} < t < z_{ik}$

for $k = 1, ..., 2^{m-1}$. Obviously, two border equivalent group strategies in $\Psi_i(r_i)$ can differ only at the border points of their common border point vector.

Combinations of rankings. A combination $r = (r_1, ..., r_n)$ of cost compatible rankings contains a cost compatible ranking r_i for each language community $i = 1, ..., n$. An *i-incomplete* combination r_{-i} of cost compatible rankings contains a cost compatible ranking for every language community j with $j \neq i$. We say than r_{-i} is *in* r if the components of r_{-i} are the corresponding components of r. The set of all cost compatible rankings of language community i is denoted by R_i; the symbol R is used for the set of all combinations r of cost compatible rankings, and R_{-i} is the set of all i-incomplete combinations r_{-i} of cost compatible rankings.

Border point systems. In the following we shall look at a fixed combination $r = (r_1, ..., r_n)$ of cost compatible rankings and the i-incomplete combination r_{-i} in r. The *monotonicity class* $\Psi(r)$ of r is the set of all strategy combinations $\varphi = (\varphi_1, ..., \varphi_n)$ with $\varphi_i \in \Psi_i(r_i)$ for $i = 1, ..., n$. Analogously, the monotonicity class $\Psi_{-i}(r_{-i})$ of r_{-i} is the set of all i-incomplete strategy combinations φ_{-i} whose components φ_j are in the corresponding monotonicity classes $\Psi_j(r_j)$. Let $\varphi = (\varphi_1, ..., \varphi_n)$ be a strategy combination in $\Psi(r)$. For $i = 1, ..., n$, let z_i be the border point vector of φ_i defined above. We call

(37) $\quad z = (z_1, ..., z_n)$

the *border point system* of φ. Analogously, the *i-incomplete border point system* z_{-i} of an i-incomplete strategy combination $\varphi_{-i} \in \Psi_{-i}(r_{-i})$ contains the border point vectors of the components of φ_{-i}. The set of all systems z of the form (37) with $z_i \in Z_i$ is denoted by Z, and the set of all i-incomplete border point systems

(38) $$z_{-i} = (z_1, \ldots, z_{i-1}, z_{i+1}, \ldots, z_n)$$

with $z_j \in Z_j$ for all components z_j of z_{-i} is denoted by Z_{-i}. The *border equivalence class* $\Lambda(z)$ of $z = (z_1, \ldots, z_n)$ is the set of all strategy combinations $\varphi \in \Psi(r)$ whose border point system is z. Similarly, the border equivalence class $\Lambda_{-i}(z_{-i})$ of an i-incomplete border point system z_{-i} is the set of all i-incomplete strategy combinations φ_{-i} whose i-incomplete border point system is z_{-i}. Two border point systems or two i-incomplete border point systems are *border equivalent* if they belong to the same border equivalence class.

Group payoffs. We continue to look at a fixed combination of cost compatible rankings $r = (r_1, \ldots, r_n)$. Let $\varphi = (\varphi_1, \ldots, \varphi_n)$ be a strategy combination in the monotonicity class $\Psi(r)$, and let $z = (z_1, \ldots, z_n)$ be the border point system of φ. The *group payoff* of a language community i for φ is defined as the integral over all individual payoffs for members of the language community. As we shall see, this group payoff is a function of z. The finitely many individual strategies at border points do not matter for the integral. Therefore, we introduce the following definition of the *group payoff* $P_i(z)$ of language community i for z:

(39) $$P_i(z) = \int_0^{a_i} H_{i+t}(\varphi_i(t), \varphi_{-i}) dt.$$

It follows by (12) and (13) that communicative benefits $U_i(C, \varphi_{-i})$ can be expressed as follows:

(40) $$U_i(C, \varphi_{-i}) = \sum_{j=1}^{n} \sum_{k=1}^{2^{m-1}} \int_{z_{j(k-1)}}^{z_{jk}} b_{ij}(t) \delta(C, C_{jk}) dt.$$

Define:

(41) $$B_{ij}(s) = \int_0^s b_{ij}(t) dt.$$

With this notation, (40) assumes the following form:

(42) $$U_i(C, \varphi_{-i}) = \sum_{j=1}^{n} \sum_{k=1}^{2^{m-1}} [B_{ij}(z_{jk}) - B_{ij}(z_{j(k-1)})] \delta(C, C_{jk}).$$

For $j = i$, the inner sum is nothing other than $B_{ii}(a_i)$, since we always have $\delta(C, C_{ik}) = 1$. Therefore, the right hand side of (42) is a function of z_{-i} in z. We introduce the symbol $U_i(C, z_{-i})$ for the right hand side of (42):

(43) $$U_i(C, z_{-i}) = B_{ii}(a_i) + \sum_{\substack{j=1 \\ j \neq i}}^{n} \sum_{k=1}^{2^{m-1}} [B_{ij}(z_{jk}) - B_{ij}(z_{j(k-1)})]\delta(C, C_{jk}).$$

For $i = 1, \ldots, n$, define

(44) $$F_i(s) = \int_0^s f_i(t)dt$$

and

(45) $$L_i(z_i) = \sum_{k=1}^{2^{m-1}} [F_i(z_{ik}) - F_i(z_{i(k-1)})]g_i(C_{ik}).$$

We refer to $L_i(z_i)$ as the *total learning cost* of language community i. Total learning cost is the integral over individual learning costs of all members of language community i. The *total communication benefit* $U_i(z)$ of language community i is defined as follows:

(46) $$U_i(z) = \sum_{k=1}^{2^{m-1}} (z_{ik} - z_{i(k-1)})U_i(C_{ik}, z_{-i})$$

for $i = 1, \ldots, n$. Obviously, $U_i(z)$ is the integral over the communication benefits of all members of language community i. The group payoff $P_i(z)$ of language community i is the difference between total communication benefit and total learning cost:

(47) $$P_i(z) = U_i(z) - L_i(z_i)$$

for $i = 1, \ldots, n$. The function P_i which assigns $P_i(z)$ to every $z \in Z$ is the *group payoff function* of language community i. The function P which assigns

(48) $$P_z = (P_1(z), \ldots, P_n(z))$$

to every $z \in Z$ is the *vector payoff function*. P_i and P are defined relative to a fixed combination $r = (r_1, \ldots, r_n)$ of cost compatible rankings. When we want to express this dependency, we shall write P_i^r and P^r instead of P_i and P, respectively.

Aggregated games. For every combination $r = (r_1, \ldots, r_n)$ of cost compatible rankings, the associated *aggregated game*

(49) $$G^r = (Z, P^r)$$

is an n-person game in normal form, where $Z_1, \ldots, Z_n$ are the strategy sets of the players $1, \ldots, s$, respectively, and where P^r is the payoff function.

Best replies and equilibrium points. We use the notation $z_i z_{-i}$ in order to denote that border point system z which contains z_i and the components z_j of z_{-i} as components. A border point vector $\tilde{z}_i \in Z_i$ is a *best reply* to an i-incomplete border point system z_{-i} in G^r if we have:

(50) $$P_i^r(\tilde{z}_i, z_{-i}) = \max_{z_i \in Z_i} P_i^r((z_i, z_{-i}).$$

We say that $\tilde{z}_i$ is a *best reply* to a border point system z in G^r if in G^r the border point vector $\tilde{z}_i$ is a best reply to the i-incomplete border point system z_{-i} in z. We also say that the border point system $\tilde{z}$ is a *best reply* to z in G^r if every component of $\tilde{z}$ is a best reply to z in G^r. An equilibrium point z^* of G^r is a border point system with the property that z^* is a best reply to z^* in G^r.

5. Existence of equilibrium points

In this section we shall prove the existence of equilibrium points for every game $\Gamma(m, n, a, b, f, g)$ whose specification satisfies assumptions 1 to 5. For this purpose we first show that an equilibrium point of the original game is associated with every equilibrium point of every aggregated game. We then turn our attention to the continuity and convexity properties of group payoffs, which later enable us to apply standard results from the literature to prove the existence of an equilibrium point for aggregated games.

Wherever we do not explicitly say anything to the contrary, statements refer to a fixed game $\Gamma(m, n, a, b, f, g)$ whose specification satisfies assumptions 1 to 5 and to a fixed combination of cost compatible rankings $r = (r_1, ..., r_n)$. The notation $G = (Z, P)$ is used for the associated game G^r. If we say that a border point vector or a border point system is a *best reply*, we mean a best reply in G.

Theorem 2 asserts the existence of equilibrium points for all aggregated games of $\Gamma(m, n, a, b, f, g)$ and therefore does not refer to a specific $r = (r_1, ..., r_n)$. Similarly, theorem 3, the final existence theorem, refers to all games $\Gamma(m, n, a, b, f, g)$ satisfying assumptions 1 to 5.

Lemma 4. Let $z_i \in Z_i$ be a best reply to $z_{-i} \in Z_{-i}$. Moreover, let φ_i be a group strategy in the border equivalence class $\Lambda_i(z_i)$, and let φ_{-i} be an i-incomplete strategy combination in the border equivalence class $\Lambda_{-i}(z_{-i})$. Then φ_i is an almost best reply to φ_{-i}. Moreover, for every $t \in [0, a_i]$ which does not coincide with one of the border points z_{ik} in z_i, the strategy $\varphi_i(t)$ is a best reply of $i + t$ to φ_{-i}.

Proof. Assume that φ_i is not an almost best reply to φ_{-i}. Then at least one of the open intervals of the form $(z_{i(k-1)}, z_{ik})$ must contain infinitely many points t where $\varphi_i(t) = C_{ik}$ is not a best reply of $i + t$ to φ_{-i}. Moreover, at least one strategy $D \in M_i$ must be a best reply of $i+t$ to φ_{-i} for infinitely many of these values of t. Let D be a strategy of this kind, and let S be the set of all $t \in (z_{i(k-1)}, z_{ik})$ such that D is a best reply and C_{ik} is not a best reply of $i+t$ to φ_{-i}. For every $t \in S$ we have:

(51) $$U_i(D, \varphi_{-i}) - f_i(t)g_i(D) > U_i(C_{ik}, \varphi_{-i}) - f_i(t)g_i(C_{ik}).$$

Essentially the same argument as has been applied to (17) and (18) in the proof of lemma 2 can be applied here, too. In this way it can be seen that S is a subinterval of $(z_{i(k-1)}, z_{ik})$. Moreover, since S contains infinitely many points, S has positive length. Let φ_i' be the following group strategy:

(52) $$\varphi_i'(t) = \begin{cases} D & \text{for } t \in S \\ \varphi_i(t) & \text{for } t \notin S \end{cases}$$

In view of (51) and the finite length of S, we obtain:

(53) $$\int_0^{a_i} H_{i+t}(\varphi_i', \varphi_{-i}) dt > P_i(z_i z_{-i}).$$

According to lemma 3, the monotonicity class $\Psi_i(r_i)$ contains a best reply Ψ_i to φ_{-i}. Let z_i' be the border point vector of a best reply $\Psi_i \in \Psi_i(r_i)$ to φ_{-i}. It follows by (53) that we have:

(54) $$P_i(z_i' z_{-i}) > P_i(z_i z_{-i}).$$

This is a contradiction to the assumption that z_i is a best reply to z_{-i}. Therefore, φ_i is an almost best reply to φ_{-i}.

We now prove the remaining assertion of the lemma. Suppose that one of the finitely many values of t, where $\varphi_i(t)$ is not a best reply to φ_{-i}, is a value t_0 which does not coincide with one of the border points. In some ε-neighborhood of t_0 all values of t with the exception of t_0 have the property that $\varphi_i(t) = C_{ik}$ is a best reply of $i+t$ to φ_{-i}, where the strategy C_{ik} is the same one for all these values of t. However, in view of lemma 2 the set of all t such that C_{ik} is a best reply of $i+t$ to φ_{-i} is an interval. It must be that t_0 belongs to the interval. This is a contradiction. Therefore, the assertion holds.

Lemma 5. Let $z = (z_1, ..., z_n)$ be an equilibrium point of the aggregated game G. Then the border equivalence class $\Lambda(z)$ contains an equilibrium point.

Proof. Let $\varphi = (\varphi_1, ..., \varphi_n)$ be an element of $\Lambda(z)$. It follows by lemma 4 that φ is an almost equilibrium point. Let Ψ be related to φ as in lemma 1. It follows by lemma 1 that Ψ is an equilibrium point. In view of lemma 4, it is clear that Ψ differs from φ only at border points in the components of z. Consequently, Ψ belongs to $\Lambda(z)$.

Lemma 6. For $i = 1, ..., n$, the group payoff $P_i(z)$ is a continuous and bounded function of z, and for every fixed $z_{-i} \in Z_{-i}$ the group payoff $P_i(z_i z_{-i})$ is a concave function of z_i.

Proof. Since definite integrals of bounded functions are continuous in their upper limits, the functions B_{ij} and F_i are continuous. Therefore, the continuity of $P_i(z)$ as a function of z is an immediate consequence of the definition of group payoffs. The boundedness of $P_i(z)$ follows by the compactness of Z.

In view of (46), for fixed $z_{-i} \in Z_{-i}$ the total community benefit $U_i(z)$ is a linear function of z_i. Therefore, it is sufficient to show that $L_i(z_i)$ is convex in z_i in order to prove that $P_i(z_i, z_{-i})$ is concave in z_i. Equation (45) can be rewritten as follows:

$$(55) \quad L_i(z_i) = \sum_{k=1}^{2^{m-1}-1} F_i(z_{ik})[g_i(C_{ik}) - g_i(C_{i(k+1)})] + a_i g_i(C_{i2^{m-1}}).$$

Since $F_i(z_{ik})$ is the integral over a monotonically increasing function, it is a strictly convex function of z_{ik}. In view of $g_i(C_{ik}) \le g_i(C_{i(k+1)})$, it follows by (55) that $L_i(z_i)$ is a convex function of z_i.

Theorem 2. For every combination $r = (r_1, ..., r_n)$ of cost compatible rankings, the associated aggregated game G^r has at least one equilibrium point.

Proof. The strategy sets Z_i are convex and compact. Lemma 6 has shown that $P_i^r(z)$ is continuous with respect to z and concave with respect to player i's strategy z_i. Therefore, the conditions for the application of well known results on the existence of equilibrium points are satisfied (see, e.g., Friedman, 1977, p. 160, theorem 7.4). We can conclude that the assertion holds.

Theorem 3. Every game $\Gamma(m, n, a, b, f, g)$ whose specification satisfies assumptions 1 to 5 has at least one equilibrium point.

Proof. The assertion is an immediate consequence of lemma 5 and theorem 2.

6. References

Breton, Albert, & Mieskowski, Peter (1975). *The returns to investment in language: the economics of bilingualism*. Toronto: University of Toronto, Institute for Policy Analysis.

Dressler, Wolfgang U. (1982). "Acceleration, retardation, and reversal in language decay?". In *Language spread*, ed. Robert L. Cooper (Bloomington, IN: Indiana University Press), pp. 321–336.

Friedman, James W. (1977). Oligopoly and the theory of games. Amsterdam: North-Holland.

Grenier, Gilles (1985). "Bilinguisme, transferts linguistiques et revenus de travail au Québec: quelques éléments d'interaction. In *Économie et langue*, ed. François Vaillancourt (Québec: Éditeur officiel), pp. 243–287.

Hočevar, Toussaint (1975). "Equilibria in linguistic minority markets". *Kyklos*, 28, 337–357.

Lang, Kevin (1986). "A language theory of discrimination". *The quarterly journal of*

economics, 100, 363–381.
Laponce, Jean (1984). *Langue et territoire*. Québec: Les presses de l'Université Laval.
Lieberson, Stanley (1982). "Forces affecting language spread: some basic propositions". In *Language spread*, ed. Robert L. Cooper (Bloomington, IN: Indiana University Press), pp. 37–62.
Sabourin, Conrad (1985). "La théorie des environnements linguistiques". In *Économie et langue*, ed. François Vaillancourt (Québec: Éditeur officiel), pp. 59–82.
Vaillancourt, François (1985). "Les écrits en économie de la langue: brève revue et introduction au recueil". In *Économie et langue*, ed. François Vaillancourt (Québec: Éditeur officiel), pp. 11–25.
Vaillancourt, François (1985a). "Le choix de la langue de la consommation". In *Économie et langue*, ed. François Vaillancourt (Québec: Éditeur officiel), pp. 209–220.
Vaillancourt, François, & Lacroix, Robert (1985). "Revenus et langue au Québec, 1970–1980: une revue des écrits". In *Économie et langue*, ed. François Vaillancourt (Québec: Éditeur officiel), pp. 221–242.

Name index

Abelson, R.P. 7
Albers, W. 17
Alger, C. 319
Allais, M. 162
Allison, G.T. 319
Arrow, K.J. 172
Aumann, R.J. 12, 14, 104, 175, 326
Avenhaus, R. 17, 319
Axelrod, R. 7, 174, 241

Berg, C.C. 9, 164
Bernholz, P. 13, 326
Bomze, I.M. 186
Boyd, R. 167, 171
Breton, A. 386
Brody, R. 319
Brown, G. 185
Büren, R. 319
Burger, E. 4, 31
Bush, R.R. 171, 172, 185, 248

Cavalli-Sforza, L.L. 167, 170, 171
Chang, P.H. 319
Cox, J.C. 264, 266, 270
Crawford, V.P. 186, 195
Cross, J. 172
Cyert, R.M. 175

Dale, P.S. 240
Darlington, R.B. 213
Deutsch, K. 295, 319
Dressler, W.U. 386
Dupuit, J. 96, 105

Eshel, I. 165, 166

Feldman, W.M. 166, 167, 170, 171
Feller, W. 341
Fisher, R. 165
Forman, R. 82
Friedman, J.W. 408

Gabisch, G. 195
Galton, F. 168
Greeno, J.G. 248
Grenier, G. 387
Guetzkow, H. 293, 319
Güth, W. 17, 110–11

Hammerstein, P. 18, 19, 166, 167, 186

Handel, M. 319
Harley, C.B. 186
Harrison, G.W. 282
Harsanyi, J.C. 4, 11–13, 16, 17, 31, 108–10, 112, 127, 129, 134–6, 343
Harstad, R.M. 264
Häselbarth, V. 6
Heimann 3
Hendrichs, H. 18
Hicks, J.R. 97
Hočevar, T. 386
Hoggatt, A. 11–13
Holsti, K.J. 319
Huber, K. 17

Isbell, J.R. 46

Jonker, L.B. 185

Kagel, J.H. 264
Kahneman, D. 9, 173, 174, 178, 319
Kalai, E. 175
Kalish, G. 5
Kalkofen, B. 17, 110, 111, 112
Kaplan, M.A. 326
Kautilya, A.-S. 326
Kendall, M.G. 258
Keser, C. 10
Kolkowicz, R. 319
Kreps, D, 15, 21, 175, 176, 241
Krischker, S. 81, 82
Kuon, B. 264, 273, 276, 277

Lacroix, R. 386
Laing, J.D. 82
Lang, K. 387
Laponce, J. 386
Lave, L.B. 240
Leopold, U. 17
Levin, D. 264
Liberman, V. 166
Lieberson, S. 386
Loomes, G. 173
Lorenz, H.W. 195
Luce, D. 335

Malawski, M. 172, 173
March, J.G. 175
Marschak, T. 13
Maschler, M. 326

Maynard Smith, J. 18, 19, 162, 165, 185
Mieskowski, P. 386
Milgrom, P. 241
Milinski, M. 186
Mitzkewitz, M. 17, 22, 264, 272, 275
Miyasawa, K. 185
Moran, P.A.P. 165
Morehous, L.G. 240
Morgenstern, O. 4, 319
Morgenthau, H. 295, 319
Mosteller, F. 171, 172, 185, 248

Nagel, R. 264, 272
Nash, J. 185
Nelson, R. 169, 175, 176
Newell, A. 319
Neyman, A. 175
Niou, E.M. 326
Noel, R. 319

Ordeshook, P.C. 326

Perlmutter, A. 13, 291, 319
Plott, C.R. 177
Potters, J. 17
Price, G.R. 18, 19, 162, 165, 185
Pruitt, P.G. 9

Raiffa, H. 335
Rapoport, A. 240
Rausch, E. 3
Ravenal, E. 319
Restle, F. 248
Ricciardi, F.M. 5
Richerson, P.I. 167, 171
Riechert, S.E. 166, 167
Roberson, B. 264, 266, 270
Roberts, R. 241
Robinson, J. 185
Rosenberg, M.J. 7
Rosenmüller, J. 185

Sabourin, C. 386
Samuelson, P.A. 97
Sauermann, H. 5, 6, 10, 319
Savage, L.H. 161, 319

Scarf, H. 12
Schelling, T.C. 80
Schumpeter, J.A. 169
Schuster, K.G. 8, 9
Selten, R. 9, 11, 17, 21, 22, 81, 82, 108–10, 112, 127, 129, 134–6, 174, 176, 177, 185, 186, 241, 264, 272, 275, 319, 343
Sen, A. 319
Shackle, G.L.S. 295, 319
Shapley, L. 50, 104
Shmida, A. 19
Siegel, S. 258
Silberberg, E. 97
Simon, H.A. 5, 175, 319
Smith, V.L 177, 264, 266, 270
Snyder, R. 319
Sorin, S. 175
Stanford, W. 176
Steinbach, U. 319
Stoecker, R. 21, 176, 185, 240, 264, 272, 319, 336
Sugden, R. 173

Taylor, P.D. 185
Tietz. R. 12
Tversky, A. 9, 173, 174, 178, 319
Tyszka, T. 162

Uhlich, G. 8, 18, 22, 275

Vaillancourt, F. 386
van Damme, E. 15, 17
van Winden, F. 17
von Clausewitz, C. 336
von Essen, B. 319
von Neumann, J. 4, 46, 319

Wagner, H. 327, 336
Walker, J.M. 264
Weber, M. 178
Williams 4
Wilson, R. 15, 21, 241
Winter, S.G. 169, 175, 176
Wolfers, A. 319

Zinnes, D.A. 326